Lecture Notes in Electrical Engineering

Volume 403

About this Series

"Lecture Notes in Electrical Engineering (LNEE)" is a book series which reports the latest research and developments in Electrical Engineering, namely:

- Communication, Networks, and Information Theory
- Computer Engineering
- Signal, Image, Speech and Information Processing
- Circuits and Systems
- Bioengineering

LNEE publishes authored monographs and contributed volumes which present cutting edge research information as well as new perspectives on classical fields, while maintaining Springer's high standards of academic excellence. Also considered for publication are lecture materials, proceedings, and other related materials of exceptionally high quality and interest. The subject matter should be original and timely, reporting the latest research and developments in all areas of electrical engineering.

The audience for the books in LNEE consists of advanced level students, researchers, and industry professionals working at the forefront of their fields. Much like Springer's other Lecture Notes series, LNEE will be distributed through Springer's print and electronic publishing channels.

More information about this series at http://www.springer.com/series/7818

Vijay Nath
Editor

Proceedings of the International Conference on Nano-electronics, Circuits & Communication Systems

Springer

Editor
Vijay Nath
Department of Electronics
and Communication Engineering
Birla Institute of Technology, Mesra
Ranchi, Jharkhand
India

ISSN 1876-1100 ISSN 1876-1119 (electronic)
Lecture Notes in Electrical Engineering
ISBN 978-981-10-2998-1 ISBN 978-981-10-2999-8 (eBook)
DOI 10.1007/978-981-10-2999-8

Library of Congress Control Number: 2016956192

Printed on acid-free paper

This Springer imprint is published by Springer Nature
The registered company is Springer Nature Singapore Pte Ltd.
The registered company address is: 152 Beach Road, #21-01/04 Gateway East, Singapore 189721, Singapore

Preface

Manufacturing plays a major role in the development of the country. However, the manufacturing industry is facing several challenges such as rapid product development, flexibility, and low to medium volume, transportation, and low cost etc. Many advanced and unconventional technologies, tools, and software are being developed worldwide to enable solutions to these challenges. Manufacturing and design of integrated circuits (ICs) comprises a large portion of research advances in manufacturing because of the need for precise work. To enable exchange of ideas on current trends in this area, the 1st International Conference on Nano-electronics, Circuits & Communication Systems (NCCS-2015) was organized by the Indian Society of VLSI Education (ISVE), Ranchi and the Institution of Electronics and Telecommunication Engineers (IETE) Ranchi at Advanced Regional Telecom Training Centre (ARTTC) near Jumar River Hazaribag Road Ranchi from 9 to 10th May, 2015. Around 300 papers were received in total, the best of which are presented in this volume. For this selection, all papers were blind reviewed by three expert reviewers and the presentation sessions were reviewed by a six-member expert committee.

There are a total of 35 papers in this volume that cover the core themes of the conference, i.e., design, simulation, verification, implementation and applications of nano-electronics, circuits and communication systems. The paper by L. Rajesh et al. describe user demand wireless network selection using game theory, with this concept unwanted power radiation will be reduced. Ankur Saxena et al. demonstrate his work on review on band notching techniques for ultra-wideband antenna. Biswajit Ghosh et al. demonstrate his research work on comparing energy efficiency of DF relay assisted cooperative and non-cooperative short-range wireless systems. Abhishek Rawat et al. describe dual band octagonal shape microstrip patch antenna at 5.70 and 8.0 GHz. Nabajyoti Mazumdar et al. explain distributed faulttolerant multi-objective clustering algorithm for wireless sensor network. Abhishek Nag et al. describe an autonomous power in clock getting technology in SRAM-based

FPGA. C. Ranjith et al. explain a hardware implementation of evolvable embedded system for combinational logic circuits using Vertex 6 FPGA. J. Dhurga Devi et al. describe design of a third order self-bias adaptive band width PLL. Srinivasulu Avireni et al. describe operational transconductance amplifier based sinusoidal oscillator with grounded capacitors. Manas Ray et al. describe the evaluation of wavelet based speech codec for VOIP applications. Trupa Sarkar et al. describe leakage reduction by test pattern reordering. R.K. Mugelan et al. demonstrate the performance enhancement of LTE HetNet using EVM based constellation combiner ECC in warp. D.C. Diana et al. describe modified PSO based equalizer for channel equalization. Abahan Sarkar et al. describe online counting of cigarette in packages as an image processing approach. Jha Vijay Kumar et al. demonstrate fuzzy min–max neural-based intrusion detection system. Amiya Sagar Das et al. describe implementation of breadth first search for storage optimization in random storage assignment of automated storage and retrieval system. Neeti Singh et al. describe efficient algorithms for removal of high-density random value impulse noises in images. Bhabani Shankar Das et al. explain prediction depth average velocity and boundary shear distribution in a single-stage channel by lateral distribution method. Kamalini Devi et al. describe the flow computation in symmetric compound channels using conveyance estimation system. Sarika Tyagi et al. describe a 21 nW CMOS operational amplifier for biomedical application. Pallavi Dutta et al. describe performance analysis of three-phase induction motor using Simulink and TMS320C6713. Abhishek Gandhar et al. describe effect of static synchronous series compensator on power balancing in wind farms. Mohini Yadav et al. describe reliability and energy benefit analysis of distribution system incorporating wind turbine generator. A.K.P. Kovendan et al. describe present context of smart girds in India: a survey. S. Sridevi et al. describe real-time simulation design for continuous process industries. M. Rajendra Prasad et al. describe system level performance analysis of embedded system for GSM application. S.M. Annapurna et al. describe robotics research visibility in BRICS country as a scientometrics study. A. Ravishankar Holla et al. describe characterization of TFT sensors for chemical sensing applications. Anurag Kumar Paliwal et al. describe design of a high-performance CMOS charge pump for phase-locked loop synthesizers. Adesh Kumar et al. describe IC packaging with 3D IC tech. method. Akshat Chitransh et al. describe analysis of a self-compensating, low-noise, low-power PLL circuit @45 nm technology node. Krishan Kumar Singh et al. describe design of comparator in sigma delta using 45 nm CMOS technology. Sudhanshu Kumar et al. describe investigation of electrostatic actuation scheme for low-voltage MEMS switch. Kumar Amit et al. describe evaluating the performance of dye-sensitized solar cell with the various key components like electrodes, dyes and electrolytes. Jitendra Yadav et al. describe modeling and simulation of the dynamic response of a generic mechanical linkage for control application under the consideration of the nonlinearities imposed by friction.

Authors and editors have taken utmost care in preparing this volume for publication, but there some errors that might have crept in despite our best efforts. Readers are requested to provide their valuable feedbacks on the quality of presentation and alert us of any inadvertent errors or omissions. We expect that the book will be welcomed by students as well as practicing engineers, researchers, and professors.

Ranchi, India

Vijay Nath

Editorial Acknowledgements

We extend our thanks to all the authors for contributing to this book by sharing their valuable research findings. We specially thank a number of reviewers for promptly reviewing the papers submitted to the conference. We are grateful to the volunteers, invited speakers, session chairs, sponsors, sub-committee members, members of international advisory committee, and members of national advisory committee, members of technical program committee, members of joint secretary, and members of scientific advisory committee for successful conclusion of the conference. The editors express their heartfelt gratitude towards Smriti Dagur, President IETE New Delhi, Sanjay Kumar Jha, Executive Engineer Government of Jharkhand & Chairman of IETE Ranchi, Prasad Vijay Bhushan Pandey, DGM ETR Jharkhand Circle Ranchi & Chairman ISVE Ranchi, A.A. Khan, Former VC Ranchi University, M.K. Mishra, VC BIT Mesra, Labh Singh, CGM BSNL Ranchi, K. Pandey, VC Ranchi University, P.K. Barhai, Former VC BIT Mesra, R. Mishra, Former CMD HEC Ranchi, Ajay Kumar, AGM(admin) ARTTC BSNL Ranchi & Secretary IETE Ranchi, P.R. Thakura, Executive member of ISVE & Professor of BIT Mesra Ranchi, Anand Kumar Thakur Treasurer, IETE Ranchi for their support, encouragement, and motivation to organize such a prestigious event that paved the way for this book on nano-electronics, circuits and communication systems (NCCS). Finally, we express our sincere gratitude towards the editorial team at Springer who helped in publishing this book.

Contents

About the Editor

Dr. Vijay Nath received his Bachelor's degree in Physics and Master's degree in Electronics from DDU Gorakhpur University, India in 1998 and 2001. He received a PGDCN from MMM Engineering College Gorakhpur (Gold Medalist) in 1999. He received his Ph.D. in VLSI Design & Technology from Dr. RML Avadh University, Faizabad in association with CEERI- Pilani in 2008. He was a member of the faculty of the Department of Electronics, DDU Gorakhpur University, Gorakhpur (2002–2006). In 2006, he joined as a faculty member in the Department of Electronics and Communication Engineering, Birla Institute of Technology Mesra, Ranchi, India. Currently he is Professor In-charge of VLSI Design Lab & Embedded System Design Lab, Department of ECE, BIT Mesra, Ranchi. His research interests include analog, digital, mixed CMOS VLSI circuits, low-power VLSI circuits, ADC, DAC, PTAT, CMOS bandgap voltage reference, piezoresistive pressure sensors, double ring capacitive pressure sensors, MEMS sensors, signal-processing circuits, ASICs, embedded systems designs, smart cardiac pacemaker, and early stage cancer detection. He has to his credit around 100 publications in international journals and conferences. He is a member of several reputed professional and academic bodies including IETE, ISVE, and IEEE.

Chapter 1
An Autonomous Power and Clock Gating Technique in SRAM-Based FPGA

Abhishek Nag and Sambhu Nath Pradhan

1.1 Introduction

Dynamic power was given the utmost importance initially for power reduction schemes in a FPGA; however, recent studies indicate the major impact of leakage power in the total power dissipation [1]. Power gating has turned out to be a popular method of leakage reduction in FPGA. All prior works in this field have used spice-based modeling of the FPGA to carry out any transistor-level operation, as in [2, 3]. Most of these prior works have been associated with the theoretical aspect of FPGA operations.

In [3], the authors have considered power-reducing techniques in a FPGA by obtaining spice net list of the architectures and varying the various parameters associated. Power gating has proven to be one of the most popular methods of reducing the leakage power of a programmable device [4]. Configuration bits were mostly used in earlier works involving power gating, to control the power. This is termed as statically controlled power gating, as in [5, 6]. The concept of dynamically controlled power gating was very recently introduced in programmable device such as in [7].

On the other hand, clock gating is an efficient technique of dynamic power reduction, by stopping the activity of clock during inactive periods of the circuit, as in [8–10]. Similar works in this field include the work done in [11, 12], considering spice precharacterization level modeling of the FPGA. However, implementing the power reduction operations in a circuit level will provide more insight into the working of the programmable device and an elaborate analysis can be carried out.

A. Nag (✉) · S.N. Pradhan
Department of Electronics and Communication Engineering,
National Institute of Technology Agartala, Agartala, India
e-mail: abhi14379@gmail.com

S.N. Pradhan
e-mail: sambhu.pradhan@gmail.com

V. Nath (ed.), *Proceedings of the International Conference on Nano-electronics, Circuits & Communication Systems*, Lecture Notes in Electrical Engineering 403,
DOI 10.1007/978-981-10-2999-8_1

In this work, an efficient and dynamic technique of power and clock gating is introduced in a SRAM-based TILE prototype, designed and developed in CADENCE tool, using a technology of 45 nm. Our TILE architecture is inspired from the work done by authors in [13]. Correct simulation results have been obtained for all the sub-blocks of the architecture. We have implemented a basic FSM circuit in the TILE and introduced a fine level gating technique in the proposed architecture. The simulation results indicate significant reduction of the total power of the TILE prototype. An in-depth delay analysis is also obtained and reported.

The following paper is presented as follows: Sect. 1.2 presents the basics of power and clock gating techniques. The transistor-level design of the TILE prototype is presented in Sect. 1.3. Section 1.4 deals with the proposed power and clock gating logic implemented in the TILE prototype. The experimental results of the power-gated FPGA are presented in Sect. 1.5 followed by conclusion in Sect. 1.6.

1.2 Power and Clock Gating

In this section, we elaborate about the basics of the gating techniques being used in our approach for power reduction of the TILE.

1.2.1 Power Gating

Power gating has proven to be one of the most effective methods of reducing leakage power of a circuit [4]. Power gating techniques were initially introduced for ASICs, which can now be implemented on programmable devices also. With the growing need for programmable devices like FPGA, power dissipation stands as a major concern. Consequently, power reduction techniques are gaining popularity. The basic of power gating is to disable/enable the functionality of a circuit by a sleep transistor or power switch. The voltage supply of a device is provided through a PMOS transistor, called the sleep transistor. Thus, the circuit can be switched "off" or "on" by switching the sleep transistor on or off. The gate input to sleep transistor is called the sleep signal or activation signal, which decides the criteria of when to turn the sleep transistor "on" or "off." The function generating the activation signal is termed as activation function. The power gating basics is explained in Fig. 1.1.

Power gating significantly reduces the overall leakage power of the circuit. However, the extra PMOS transistor results in some resistance, which may affect the performance of the circuit. By appropriately sizing the transistor, an acceptable trade-off between power and performance can be obtained. Again, the excess area

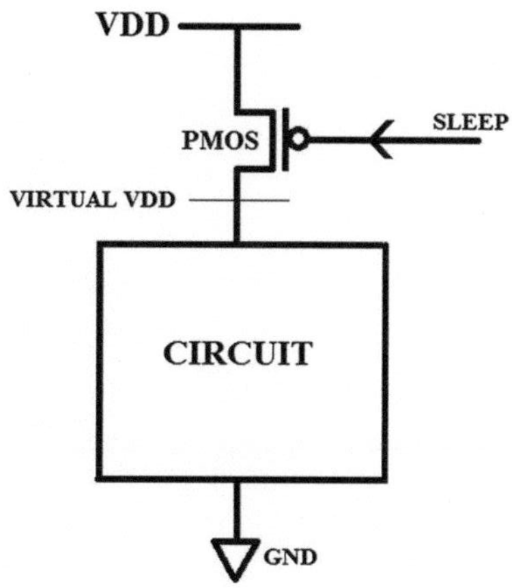

Fig. 1.1 Basic of power gating

overhead will be minimal when the same transistor is used to power gate a large portion of the circuit.

1.2.2 Clock Gating

Clock gating has turned out to be a popular technique of dynamic power reduction in a sequential circuit [8]. The clock gating logic is to stop the activity of the clock, when no useful computation is being performed by the circuit. In other words, when a sequential circuit does result in any state transition, the clock is being disabled to reduce the unnecessary power dissipated by the clock during idle states. The activation signal for clock gating is kept low when the necessary clock gating condition occurs, which is then AND gated with the global clock signal to obtain the gated clock. By knowing the inputs and states of a circuit, activation signal can be generated. The gated clock signal is then finally provided to a sequential block, which needs to be gated. The basic idea of clock gating in a sequential circuit is depicted in the following Fig. 1.2.

1.3 Transistor-Level Design

The FPGA prototype developed in this work is a SRAM-based TILE-oriented FPGA. Static memory cells are the basic building blocks of the FPGA. The SRAM cells are arranged throughout the entire FPGA architecture for providing configuration. SRAM-based programming technology is the most popular approach for FPGA due to its reprogrammable nature and use of standard CMOS process

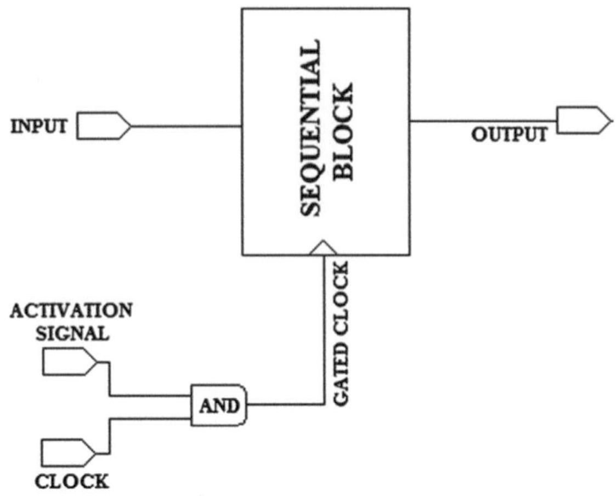

Fig. 1.2 Basic of clock gating

technology. This leads to increased integration, higher speed, and lower dynamic power consumption.

FPGA, being a regular architecture, is divided into TILEs, where each TILE compromises of logic element and routing resources. The basic building blocks of a TILE-based FPGA in its highest level are logic element (LE), which consists of a lookup table (LUT), flip-flop, and a multiplexer that chooses whether to forward the output of the LUT or the flip-flop outside the LE. A k-LUT consists of a set of multiplexers that can implement any function of *k* inputs by forwarding one of the 2 k configuration bits to the output of the LUT. LEs are interconnected with routing resources, which are also configurable. Configuration of routing resources is achieved using pass gates controlled by a routing configuration bits. The desired configuration is stored in SRAM cells. One SRAM cell is required per pass gate, while a k-LUT requires 2 k SRAM cells.

The routing resources include connection blocks (C) and switch blocks (S), where the purpose of connection block is to interconnect the logic elements, whereas the routing wires (running horizontally and vertically) are connected inside the switch block. The FPGA developed in this work is based on the architecture developed in [13]. The top view of a TILE-based FPGA architecture is shown in Fig. 1.3 [13].

The structure of a TILE with the corresponding programming and routing resources used in this work is shown in Fig. 1.4. The entire architectural description of the internal structures is explained in [13].

1.3.1 Logic Element

The logic element is the computational block of an FPGA. This is where an arbitrary function can be applied to the input signals to produce a specific result.

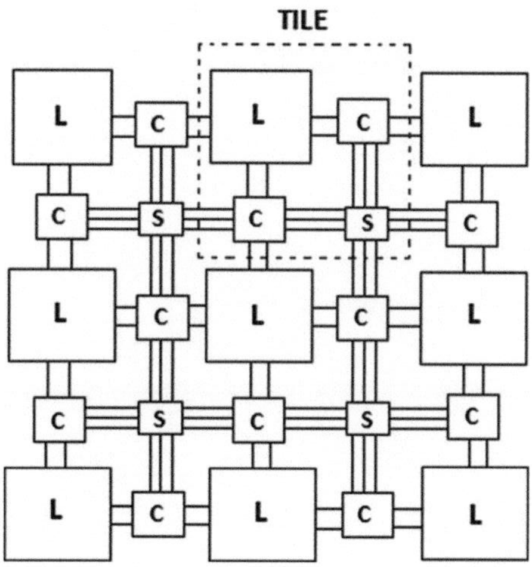

Fig. 1.3 Top view of FPGA architecture

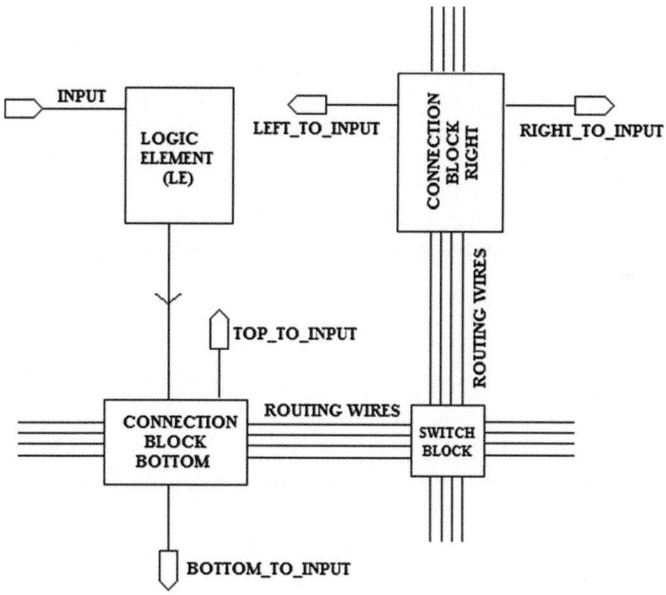

Fig. 1.4 Outline of a basic TILE structure

Figure 1.5 shows the top level view of the logic element. The logic element used in this work consists of three sub-blocks: a 4 input LUT, a register, and a 2:1 MUX.

The LUT is used to create the result of the desired arbitrary function. LUTs are chosen because at the time of fabrication of the FPGA, it is unknown what functions

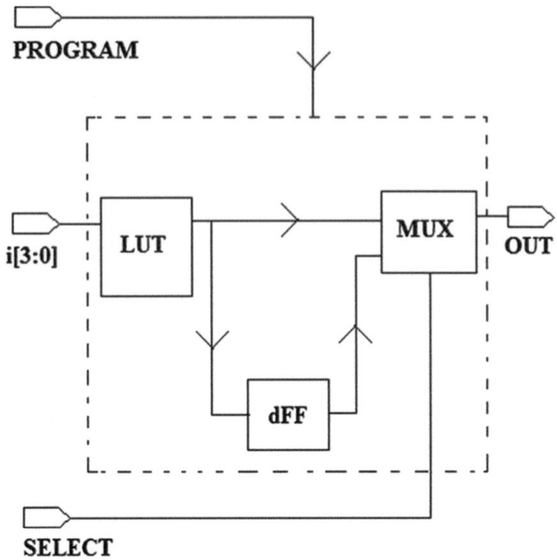

Fig. 1.5 Logic element schematic (LE)

the end user of the FPGA wants. LUT is the simplest way to provide the functionality for allowing arbitrary function to be programmed. In this work, we use 4 input LUTs, and hence, each LUT needs 16 (24) SRAM cells (it is 24 because each input can have two values [0 or 1], and there are 4 inputs). The SRAM cells are to be programmed with the 16 possible results of the arbitrary functions. The inputs to the LUT are provided through the select lines.

1.3.2 TILE

The design and simulation of the TILE architecture in Fig. 1.4 has been carried out in CADENCE Virtuoso spectrum at 45 nm technology. Figure 1.6 shows the input and output terminals of a single TILE.

The description and configuration of the various input and the output pins of the TILE are presented in Table 1.1.

As mentioned earlier, using an array of rows and columns of the above TILE, an entire FPGA architecture can now be established [13]. In our architecture, we have not considered the impact of I/O blocks associated with FPGAs. Also, since our aim is just to analyze the effectiveness of our proposed gating technique in the TILE, we have not considered the resistive and capacitive effect after layout generation of the TILE prototype. All these points are kept in considerations for future scope.

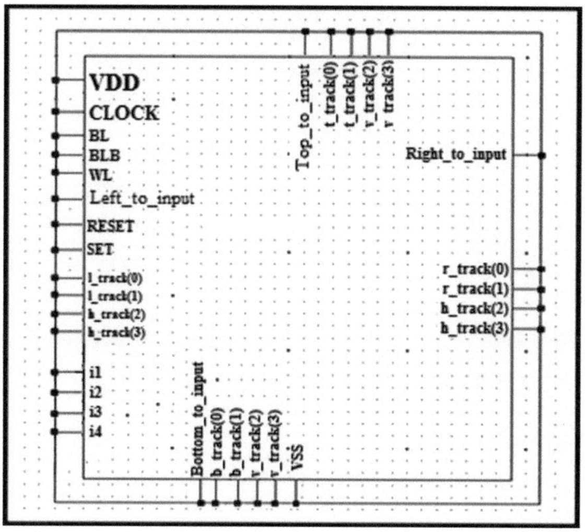

Fig. 1.6 I/O terminals of a TILE

Table 1.1 Description of I/O terminals of a TILE

Sl. no.	PIN name	Description	Direction
1	VDD	Supply voltage	Input
2	Clock	DFF clock signal	Input
3	BL, BLB	Bit lines (program data)	Input
4	WL	Word line	Input
5	Set, reset	Set/reset signal of DFF	Input
6	l_track(0,1), r_track(0,1), b_track(0,1)	Routing tracks	I/O
7	t_track(0,1), v_track(2,3), h_track(2,3)	Routing tracks	I/O
8	i1, i2, i3, i4	LE inputs	Input
9	Bottom_to_Input, Top_to_Input	Outputs from connection block bottom	Output
10	Right_to_Input, Left_to_input	Output from connection block right	Output
11	VSS	Ground terminal	Input

1.4 Proposed Power and Clock Gating Technique

In order to carry out gating in the TILE, we have first implemented an arbitrary 4 input FSM circuit “EXCKT” in the TILE. A finite state machine (FSM) is an abstract model of a system (physical, biological, mechanical, electronic, or software). FSM is modeled as a useful abstraction for sequential circuits containing

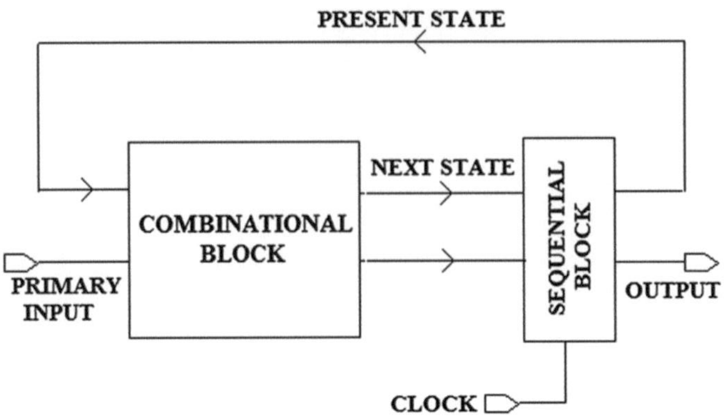

Fig. 1.7 Block diagram of a FSM

Table 1.2 State table of a 4 input FSM circuit

I(2)	I(1)	I(0)	PS	O	NS
0	0	0	0	0	0
0	0	0	1	1	0
0	0	1	0	1	0
0	0	1	1	1	0
0	1	0	0	0	1
0	1	0	1	1	1
0	1	1	0	1	1
0	1	1	1	0	0
1	0	0	0	1	0
1	0	0	1	0	0
1	0	1	0	0	1
1	0	1	1	0	0
1	1	0	0	0	1
1	1	0	1	1	1
1	1	1	0	1	1
1	1	1	1	0	1

centralized “states” of operation. At each clock edge, combinational logic generates outputs and next state as a function of inputs and present state. The block diagram of a mealy FSM is shown in Fig. 1.7.

The description of an arbitrary FSM circuit “EXCKT” is presented in the following Table 1.2. Here, “PS” stands for present state and “NS” stands for next state. The implementation of the FSM circuit in the FPGA is done by simple binary coding technique, where all the output and next state bits are generated in an individual TILE. Based on the description of the FSM circuit EXCKT in Table 1.2, we require two TILEs to generate the output (O) and next state (NS). The

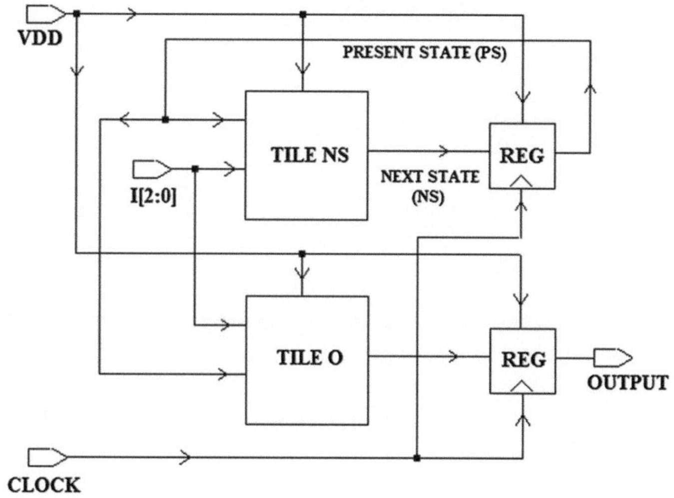

Fig. 1.8 Implementation of EXCKT in FPGA

architecture of the TILEs implementing the FSM EXCKT is shown in the Fig. 1.8. A register (REG) is being utilized in the architecture to generate the present state of the FSM as FPGA has scope for inbuilt memory devices in their architecture. We have developed the register using D-F/Fs, in the same 45 nm technology.

1.4.1 *Method of Gating*

The idea of gating the TILE is obtained from the logic used in [14]. The logic of gating the circuit in [14] was used for reducing the dynamic power, by implementing clock gating in ASICs. However, we have extended that approach in our architecture for implementing concurrent clock and power gating, to achieve total power reduction of a programmable TILE.

The basic logic behind gating the architecture is to stop the functionality of a circuit when the present state and next state of the FSM become equal, referred to as "self-loops." During that instant, the TILE implementing the next state function is no longer required to operate and compute. To avoid the functioning of the TILEs during self-loops, the architecture is power gated, depending on the sleep signal generated. The sleep signal, also referred to as the activation signal, is generated from the knowledge of the present state and primary inputs of the circuit, and the function generating the sleep signal is termed as activation function. The activation signal becomes high when a self-loop occurs and restricts the power supply to the TILE, through a PMOS (sleep transistor). The activation function is implemented as a combinational block which extracts information from the primary inputs and state inputs of the circuit.

$$\text{Activation signal} = \begin{matrix} 1, & \text{if PS} = \text{NS} \\ 0, & \text{else} \end{matrix} \qquad (1.1)$$

Based on the state table presented in Table 1.2, the activation function (F) generated for the FSM EXCKT is,

$$F = [\sim I(0).\sim I(1).\sim \text{PS}] + [\sim I(1).\sim I(2).\sim \text{PS}] + [I(1).\sim I(2).\text{PS}] \qquad (1.2)$$

Here "$\sim$" stands for NOT operation. Concurrently, with the power gating of the TILE, the register (REG) clock signal in Fig. 1.8 is disabled during self-loops. Disabling the activity of clocks during periods of inactivity (self-loops in this case) results in significant dynamic power savings of the FPGA. The clock gating of the registers is achieved by AND gating the global clock signal with the activation signal. The clock gating condition called activation signal is so generated such that it becomes low whenever a self-loop occurs. Thus, the gated clock signal will be low during self-loops irrespective of the state of the global clock.

1.4.2 Power and Clock-Gated TILE

The TILE architecture displaying the concurrent power and clock gating is shown in Fig. 1.9. The activation function is same for both clock gating and power gating. The power gating is carried out through a PMOS transistor, where the activation signal is fed as the gate input and the power supply to its drain input. Based on the activation signal, the PMOS gets switched on and off, thus controlling the supply voltage from entering the TILE. The activation signal becomes high, whenever a self-loop appears and hence shuts down the pMOS transistor. On the other hand, during clock gating, the activation signal should be low whenever a self-loop appears. For this reason, we have used a NOT gate to generate the activation signal for clock gating. The inverted activation signal is now passed through a AND gate, whose other input is the clock signal. This generates the gated clock signal, which is fed to the external register.

One important thing to note here is that the gating is being implemented only for the TILE implementing the next state. The output TILE is not being gated. This is because, in several FSMs, there may arrive a situation where even during self-loops the output changes. The TILE architecture displaying the proposed gating logic in the next state function of the FSM EXCKT is shown in the Fig. 1.9.

The proposed architecture has successfully resulted in the total power reduction of the TILE with a minimal increase in delay. One of the limitations of our approach is the excess area overhead due to the introduction of new circuit elements

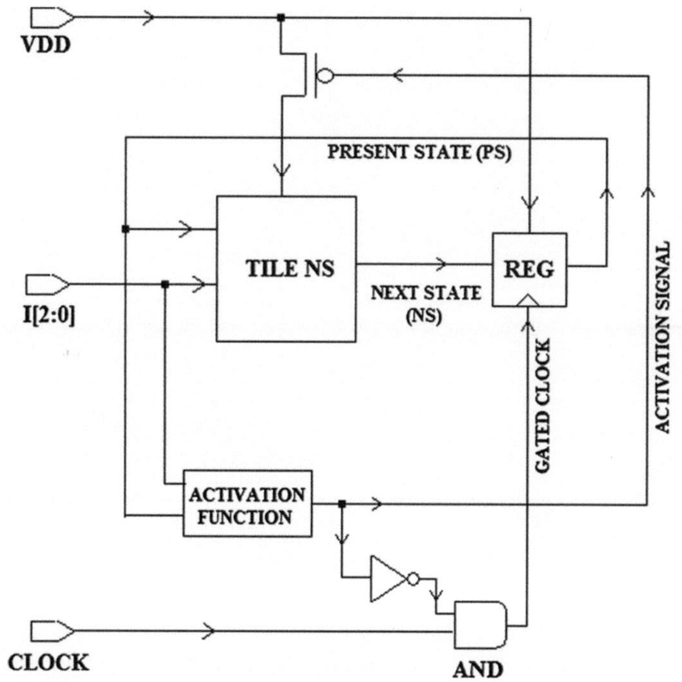

Fig. 1.9 Proposed power and clock gating in a TILE

for gating. However, by appropriately sizing the transistors in the activation function circuit, a proper trade-off between area and power can be maintained.

1.5 Results

The power analysis and delay analysis of the TILE are presented in this section. The simulation of the TILEs has been carried out in CADENCE tool, using a technology of 45 nm and power supply (VDD) of 1 V. The simulation waveform of the TILEs implementing the gated FSM circuit EXCKT is shown in the following Fig. 1.10.

The power, delay, and area analyses for both with power and clock gating (PCG) and without power and clock gating (WPCG) for the circuit EXCKT are shown in the Table 1.3.

Here, "*k*" is the width of a minimum width transistor and "λ" stands for the feature size of the concerned technology. The power results in Table 1.3 show a reduction of approximately 19% of the total power dissipation of the TILEs after gating. This number is, however, likely to increase as the number of self-loops in a FSM increases. However, power and clock gating results in an increase in delay due to the wake up time associated with the logic block for power supply adjustments as

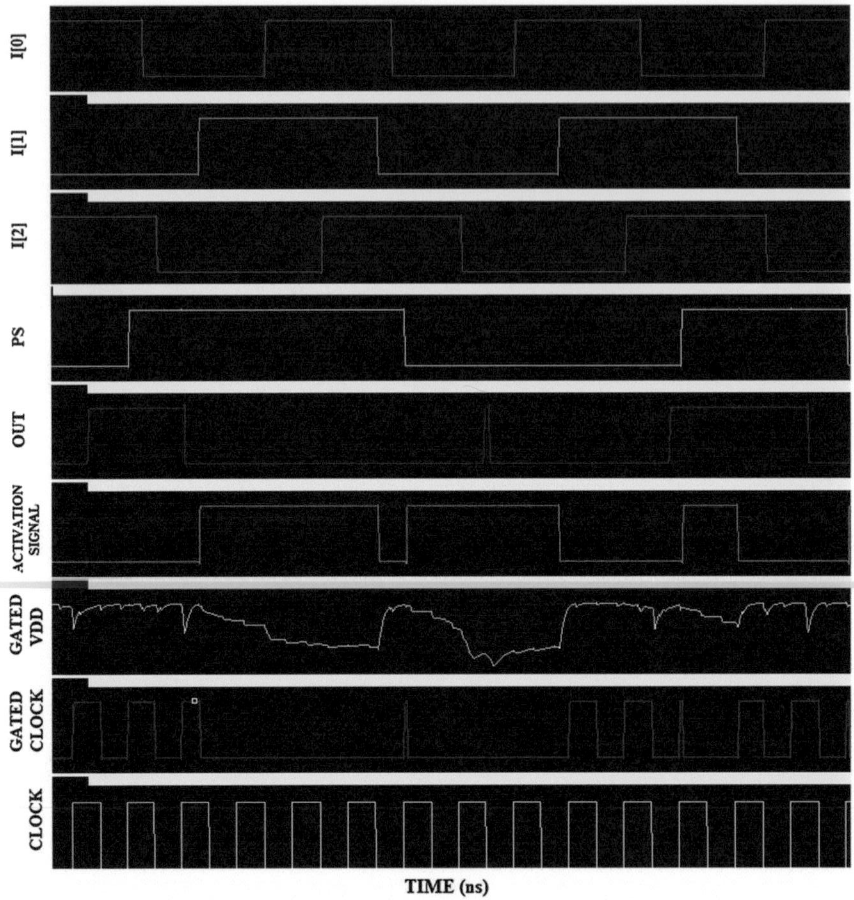

Fig. 1.10 Simulation waveform of gated EXCKT

Table 1.3 WPCG (without power and clock gating) and PCG (power and clock gating) results for "EXCKT"

	Power (uW)	Delay (ps)	Area (λ^2)
WPCG	17.81	187.9	1284 k
PCG	14.49	189.9	1332 k

well as for the introduction of new circuit elements. The increase in area after gating is only 3.7%.

The proposed gating technique in our work works at a fine level of granularity, considering power reduction of an individual TILE. Majority of the previous works in this field considered coarse-level power reduction, shutting down the activity of some portion of the FPGA at a time. Power reduction in individual TILE level can also result in significant power savings of a practical FPGA, which consists of several thousands of TILEs.

1.6 Conclusion

In this work, an efficient and autonomous power and clock gating technique is being proposed and analyzed in a transistor-level TILE architecture, designed and developed in CADENCE tool, using a technology of 45 nm. FPGA, being a regular architecture, is divided into TILEs, where each TILE compromises of a logic element and routing resources. Using an array of rows and columns of the TILEs, an entire FPGA architecture can be established. The prototype designed in this work results in correct functionality and resembles the working principle of an actual TILE. The gating logic implemented in this work is based on the knowledge of self-loops in a finite state machine and works at a fine level of granularity. The results show a reduction of approximately 19% of the total power dissipation of the TILE after implementing power and clock gating with a minimal increase in delay and area. Our proposed gating method has proven to be effective and can be generally implemented to any FSMs with sufficient self-loops. In an extension to this work, we are focusing on introducing the proposed gating logic in a larger module consisting of multiple TILEs.

Acknowledgments This work was supported by Fast Track Project (Ref. No. SB/FTP/ETA-341/2012) sponsored by Science and Engineering Research Board (SERB), Department of Science and Technology (DST), Govt. of India, New Delhi—110016.

Reference

1. "40-nm FPGA Power Management and Advantages", Link: http://www.altera.com/literature/wp/wp-01059-stratix-iv-40nm-power-management.pdf
2. Smith AM, Constantinides GA, Cheung PYK (2010) FPGA architecture optimization using geometric programming. Comput-Aided Des Integr Circ Syst IEEE Trans 29(8):1163–1176
3. Kapre N, DeHon A (2009) Accelerating SPICE model-evaluation using FPGAs. In 17th IEEE Symposium on Field Programmable Custom Computing Machines, pp 37–44
4. Henzler S (2007) Power management of digital circuits in deep sub-micron CMOS technologies. Springer, Berlin Springer Series in Advanced Microelectronics, Secaucus, NJ, USA
5. Lin Y, Li F, He L (2005) Routing track duplication with fine-grained power-gating for FPGA interconnect power reduction. In Proceedings of the 2005 Asia and South Pacific design automation conference, 2005, pp 645–650
6. Gayasen A, Tsai Y, Vijaykrishnan N, Kandemir M, Irwin MJ, Tuan T (2004) Reducing leakage energy in FPGAs using region-constrained placement. In Proceedings of the 12th ACM/SIGDA international symposium on field-programmable gate arrays, 2004, pp 51–58
7. Bsoul AAM, Wilton SJE (2010) An FPGA architecture supporting dynamically controlled power gating. In International conference on field-programmable technology (FPT), pp 1–8

8. Dev MP, Baghel D, Pandey B, Pattanaik M (2013) Clock gated low power sequential circuit design. In IEEE Conference on Information & Communication Technologies (ICT), 2013, pp 440–444
9. Kuo YM, Weng SH, Chang SC (2008) A novel sequential circuit optimization with clock gating logic. In International conference on Computer-Aided Design, ICCAD 2008, IEEE/ACM, pp 230–233
10. Man X, Kimura S (2011) Comparison of optimized multi-stage clock gating with structural gating approach. In IEEE Region 10 conference on TENCON 2011, pp 651–656
11. Li F, Chen D, He L, Cong J (2003) Architecture evaluation for power-efficient FPGAs. In Proceedings of the 2003 ACM/SIGDA eleventh international symposium on field programmable gate arrays, pp 175–184
12. Poon KKW, Yan A, Wilton SJE (2002) Flexible power model for FPGAs. 12th Int Conf Field-Program Logic Appl 2438:312–321
13. Fort B, Paladino D, Plavec F (2004) Full custom layout of an SRAM-based FPGA. University of Toronto, Toronto
14. Nag A, Pradhan S (2014) An efficient clock gating approach based on registers splitting. In International conference on VLSI and signal processing (ICVSP). IIT Kharagpur

Chapter 2
A Hardware Implementation of Evolvable Embedded System for Combinational Logic Circuits Using Virtex 6 FPGA

C. Ranjith and S.P. Joy Vasantha Rani

2.1 Introduction

The traditional method of designing digital systems was in the form of schematic or programming by hardware description languages (HDL). The above two methods require an in-depth analysis of the system under design. The manual design through schematic or programming can be sometimes complex and may lead to monotonous solutions. Another, novel method of circuit design is by evolvable hardware, wherein the circuits is evolved automatically based on some optimisation algorithm. The mode of digital design is now getting due importance from the research community and the industry. Research in this field received a boom with major chip industries innovating novel tools and chips for the application and testing of "evolvable" circuits.

Evolvable hardware (EHW) was a term coined by Hugo De Garis in the year 1992 for circuits which could configure its own hardware structure dynamically depending on the changes in the environment or on design parameters [1]. This capability of configuring the hardware was achieved by employing efficient search algorithms, like genetic algorithms (GA) [2, 3]. Evolvable systems for digital domain are mainly designed using field programmable gate array (FPGA) chips. FPGA-based EHW can be classified based on the evaluation of the solutions. The first method is recognized as *extrinsic evolution*, where the development of circuits uses a simulation approach of determining the best evaluation; such solutions are then implemented in the device. The second approach is *intrinsic evolution*, where

C. Ranjith (✉) · S.P. Joy Vasantha Rani
Electronics Engineering Department, M.I.T. Campus,
Anna University, Chennai, Tamil Nadu, India
e-mail: ranjith.kmct@gmail.com

S.P. Joy Vasantha Rani
e-mail: joy_mit@annauniv.edu

V. Nath (ed.), *Proceedings of the International Conference on Nano-electronics, Circuits & Communication Systems*, Lecture Notes in Electrical Engineering 403,
DOI 10.1007/978-981-10-2999-8_2

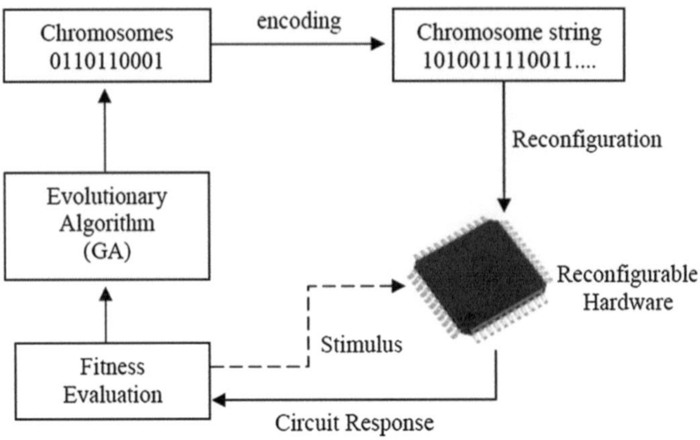

Fig. 2.1 Basic structure of EHW

each candidate solution is physically tested on the hardware device. The latter offers better accuracy and operation of self-evolved circuits [4]. The basic concept of the combination of GAs with FPGA in EHW is with regard that the configuration bit strings to the FPGA is chromosomes of the GA. The fitness function is designed, such that the GA can autonomously find the best hardware for the design to be implemented in the FPGA. Figure 2.1 depicts a simple example of the EHW concept [5]. In conventional works on EHW, the GA operations were conducted in computer or workstations, which would make the system robust and slow. In recent works, the GA operation is performed on the same FPGA chip either by hardware or by software in a dedicated core.

This paper describes the implementation of the evolvable embedded system which uses the evolutionary algorithm to dynamically modify some of the system components or parameters in order to adapt to the changing environment [2]. The use of a soft processor (MicroBlaze) for computing the optimal search algorithm (GA) and hardware architecture for evolution are integrated in a single FPGA, therefore termed *evolvable embedded system*. An evolvable embedded system architecture of combinational circuits using the intrinsic mode of evolution is discussed. The hardware was changed based on the fitness evaluation of the GA. The evolvable embedded hardware architecture was implemented on Virtex 6 (XC6VLX240T-1FFG1156) ML605 Evaluation Kit. Combinational circuits with 8 inputs and outputs could be evolved in this architecture with GA results displayed onto a PC. A 2-bit adder and multiplier circuits were evolved as an example using this architecture.

The paper is organized as follows: Section 2.2 describes the structure of EHW system, with Sect. 2.3 giving a complete view of the modeled system and its design. Section 2.4 describes the specifications and implementation process. Section 2.5 analyses the results, and final conclusion of the paper is discussed in brief.

2.2 Structure of Evolvable Embedded System

The architecture of the EHW uses the concept of the virtual reconfigurable architecture (VRA) [6]. A hardware description of the architecture was layered over the reconfigurable chip, to implement the evolutionary structure. The GA program was fused in the MicroBlaze soft processor where the computed fitness measure was displayed on the PC through a configured UART peripheral. The complete structure is as shown in Fig. 2.2.

2.2.1 *Study of VRA*

VRA is modeled in HDL and is taken as a second reconfigurable layer on the FPGA. The primary advantage of this concept is to provide a simple and efficient mode of intrinsic evolution [6]. Figure 2.3 shows the VRA structure implemented for the design of evolvable combinational circuits. An array of HDL-defined, configurable cells are arranged in rows and columns, where each cell's input is connected to the outputs of the two previous columns with the exception of the first-array column which is connected to the inputs and its invert as shown in Fig. 2.3. The number of rows and columns selected depends on the complexity of the problem. Configuration bits from the GA provide the connectivity and logic based on the 16-bit input combinations to the cell. The outputs are checked for problem logic (truth table conditions) to determine the fitness criteria. The VRA approach is widely employed in the implementation of EHW systems as the configuration bit formats of the FPGA are company proprietary. The VRA concept is similar to the Cartesian genetic programming (CGP) and provides other benefits, including (1) array of configuration cells directly connected to the hardware of the GA in the same FPGA, making the communication faster; (2) the VRA is modeled as an HDL source code which makes it easier to modify and synthesize in other

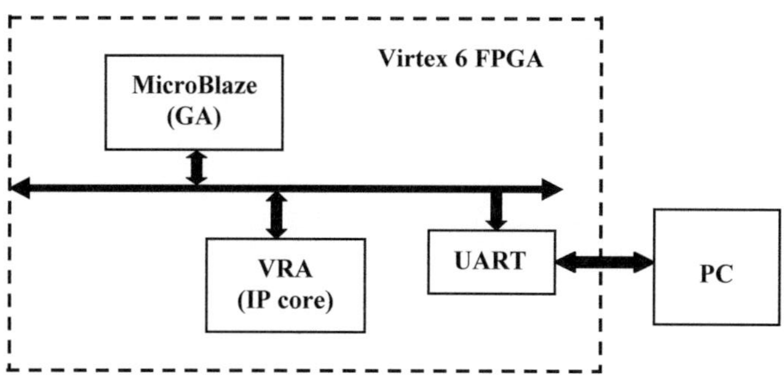

Fig. 2.2 Structure of evolvable embedded system

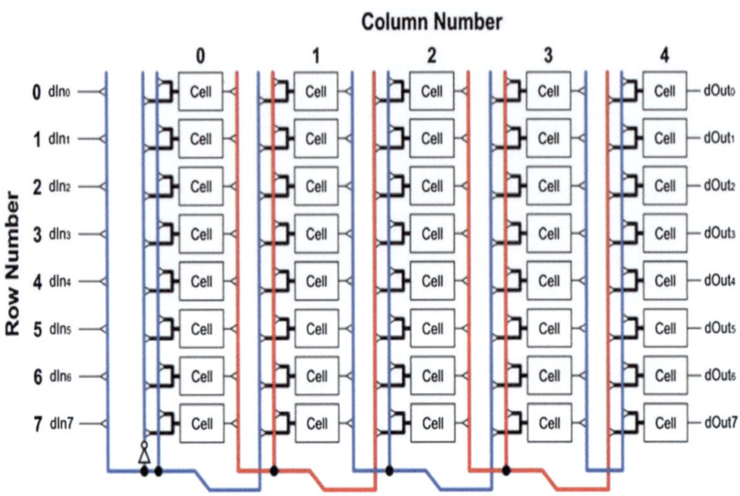

Fig. 2.3 An 8 × 5 VRA structure

target platforms; and (3) The VRA architecture modeled can be utilized for similar set of problem definitions.

2.2.2 *Genetic Algorithm Flow*

In this paper, we make use of a simple GA to search for the best solution [7]. The algorithm runs on the principle of population size of individuals (candidate) solutions to the optimization problem. These individuals consist of a string (chromosomes) of genes (genotype). The genotype is encoded to produce the configuration set (phenotype). The operations of GA include selection, reproduction, crossover, and mutation on the individuals to get a better solution [3]. The initial step is to prepare a chromosome format for the given problem, the length of which corresponds to the total number of decision variables in the search. Each of these chromosomes is passed on their fitness probability and the best fit chromosomes are selected. The best chromosome strings from the pool of populations undergoes genetic operations like crossover, mutation, and selection. New populations are generated at each iteration and checked for their fitness to produce new solutions. The cycle repeats till an optimized and best solution is found. The complete flow diagram of GA process is shown in Fig. 2.4.

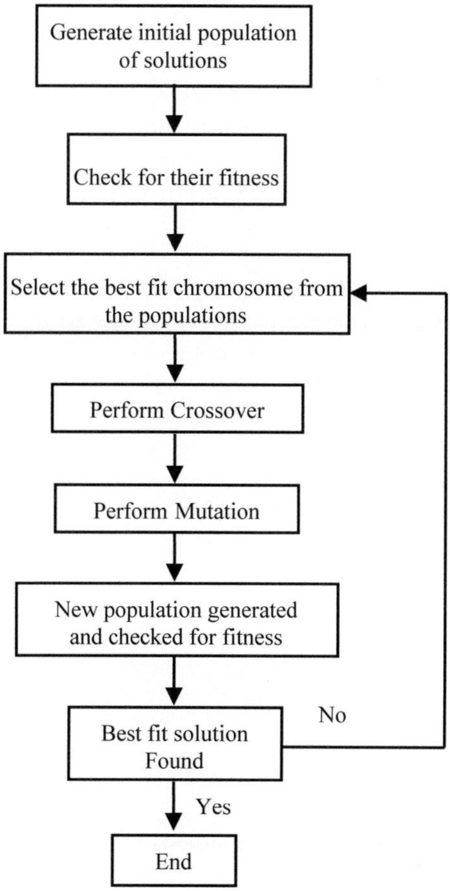

Fig. 2.4 Flow graph of simple GA structure

2.2.3 *Reconfigurable FPGA Chip*

The evolvable embedded architecture is implemented on a ML605 Evaluation board with Virtex 6 FPGA. The Virtex-6 family is built on a 40-nm process for high computational electronic systems. This series of FPGAs have integrated features that include DSP blocks, PCI-Express controllers, Ethernet MAC blocks, and high-speed transceivers [8]. The use of Virtex 6 FPGA was mandatory due to simple and flexible implementation of a 32-bit soft core MicroBlaze processor [9]. This processor computes the GA, therefore faster evaluation time can be achieved. The main advantage is that the complete EHW process can be implemented on a single chip, thereby making area efficient and flexible.

2.3 Evolvable Embedded System Design

This section gives an overall view of the complete evolvable embedded system design. The VRA architecture of combinational circuits is modeled and coded by a hardware description language. The architecture of the VRA is modeled as shown in Fig. 2.5. The heart of the architecture unit is named as configuration *cell*, consisting of three 16:1 multiplexers and an 8 × 1 bit RAM. This unit offers logical function and interconnectivity for the system under design. Three multiplexers are used to select the inputs to the lookup table (LUT), which are driven from one of the set of sixteen inputs. The configuration cell is driven by 20 configuration bits (3 * 4 select lines of MUX + 8 selectable lines to RAM). The configuration cells are interconnected in '*m*' rows and '*n*' columns to form a *Cell Array*. A simple interconnection principle is followed where each cell input is connected to the output of the two columns, except for the first cell array column, which are connected by the inputs and its invert. This interconnection principle simplifies the cell-to-cell routing. For the design, an 8 × 5 cell array was selected, with a prospect of evolving simple combinational circuits with a maximum of 8 inputs and outputs. Here, a 400 configuration bits (40 cells * 20 configuration bits) were required to perform the complete logical functions and interconnection between cells. The configuration bits are provided by the GA taking the fitness function criteria. A reconfigurable hardware is tested using a hardware setup of simple logical gates to check the functionality of the design.

The major advantage of performing fitness evaluations in hardware was that multiple evaluations in parallel could be achieved easily. In this design, we have combined 8 cell arrays into a single block. The block contains extra circuitry to help determine the fitness of an individual. Figure 2.6 shows one such cell array with the additional circuits to perform a fitness evaluation. The RAM is used to store the

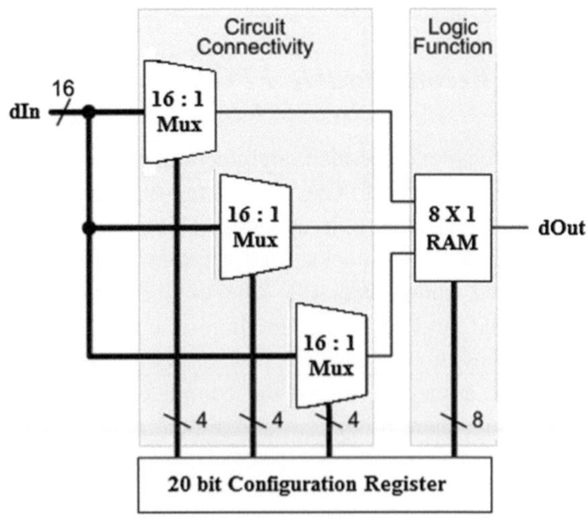

Fig. 2.5 Architecture of the cell

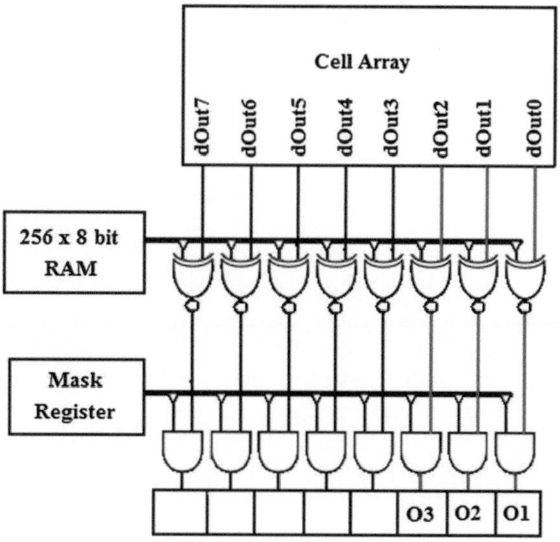

Fig. 2.6 Single cell array with fitness check connections (showing the outputs for a 2-bit adder circuit)

truth table of the target circuit (in this case, a 2-bit adder/multiplier). The output bits of the cell array after configuration bit selection are compared with the correct values of the 8-bit RAM outputs, using XNOR gate (last 3 bits are the truth table entries of Cout, S0, S1 for a 2-bit adder), and final values are masked by keeping one input of AND gate to high depending on the number of outputs of the evolved circuit. Match outputs are obtained (O1, O2, O3) as shown for a 2-bit adder circuit. Here, O1, O2, O3 symbolizes Cout, S0, S1 of a two-input adder. The same could be employed for a 2-bit multiplier where 4 outputs are valid, so the last 4 AND gates input are held high.

The evolutionary system was developed using Xilinx Platform Studio (XPS) tool. An embedded system with both hardware and software elements was created in the EDK (Embedded Development Kit) and SDK (Software Development Kit), respectively; a 32-bit MicroBlaze soft core processor (for performing the GA) with a clock frequency of 50 MHz and an 8-cell array VRA core structure (for evaluating candidate solutions). The system also includes a configured serial UART (universal asynchronous receiver/transmitter) to send GA computations to the PC. The cell array VRA block was imported as an IP core (intellectual property) and integrated as a peripheral to the Microblaze processor. The IP core was plugged into the processor bus using the generated macros in situ with an interface program [9]. The system netlist (ngc files) and bitstream (bit files) are generated from the EDK. The C program for the GA optimization was hand-coded, encoding the inputs and outputs of the cell array. The input and output cell array configuration for a 2-bit adder and 2-bit multiplier is as shown Fig. 2.7a, b. The GA parameters such as crossover rate, mutation rate, and number of selections are assigned by trial and error. The evaluation is based upon minimum iterations to converge to a better fit for the optimization of the circuit. The program calls for

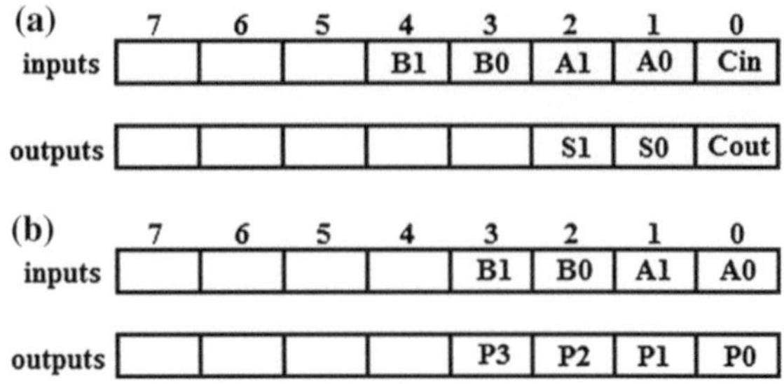

Fig. 2.7 Input and output cell array configuration **a** 2-bit adder, **b** 2-bit multiplier

additional header files, to hold parameters required to communicate the application program with the devices to the embedded system. This application program creates an executable and linkable format file (elf file) for the program to be realized by the FPGA. This file along with the bit file is integrated to be programmed into the FPGA (ML605) through a JTAG cable.

2.4 Implementation

The VRA architecture was coded in VHDL and synthesized in Xilinx ISE Design Suite 14.6. An inbuilt MicroBlaze soft processor having 32 bit, Reduced Instruction Set Computer (RISC) architecture is utilized for modeling the system [9]. The MicroBlaze processor with its peripherals for the design include UART (for communication with PC), 4 seven-segment LEDs, push buttons for inputs and reset options, realized from XPS 14.6 EDK platform tool. The peripherals are linked to the processor via the processor local bus (PLB). The synthesized VRA module files are exported to realize as a user IP core. This IP core is connected to the processor via the PLB. The generated macros are modified to interconnect the user IP core to the PLB bus. The complete embedded processor structure with its peripherals realized as block schematic is presented in Fig. 2.8. The microprocessor hardware specification file (mhs file) and peripheral analysis order file (pao file) has to be updated to include the changes accommodated by the addition of new IP core, along with the creation of black box definition file (bbd file). Once the following peripherals are connected, new netlist files (*system.ngc*) and bit files (*system.bit*) are generated.

The hardware system created in EDK is exported to SDK tool to develop the GA program to operate in association with the hardware. A simple GA program with

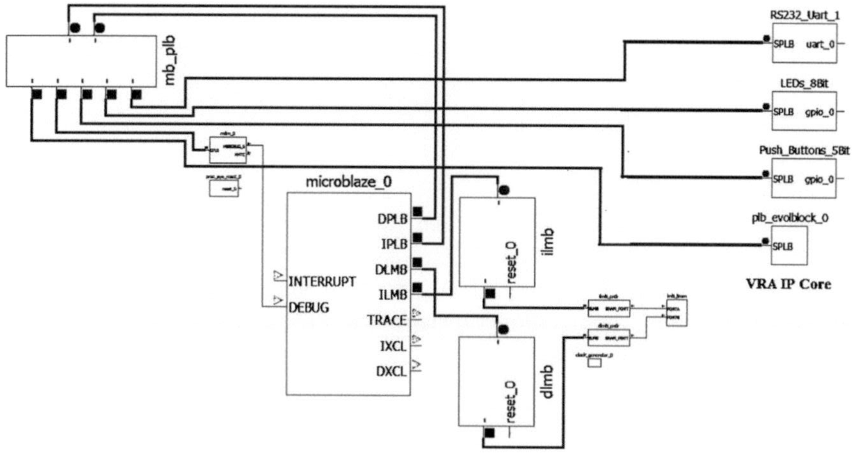

Fig. 2.8 MicroBlaze processor with peripheral interconnections (viewed through XPS tool)

optimization for the evolution of circuits is coded in C language. The GA program is developed with the following subroutines:

(i) The truth table of the target circuit to be stored in the 256 × 8 bit RAM. Here, 2-bit adder/multiplier truth tables are initially stored occupying the first 16 locations of the RAM, whereas the rest is default taken to be nil.
(ii) Configure the input and output registers of the cell array with the bit placing as shown in Fig. 2.7a, b. For a 2-bit adder, the first 5 bits of the input register represent the 2-bit input, and a carry (Cin, A0, A1, B0, B1), whereas the first 3 bits of output register represent the outputs (C0, S0, S1) of the 2-bit adder circuit. Similar logic holds for a 2-bit multiplier, having of 4 inputs (A0, A1, B0, and B1) and generating 4 outputs (P0, P1, P2, and P3).
(iii) A simple GA program to find a best fit configuration is to be applied to the cell. The fitness function is the truth table of the 2-bit adder/multiplier circuit. The fitness function can be calculated as:

$$\text{fitness} = 2^i, \quad \text{where } i = \text{inputs of the system} \tag{2.1}$$

From the criteria, the fitness of the 2-bit adder circuit is summed to 96 (sum of the fitness of C0, S0, S1), and that of 2-bit multiplier to be 64. The GA program is developed and tested for the following parameters:

No. of iterations: 500
Population size: 128 with 16 sets of population for each cell array (8 cell arrays in parallel)
Crossover rate: 55%
Mutation rate: 0.1%
Selection method: Tournament selection.

The GA program communicates with the devices of the embedded system through the header files provided by the vendor. Two header files, namely *xparameters.h* and *xgpio.h,* holding parameters for the communication are added to the application program. The *.elf* file (software) from the SDK and the *system.bit* file (hardware) from the EDK are combined to the Virtex 6 FPGA (XC6VLX240T-1FFG1156). The combinations of these two files integrate to form *download.bit.* This bit files are used to program the FPGA through the JTAG cable.

2.5 Results

The results are monitored on a PC connected through the UART peripheral. The results display the GA program evaluating the fitness calculations for a 2-bit adder/multiplier circuits. The program is terminated once the fitness is evaluated and displays the number of generations and configurations for each cell array as shown in Fig. 2.9a, b for 2-bit adder and multiplier, respectively. From the results, it can be analyzed that a 2-bit adder was evolved over 88 generations, and 135 generations were required to evolve a 2-bit multiplier. The number of generations differs with the number of inputs and outputs of the system. The experiments were repeated for different values of crossover rate, mutation rate, and selection criteria.

Optimized results were obtained for crossover rate = 55%, mutation rate = 0.1%, and tournament selection of 5 individuals at a time. The hardware interconnections could be analyzed from the configurations of the individual cells. The complete floor plan of the evolvable embedded system for the combinational circuit using the VRA concept is shown in Fig. 2.10 using the PlanAhead tool. From the figure, it can be seen that around 40% of the implemented chip area were occupied by the MicroBlaze processor and its peripherals and the rest, by the VRA IP core. From the total chip area, around 6% of LUTs, 4% IOBs and 3% of registers were used from the available resources as tabulated in Table 2.1. The timing criteria were met with an implementation time of 17.063 ns (maximum frequency of 58.606 MHz).

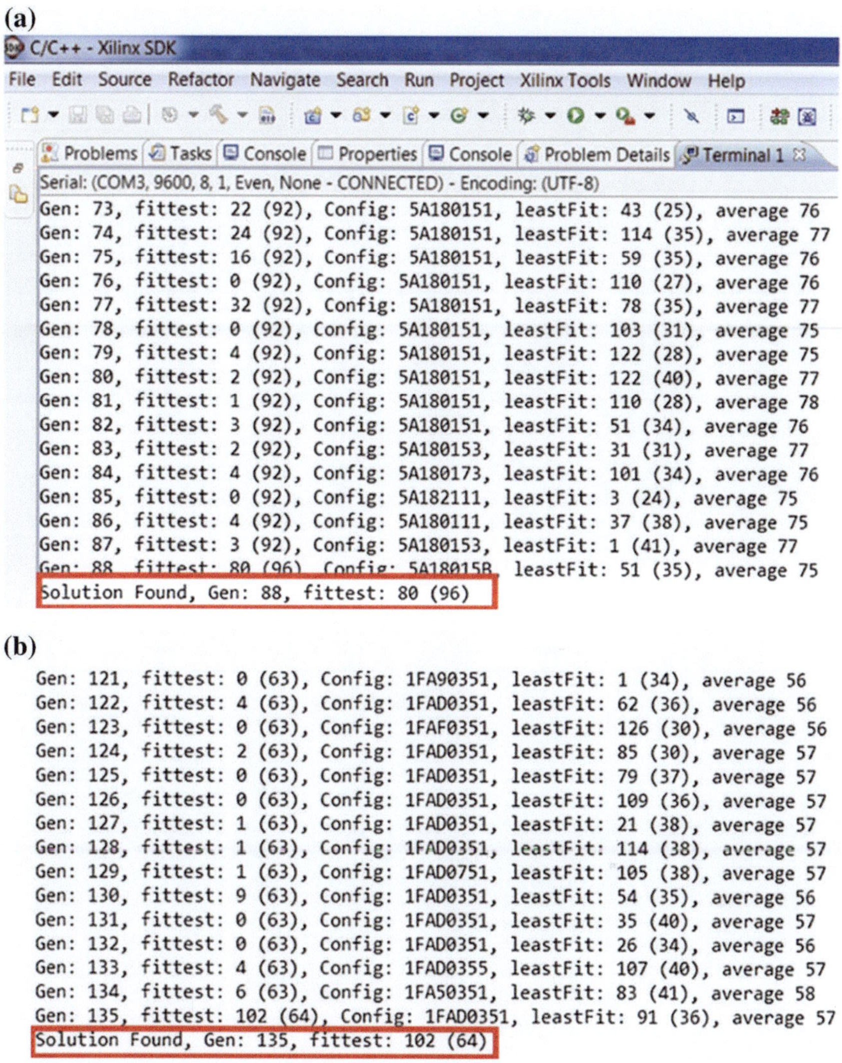

Fig. 2.9 GA fitness evaluation results **a** 2-bit adder **b** 2-bit multiplier

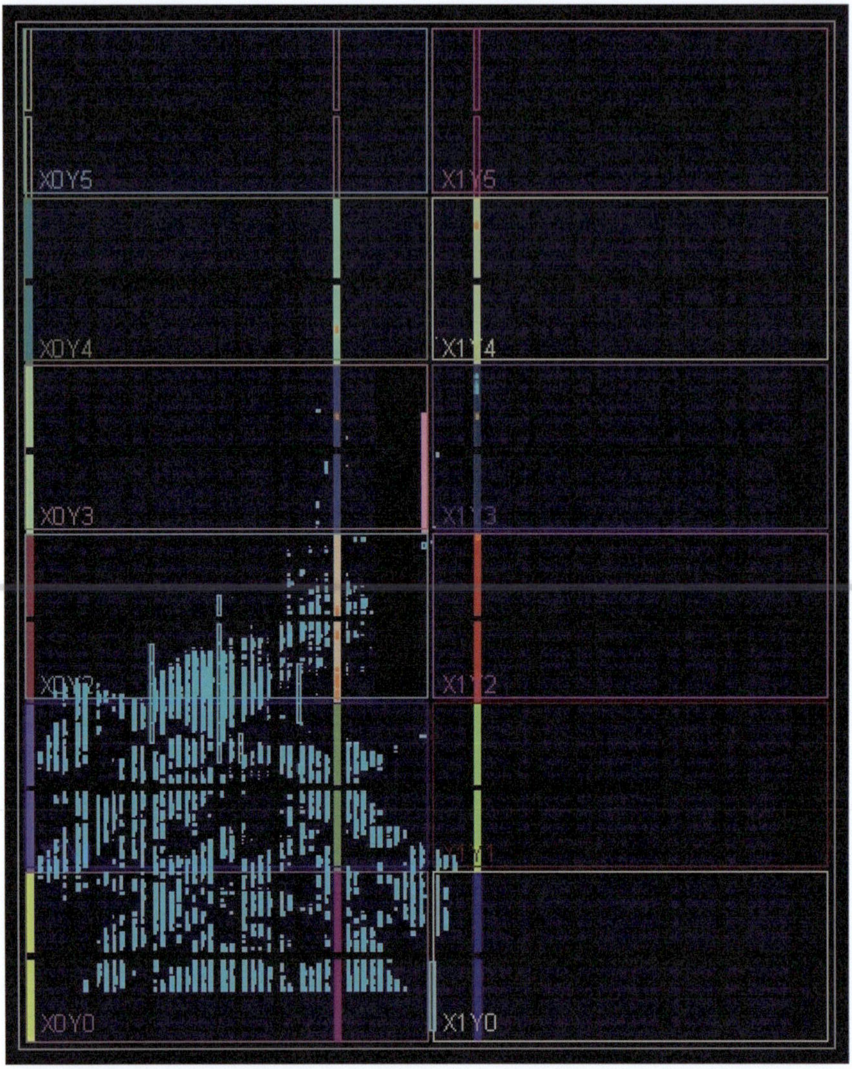

Fig. 2.10 Complete floor plan of the evolvable embedded system

2.6 Conclusions

The concept of the evolvable embedded systems is introduced through this paper which could provide automation of complex circuits like filters, fault tolerance circuits, and adaptive circuits. The complete evolvable hardware process can be conceptualized in a single FPGA, in which evaluation and evolution are performed in parallel in the same FPGA. The MicroBlaze soft processor is used for the

Table 2.1 Total resource utilization of Virtex 6 (XC6VLX240T-1FFG1156) for evolvable embedded system structure

Description	Available	Used	Utilization (%)
No. of slice registers	301,440	8413	3
No. of slice LUTs	150,720	8496	6
No. of bounded IOBs	600	22	4
No. of RAMB36E1	416	16	4
No. of RAMB18E1	832	1	1
No. of DSP48E1	768	3	1

computation of genetic algorithm and communication link between the PC and FPGA system. Optimized utilization of FPGA resources was created by selecting optional peripherals for the CPU. The computation of the GA program in the processor has speeded up the evaluation process compared to the conventional use of evaluating through the PC. The evaluation time could still be accelerated up by programming the FPGA through a Compact Flash card rather than a JTAG cable. Here, the bitstreams are stored on a Compact Flash and interconnected to the processor using the SysACE peripheral.

Here, an evolvable embedded system was implemented on Virtex 6 (XC6VLX240T-1FFG1156) ML605 evaluation kit. The GA performed 88 and 135 generations, to evaluate the fitness of the 2-bit adder and multiplier circuits, respectively. The optimized generations were obtained after a manually varying the GA parameters. The system evolved through this process give novel or optimized systems. The advantage of this architecture is that any combinational circuit whose inputs and outputs not exceeding 8 bits could be evolved from the same architectural model. The GA program can also be programmed to adaptively change the parameters so as to optimize the number of generations effectively. This could solve the problem of manually varying the GA parameters for optimization. Memetic algorithm (MA) is another optimization algorithm which could be employed in the evolutionary process to obtain faster and better convergence [10]. This optimization process is still under research, but able optimize circuits with large number of inputs or circuits which require large search space.

References

1. Lambert C, Kalganova T, Stomeo E (2007) FPGA-based systems for evolvable hardware. World Acad Sci Eng Technol 1:743–749
2. Sekanina L, Drabek V (2004) Theory and applications of evolvable embedded systems. In: Proceedings of the 11th IEEE computer-based systems (ECBS'04). IEEE Computer Society Press, Aug 2004
3. Stomeo E, Kalganova T, Lambert C (2006) A novel genetic algorithm for evolvable hardware. In: IEEE congress on evolutionary computation, Vancouver, Canada, July 2006, pp 134–141

4. Bartolini DB, Cancare F, Carminati M, Sciuto D (2011) HERA: hardware evolution over reconfigurable architectures. In: 1st international workshop on computing in heterogeneous, autonomous 'N' goal environments (CHANGE 2011), March 2011, pp 1–6
5. Vasicek Z, Sekanina L (2007) An evolvable hardware system in Xilinx Virtex II Pro FPGA. Int J Innov Comput Appl 1(1):63–73
6. Wang J, Chen QS, Lee CH (2008) Design and implementation of a virtual reconfigurable architecture for different applications of intrinsic evolvable hardware. IET Comput Digital Tech 2(5):386–400
7. Torreson J (2004) An evolvable hardware tutorial. In: Field programmable logic and applications. Lecture notes in computer science, vol 3203, pp 821–830
8. Xilinx (2012) ML605 hardware User Guide, Application UG534, Oct 2012
9. Jesman R, Vallina FM, Saniie J (2006) MicroBlaze tutorial creating a simple embedded system and adding custom peripherals using Xilinx EDK software tools. Embedded Computing and Signal Processing Laboratory, Illinois Institute of Technology
10. Mo H, Meng L (2013) Research on evolution hardware design based on memetic algorithm. In: IEEE workshop on memetic computing, April 2013, pp 32–36

Chapter 3
Evaluation of Wavelet-Based Speech Codecs for VoIP Applications

Manas Ray and Mahesh Chandra

3.1 Introduction

Voice over Internet Protocol (VoIP) is a pioneer technology in the modern communication world that allows delivery of voice calls over packet-switched network like broadband Internet connectivity. This technology utilizes the existing data communication infrastructure to deliver the voice packets. The current challenges faced by VoIP includes relatively high bandwidth requirement, traffic congestion leading to propagation delay, network delay variations, and excessive delay [1]. To minimize these limitations of VoIP, an efficient speech compression technique is desired. The speech signal is required to be compressed so as to conserve the precious resource of bandwidth.

In the recent years, wavelet transforms and its applications are being extensively studied in the field of signal processing. Wavelet transform provides excellent resolution in frequency as well as in time domain [2]. The wavelet transform represents the signal with very high precision and limited storage requirements [3]. The wavelet is defined as limited waveform having zero average value. It is finite in nature. The multi-resolution capability of the wavelet provides us with dilate and translate versions of the wavelet [4]. The resolution of the analysis is determined by the scaling function, and the analysis is performed by the mother wavelet function $\Psi(k)$ [3]. Wavelet transform is calculated by the convolution of original signal $s(k)$ and the mother wavelet function $\Psi(k)$ as defined as follows [5]:

M. Ray (✉) · M. Chandra
Department of Electronics and Communication Engineering, Birla Institute of Technology Mesra, Ranchi 835215, India
e-mail: manas_ray.me@rediffmail.com

V. Nath (ed.), *Proceedings of the International Conference on Nano-electronics, Circuits & Communication Systems*, Lecture Notes in Electrical Engineering 403,
DOI 10.1007/978-981-10-2999-8_3

$$\begin{aligned} W_{\Psi}(m,n) &= \int_{-\infty}^{\infty} s(k)\Psi'_{mn}(k)\mathrm{d}k \\ &= \frac{1}{\sqrt{m}} \int_{-\infty}^{\infty} s(k)\Psi\left(\frac{k-n}{m}\right)\mathrm{d}k \end{aligned} \tag{3.1}$$

where $s(k)$ is the original signal, 'm' is the scaling factor, 'n' is the translation parameter, and $\Psi(k)$ is the mother wavelet. The wavelet function is given by

$$\Psi_{m,n}=\frac{1}{\sqrt{m}}\Psi\left(\frac{k-n}{m}\right) \tag{3.2}$$

The discrete version of the continuous wavelet transform (CWT) with dyadic grid parameters of translation $n = p$ and the scale $m = 2^j$ and the mother wavelet is defined by

$$\Psi(x) = 2^{j/2}\Psi(2^j x - p) \tag{3.3}$$

Similarly, the scaling function is defined as follows:

$$\phi(x) = 2^{j/2}\phi(2^j x - p) \tag{3.4}$$

The original function $f(x)$ can be obtained from the scaling and the wavelet functions from the [6]:

$$f(x) = \sum_{p=-\infty}^{\infty} c_p\phi_p(x) + \sum_{p=-\infty}^{\infty} d_{j,p}\Psi_{j,p}(x) \tag{3.5}$$

where C_p are the average coefficients, and $d_{j,p}$ are detail coefficients.

3.2 Various Wavelet Families

In this paper, we evaluate the following wavelet families Haar, Daubechies, Discrete approximation of Meyer wavelet (dmey) and Coiflets. Each of these wavelet families is defined as follows:

(a) Haar Wavelet (Haar)

Haar wavelet is the simplest possible wavelet [7]. For a signal represented by 2^t values, the wavelet transform recursively provides the difference and forwards the sum to the next level, resulting in $2^t - 1$ differences and one total summation. It is not continuous. The wavelet function $\Psi(x)$ is defined as follows

$$\Psi(x) = \begin{cases} 1 & 0 \le t \le \frac{1}{2} \\ -1 & \frac{1}{2} \le t \le 1 \\ 0 & \text{Otherwise} \end{cases} \tag{3.6}$$

The scaling function is defined as follows

$$\Phi(x) = \begin{cases} 1, & 0 \le t < 1 \\ 0, & \text{Otherwise} \end{cases} \tag{3.7}$$

(b) Daubechies Wavelet

Daubechies wavelets are orthogonal wavelets, having largest number of vanishing moments for some support and are commonly used for the analysis of a signal. Here, the scaling and the wavelet functions are not defined [8]. The number of coefficients generated is defined by the index N of the coefficients, and the number of vanishing moments is $N/2$ [3].

(c) Discrete Approximation of Meyer Wavelet (dmey)

The discrete format of the Meyer wavelet function is defined as follows

$$G_o\left(e^{j\omega}\right)\sqrt{2}\sum_{K} \Phi(2\omega + 4k\Pi) \tag{3.8}$$

Given the basis function 'Φ', DTFT techniques are employed to obtain the scale coefficients [9].

(d) Coiflet Wavelet

Coiflets are wavelets having scaling functions with vanishing moments. The wavelet is near symmetric and has $N/3$ vanishing moments, and scaling function has $N/3 - 1$ vanishing moments [3]. If the taps $N = 6p$, then $2p$ number of vanishing moment conditions are imposed on wavelet function and $2p - 1$ on scaling function and the remaining on normality and orthogonality conditions.

Thus, the conditions imposed are as follows [10]:

$$\int \phi(k)\mathrm{d}k = 1 \tag{3.9}$$

$$\int \phi(k)\phi(k-l)\mathrm{d}k = \delta_{0,l} \tag{3.10}$$

$$\int k^n \Psi(k)\mathrm{d}k = 0 \quad \text{for } n = 0, 1, 2, \ldots 2p - 1 \tag{3.11}$$

$$\int k^n \phi(k)\mathrm{d}k = 0 \quad \text{for } n = 0, 1, 2, \ldots 2p - 1 \tag{3.12}$$

3.3 Speech Signal Processing Using Wavelet Transform

The speech signal processing or compression by wavelet transform is performed by choosing a particular wavelet function. The speech quality requirements of the codec govern the selection of the wavelet function for the analysis. The objective of the processing is to maximize the signal quality and minimize reconstructed error variance [11]. Wavelets decompose a signal into components of different frequency bands called as resolution. The signal compression is achieved by reconstructing the signal by considering a limited set of approximation coefficients and some detail coefficients. This is done by the process of thresholding, wherein coefficients falling below a threshold value are ignored and made equal to zero [11]. The signal is reconstructed by performing inverse wavelet transform using the coefficient values which are above the threshold values. Generally, 5-level decomposition is adequate for speech signals [12]. Figure 3.1 [13] shows the process of the speech signal processing for the purpose of compression using the wavelet transform technique.

3.4 Performance Evaluation Parameters

The speech codecs based on the above-defined families of wavelets are implemented in MATLAB for the simulation purpose. The acceptability of the performance of the wavelet-based speech codec for VoIP application is gauged by the subjective testing of mean opinion score (MOS), wherein the original signal and re-constructed signal are presented to a user, who then provide a performance rating between 1 and 5, where 5 is excellent grade [14]. Further, the performance evaluation of the wavelet-based codec is carried out by objective testing of the speech samples. The tests were carried out by comparing the performance in terms of compression ratio (CR), SNR, NRMSE [12, 13], and retained signal energy (RSE). The expressions of these parameters are given below.

$$\mathrm{CR} = \frac{\text{Length of } (o(k))}{\text{Length of } (p(k))} \tag{3.13}$$

where

$o(k)$ is the input signal
$p(k)$ is the re-constructed signals, respectively

$$\mathrm{SNR} = 10\log_{10}\left(\frac{\sigma_x^2}{\sigma_e^2}\right) \tag{3.14}$$

where σ_x^2 and σ_e^2 are mean square of the input signal and the mean square difference between the input and re-constructed signal, respectively.

Normalized root mean square error (NRMSE) is given by

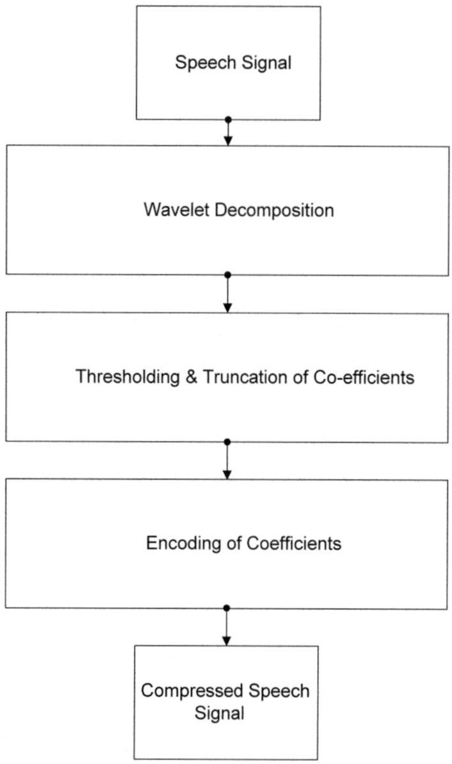

Fig. 3.1 Speech compression using DWT

$$\text{NRMSE} = \sqrt{\frac{(o(n) - p(n))^2}{(o(n) - \mu o(n))^2}} \tag{3.15}$$

where

$o(n)$ is the original input signal,
$p(n)$ is the signal, re-constructed and
$\mu o(n)$ is the mean of the original signal.

Retained signal energy (RSE) [15] is defined as follows

$$\text{RSE}\,(\%) = \frac{\|o(n)\|^2}{\|p(n)\|^2} \times 100 \tag{3.16}$$

where

$\|o(n)\|$ is the original signal norm
$\|p(n)\|$ is the norm of the re-constructed signal.

Table 3.1 Details of sample sentences used in the experiment

Sample no.	Language	Speaker
1	Hindi	Male
2	Hindi	Male
3	English	Male
4	English	Female
5	Hindi	Female
6	Hindi	Female

3.5 Results

Discrete wavelet transform-based codec is simulated in MATLAB based on the speech compression principle adopted in wavelet transforms. The test sentences as presented in Table 3.1 are iterated against each of the set of 4 different wavelet families, viz. Haar, Daubechies, dmey, and Coiflet wavelets.

The speech signal is decomposed into 5-level approximation and detail coefficients. A global threshold value is used for the decomposition of signal. The quality of the signal was measured based on MOS, SNR, RSE, and compression ratio.

The results are shown in the following figures. Figure 3.2 shows the comparison of the wavelets in terms of the MOS, Fig. 3.3 compares the wavelets in terms of the compression ratio, Fig. 3.4 compares the wavelets in terms of SNR, and Fig. 3.5 shows the comparison in terms of the RSE %.

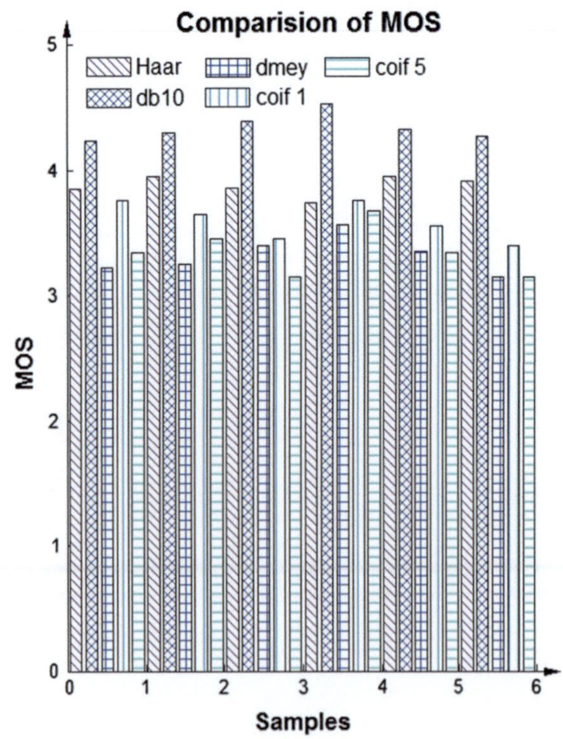

Fig. 3.2 Comparison of MOS of wavelets

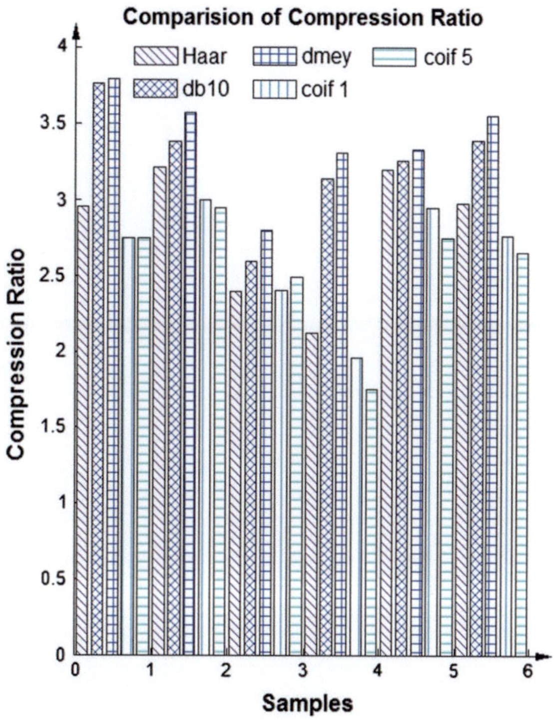

Fig. 3.3 Comparison of compression ratio of wavelets

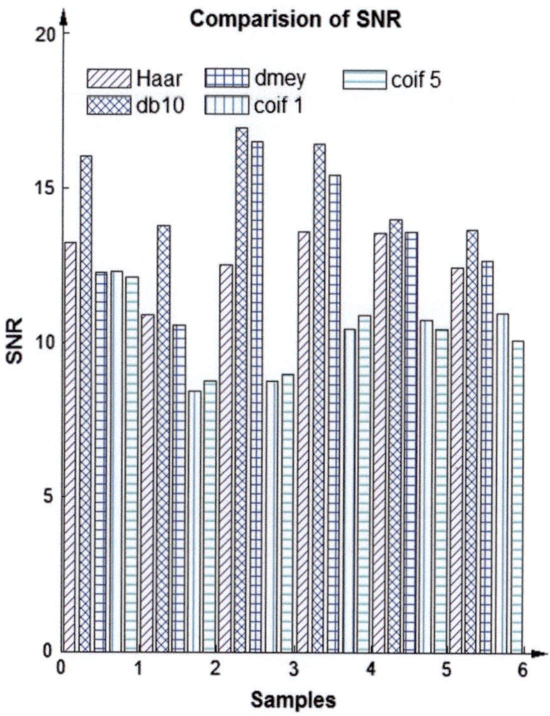

Fig. 3.4 Comparison of SNR of wavelets

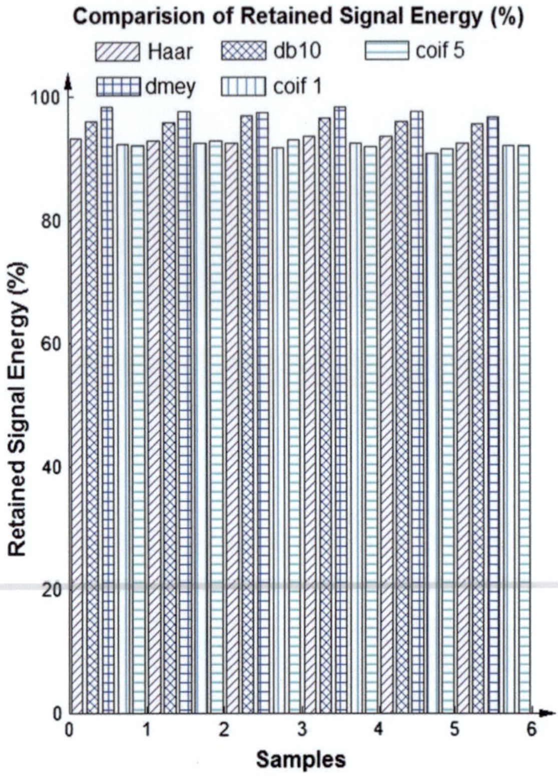

Fig. 3.5 Comparison of retained energy

It can be seen from the above that dmey wavelets provide excellent results in terms of % energy retention and compression ratio, followed by Daubechies family. Further wavelets of Daubechies family provide better degree of performance in terms of the MOS and SNR of the signal.

3.6 Conclusions

Wavelet-based speech coding, in general, offers a good degree of compression of the speech signal, whose magnitude can be varied easily. The Haar wavelet transform is the straight forward and fastest transform to be used for speech compression. However, due to its discontinuity, it is not advantageous for the simulation of speech signals. Daubechies wavelet has shown its superiority over other families of wavelet for speech compression in terms of all parameters such as % compression and SNR value and hence extensively used in various speech processing applications. Further, it can be inferred from the above that the average MOS of wavelet-based speech codec is in the range of 3.9–4.5, which is near toll quality; hence, they compare well with the currently deployed speech codec in the VoIP applications. The results further reveal that the performance of codec under study remains unaffected with change in language or speakers.

References

1. Ray M, Chandra M, Patil BP (2012) Evaluation of CDMA microwave links at different environments for VoIP applications. Int J Adv Res Comput Commun Eng (IJARCCE) 1 (8):508–512
2. Karam J (2007) Various speech processing techniques for speech compression and recognition. In: Proceeding of world academy of science, engineering and technology, vol 26, pp 209–213, ISSN 1307-6884
3. Somani KP, Ramchandran KI, Resmi NG (2009) Insights into wavelets from theory to practice. Prentice Hill International
4. Daubechies I (1998) Ortho-normal bases of compactly supported wavelets. IEEE Commun Pure Appl Math 41:906–996
5. Debdas S, Jagrit V, Chandrakar C, Quereshi MF (2011) Application of wavelet transform for speech processing. Int J Eng Sci Technol (IJEST) 3(8):6666–6670
6. Agbinya JI (1996) Discrete wavelet transform techniques in speech processing. IEEE TENCON-Digital Signal Process Appl 514–519
7. http://en.wikipedia.org/wiki/Haar_wavelet
8. http://en.wikipedia.org/wiki/Daubechies_wavelet
9. Kornsing S, Srinonchat J (2012) Enhancement speech compression technique using modern wavelet transforms. IEEE Int Symp Comput Consum Control 393–396
10. http://en.wikipedia.org/wiki/Coiflet
11. Kinsner W, Langi A (1993) Speech and image signal compression using wavelets. In: IEEE Wescanex Conference Proceedings, pp 368–375
12. Ambika D, Radha V (2012) A comparative study between discrete wavelet transform and linear predictive coding. In: IEEE world congress on information and communication technologies, pp 965–969
13. Joseph SM (2010) Spoken digit compression using wavelet. IEEE international conference on signal and image processing, pp 255–259
14. Ray AK, Acharya T (2004) Information technology: principles and applications. Prentice-Hill of India Private Limited, New Delhi
15. Nan L, China J (2009) A new adaptive threshold algorithm to speech enhancement based on minimum description length criterria. In: IEEE international conference on information engineering and computer science, pp 1–9

Chapter 4
User Demand Wireless Network Selection Using Game Theory

L. Rajesh, K. Boopathy Bagan and B. Ramesh

4.1 Introduction

In the current generation and future generation, the usage of the Internet for various applications increases, which in turn increases the data traffic. To provide a continuous connection to the users, different operators started to implement radio access technology in the overlapping coverage area of different radio access networks such as WLAN, WiMAX, and UMTS. So that user can get better seamless connectivity and QoS for various applications and the operators can increase their revenue. These radio access networks differ in the available bandwidth, operating cost, coverage area, and capacity. The main challenge is to provide seamless connectivity for various multimedia applications with the highest quality of service to the user and to select the best radio access network available in a heterogeneous network environment. The access network selection algorithm helps to select best available network based on a utility function score calculated with energy, quality, and cost which are considered as the main parameter. In order to ensure a profit for both the user and network, a repeated game theory is used in which the user–network interaction is considered. The wireless networks considered here are WLAN and WiMAX. WLAN provides good throughput, and more reliability provides a throughput of 54 Mbps. WiMAX provides a higher coverage area, provides mobility support, and achieves data rate of 1 Gbps for stationary user and 100 Mbps for mobile user. In this paper, a decentralized network selection method is used; that is, the decision criteria are mainly user preference. The user preference decision criteria are mainly based on the energy conservation needs, cost, and the

L. Rajesh (✉) · K. Boopathy Bagan · B. Ramesh
Department of Electronics Engineering, MIT Campus,
Chromepet, Chennai, India
e-mail: virudhairaj@gmail.com

B. Ramesh
e-mail: rameshb089@gmail.com

V. Nath (ed.), *Proceedings of the International Conference on Nano-electronics, Circuits & Communication Systems*, Lecture Notes in Electrical Engineering 403,
DOI 10.1007/978-981-10-2999-8_4

service quality expectations of the user. The user preference plays an important role in the network selection decision by giving certain weight for each decision criteria. Game theory is a mathematical tool used to evaluate the competitive situations which imply the interaction of rational decision makers with mutual and possibly conflicting interests.

For network selection mechanism, different methods have been proposed by the researchers. The existing solutions are divided as multiattribute decision-making (MADM) methods and game theory-based network selections. Wang et al. [1] proposed SAW method in which the network selection score function is defined as the sum of the weighted normalization of the parameters considered. One of the main drawbacks of the SAW method is a low value of the parameter that can outweigh the high value of another parameter. Bakmaz et al. [2] proposes a TOPSIS method for network selection algorithm, in which the networks are ranked on the basis of the values close to the solution of TOPSIS method. Song et al. [3] proposed a mathematical technique-based network selection, such as AHP and GRA [4]. AHP calculates the relative weight of each parameter considered, and GRA is used to rank the networks. Game theory is modeled as cooperative game theory and noncooperative game theory. Most of the related works model the network selection problem as noncooperative game theory. In this paper, we model the network selection problem as a cooperative game theory.

The organization of the paper is as follows. Sect. 4.2 presents about game theory and network selection. Section 4.3 presents about proposed access network selection algorithm and the game theory for user–network interaction. Section 4.4 presents about the simulation results and their discussions. Section 4.5 presents about the conclusion of the proposed work and future work.

4.2 Game Theory and Network Selection

Game theory is a mathematical tool used to evaluate the competitive situations which imply the interaction of rational decision makers with mutual and possibly conflicting interests. Nowadays, game theory is applied in various areas such as economics, politics, and engineering. Game theory is used in wireless environment such as wireless sensor networks [5] and cognitive radio networks [6]. Game theory is also used as a tool for studying and modeling and analyzing the interactions between individuals using various strategies. It is also used in wireless environments to study about the resource allocation problems, dynamic pricing problems [7–9], and distributed power control [10, 11] problems. Charilas et al. [12] describe about the survey on the various game theory applications in wireless networks categorizing the applications of each layer in OSI layer.

The main components of game theory are the set of players, the set of actions, and the set of payoffs. The players try to maximize their payoffs by using various strategies and utilize actions depending upon the available information at that moment. Each player chooses different strategies available to maximize their payoff. The combination of the best strategy for each player is known as equilibrium. The utility is defined as the payoff that each player gets from the strategies used by them.

The game theory components can be mapped with network selection as shown in Table 4.1. The players in the game are mobile users or networks. The strategies chosen by the players to maximize their payoffs are available bandwidth and subscription plan. The payoffs can be calculated using the utility function based on various decision criteria such as monetary cost, energy conservation, and network load. The game can be modeled to use for maximizing or minimizing different resources such as bandwidth and power.

There are two types of game theoretic approaches: a) cooperative approaches, in which both the players have the same strategy and cooperate each other to maximize their payoff, b) noncooperative game theory in which each player selects his own strategies. Game theoretic approach for network selection is shown in Table 4.2.

Khan et al. [13] proposed the network selection problem as a noncooperative auction game, in which the game components are bidders, sellers, and auctioneer. The winning bid is computed as such, and it will maximize their utility. Charilas et al. [14] propose a congestion avoidance control mechanism, which models the game as a noncooperative two player game. Antoniou et al. [15] proposed the user network interaction as a cooperative repeated game where the users have four strategies.

4.3 Proposed Work

Due to a large increase in the Internet data, the mobile operators are facing a large increase in the data traffic. The increase in the quality of the video content had created a challenge for the network operators to provide a seamless connectivity for multimedia applications to the end users with the required quality level utilizing the available network resources. The challenge is to provide a seamless connectivity for multimedia applications in a heterogeneous wireless network environment.

Table 4.1 Mapping of game components to network selection

Game component	Network selection environment correspondent
Players	The agents who are playing the game: User and/or networks
Strategies	Available/requested bandwidth, subscription plan, offered prices, and available APs
Payoffs	The motivation players represented by profit and estimated using utility functions based on various parameters: monetary cost, quality, network load, and QoS
Resources	The resources for which the players involved in the game are competing: bandwidth and power

Table 4.2 Game theoretic approaches for network selection

Players Interaction	Game theoretic approach	Objective
Users versus users	Noncooperative	User competes against each other seeking to maximize their own utility
	Cooperative	Users cooperate in order to obtain mutual advantage
Networks versus users	Noncooperative	Users compete against networks, each seeking to maximize their own utility. On the one side, the users try to maximize their cost-benefit performance. On the other side, networks aim to maximize the profit for the provided services
	Cooperative	Both sides cooperate in order to achieve mutual satisfaction
Networks versus networks	Noncooperative	Networks compete against each other seeking to maximize their individual revenues
	Cooperative	Networks cooperate in order to achieve global welfare maximization

The access network selection mechanism proposed will calculate the utility score for each network and select the network, which is having the highest score, that is, the targeted network. The access network selection algorithm considers user preference and application requirements. The radio access networks considered are WLAN and WiMAX networks. The proposed algorithm helps the user to select the best network available in a heterogeneous wireless network environment in Fig. 4.1.

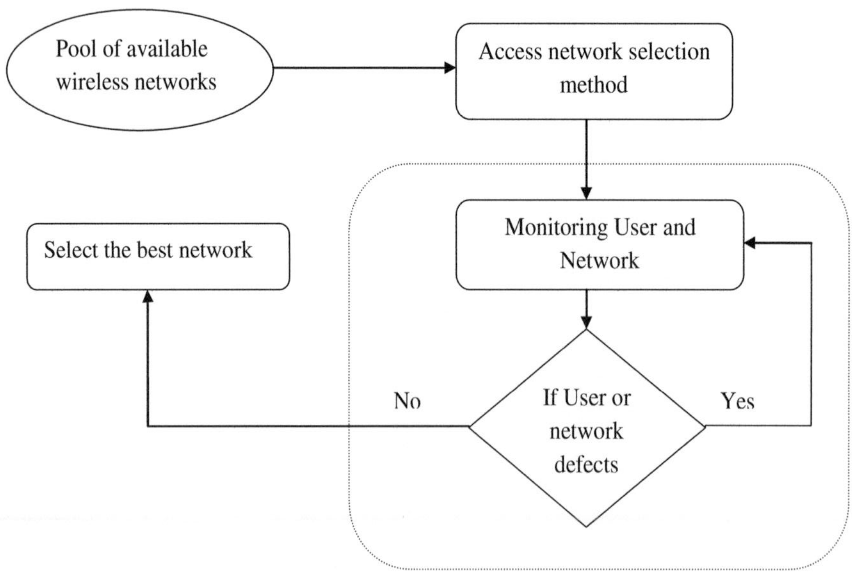

Fig. 4.1 Flowchart of the proposed algorithm

The user preference plays an important role in the network selection decision. The decision making considered in access network selection algorithm mainly depends on three criteria: quality, energy, and cost. The user can specify his preference by giving importance to the each criterion. For example, if the user is concerned about strict budget, then he can give more weightage to the criteria cost. If the user is more concerned about the energy consumption of the device, then the user can give more importance to criteria energy; if the user is more quality oriented, then he can give more weight to the criteria cost.

The access network selection score can be calculated using the multiplicative exponential weighted [MEW] method. The score depends on three criteria mainly: energy consumption, multimedia stream quality, and monetary cost. The criteria can be divided into two classes: (1) the smaller the better, the smaller values of the criteria are considered to better than the higher values (e.g., cost and energy) and (2) the larger the better, the larger values of the criteria are considered to be better than the smaller values. The criteria should be scaled because each criterion value will be in different range and different units. The scaling process will help to make the values within the range [0,1] and make it comparable. Each criterion is scaled with the help of using utility functions.

$$U_{\mathrm{i}} = u_{\mathrm{ei}}^{w_{\mathrm{e}}} \cdot u_{\mathrm{qi}}^{w_{\mathrm{q}}} \cdot u_{\mathrm{ci}}^{w_{\mathrm{c}}} \tag{4.1}$$

where U_{i} is the utility function score for each $\mathrm{RAN_i}$, where u_{e}, u_{q}, and u_{c} are the utility function of energy, quality in terms of receiving bandwidth, and monetary cost for the $\mathrm{RAN_i}$ respectively. w_{e} and w_{q} represent the weights for each criterion based on the importance of each criterion in the decision algorithm such that $w_{\mathrm{e}} + w_{\mathrm{q}} + w_{\mathrm{c}} = 1$. The utility function score has values in [0,1] interval unit. Each utility is described as follows.

4.3.1 *Energy Utility*

The energy utility u_{e} follows the "smaller the better" principle, that is, the smaller the value of energy consumption, the larger the value for $u_{\mathrm{e},}$ and for high value of energy consumption, $\mathrm{u_e}$ has smaller value. It has value in [0,1] interval and has no unit. The estimated energy for the real-time application is calculated using Eq. (4.2) [16],

$$E_{\mathrm{i}} = t(r_t + \mathrm{Th_i} \cdot r_{\mathrm{d}}) \tag{4.2}$$

where E_{i} is the energy consumption (joule) for the radio access network "*i*", "*t*" represents the transaction time of the application in seconds, r_t is the mobile device energy consumption per unit time in watts, $\mathrm{Th_i}$ is the available throughput in kbps for the radio access network "*i*", and r_{d} is the energy consumption rate for data/received stream (joules/kb). The transaction time can be predicted from the multimedia application duration. r_d and r_t will be specified by the device manufacturer.

$$U_e(E) = \begin{cases} 1, & E < E_{min} \\ \frac{E_{max} - E}{E_{max} - E_{min}} & E_{min} <= E < E_{max} \\ 0, & \text{otherwise} \end{cases} \quad (4.3)$$

where E_{min} is the minimum energy consumption (joule), E_{max} is the maximum energy consumption (joule), and E is the energy consumption of the current network (joule). E_{min} and E_{max} can be calculated for Th_{min} (minimum throughput) and Th_{max} (maximum throughput), respectively. The energy consumed is calculated using Eq. (4.2).

4.3.2 *Quality Utility*

In this, a zone-based sigmoid quality utility function is defined. The quality utility is based on the minimum throughput (Th_{min}) required to maintain the multimedia service at the minimum acceptable quality (any values below this threshold will result in an unacceptable quality level, that is, zero utility), the required throughput (Th_{req}) is to ensure high-quality level for multimedia services, and maximum throughput (Th_{max}) human cannot distinguish well the quality level above this threshold values. The quality utility function can be computed as follows:

$$U_q(Th) = \begin{cases} 0, & Th < Th_{min} \\ 1 - e^{\frac{-\alpha * Th^2}{\beta + Th}}, & Th_{min} <= Th < Th_{max} \\ 1, & \text{otherwise} \end{cases} \quad (4.4)$$

where α and β are two parameters which determine the shape of the utility function (no unit). Th is the average predicted throughput for each RAN. Th_{min} and Th_{max} are the minimum and maximum throughput, respectively. The quality utility has the values in [0,1] interval and has no unit in Fig. 4.2.

4.3.3 *Cost Utility*

The cost utility [17, 18] also follows the "smaller the better" principle. This means that for small values of the monetary cost, the cost utility has high value, and for large values of monetary cost, the cost utility has low values. Cost utility is calculated as follows:

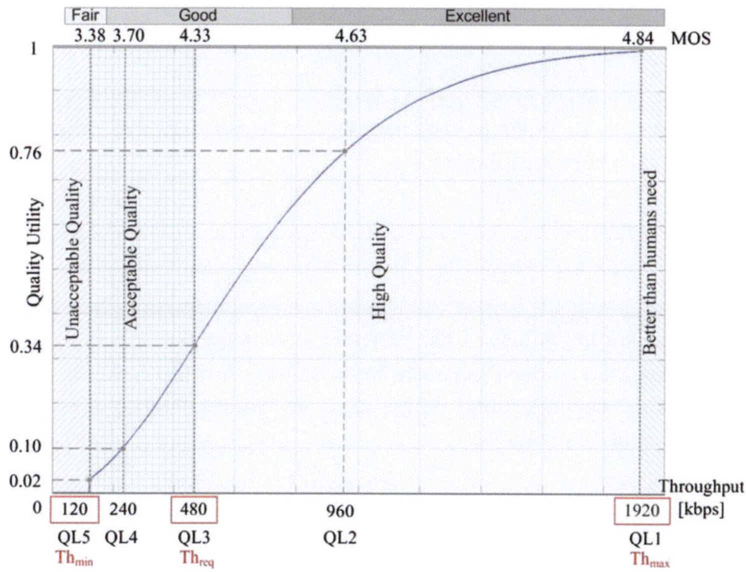

Fig. 4.2 Sigmoid quality function

$$U_c(C) = \begin{cases} 1, & C < C_{min} \\ \frac{C_{max} - C}{C_{max} - C_{min}}, & C_{min} <= C < C_{max} \\ 0, & \text{otherwise} \end{cases} \quad (4.5)$$

where C is the monetary cost for the current network, C_{min} is the minimum cost that the user is willing to pay, and C_{max} is the maximum cost that the user can afford to pay. The cost utility has value in [0,1] interval and has no unit.

4.3.4 Ranking Methods

Each ranking method will assign a score to each network and for each quality level. The network that has the highest score for a certain quality level will be selected as the target network. The simple additive weighting [SAW] method uses additive operation and is computed as follows:

$$\mathrm{SAW_i} = w_e \cdot u_{ei} + w_q \cdot u_{qi} + w_c \cdot u_{ci} \quad (4.6)$$

The multiplicative exponential weighted [MEW] method uses multiplicative operations and can be computed as follows:

$$\mathrm{MEW_i} = u_{\mathrm{ei}}^{w_e} \cdot u_{\mathrm{qi}}^{w_q} \cdot u_{\mathrm{ci}}^{w_c} \tag{4.7}$$

The Gray relational analysis [GRA] method makes use of the best reference network parameters in order to describe the similarity between each candidate network. It is calculated as follows:

$$\mathrm{GRA_i} = \frac{1}{w_e \cdot \left|u_{e1} - u_e^b\right| + w_e \cdot \left|u_{q1} - u_q^b\right| + w_e \cdot \left|u_{c1} - u_q^b\right| + 1} \tag{4.8}$$

where u_e^b, u_q^b, u_c^b are the values of the best reference networks.

The Technique for Order Preference by Similarity to Ideal Solution [TOPSIS] method orders the network based on the best and worst reference networks score which are calculated as follows:

$$\mathrm{TOPSIS_i} = \frac{D_{w,i}}{D_{b,i} + D_{w,i}} \tag{4.9}$$

$D_{w,i}$ and $D_{b,i}$ are the Euclidean distance from the worst and best, respectively, reference values, and their values are given by the following,

$$D_{w,i} = \sqrt{w_e^2\left(u_{e1} - u_e^w\right)^2 + w_e^2 \cdot \left(u_{q1} - u_q^w\right)^2 + w_c^2 \cdot \left(u_{c1} - u_c^w\right)^2} \tag{4.10}$$

$$D_{b,i} = \sqrt{w_e^2\left(u_{e1} - u_e^b\right)^2 + w_e^2 \cdot \left(u_{q1} - u_q^b\right)^2 + w_c^2 \cdot \left(u_{c1} - u_c^b\right)^2} \tag{4.11}$$

where u_{ew}, u_{qw}, and u_{cw} are the reference values of the worst network.

Game theory is formulated as a Prisoner's dilemma game in which the user and network cooperate each other in order to achieve Nash equilibrium; if the equilibrium is achieved, the current network is selected. The game can be modeled as follows:

Players: The players in the game are users and networks.

Strategies: For a Prisoner's dilemma game, two types of strategies are defined for each player.

The user strategies are as follows:

(a) Cooperate: The user accepts the networks offer and stay.
(b) Defect: The user decides to leave the network if the network does not offer minimum requested QoS.

The network strategies are as follows:

(a) Cooperate: The network accepts to maintain the QoS at the required level of the user.

Table 4.3 User network payoff table

User	Network	
	Cooperate	Defect
Cooperate	(B1,B2)	(D1,A2)
Defect	(A1,D2)	(C1,C2)
	Payoff (player 1, player 2)	

(b) Defect: The network does not fulfill the QoS requirements of the user and selfishly trying to increase its revenue by adding more number of users in Table 4.3.

Observing the payoff table, it can be found that the user gets the highest payoff when the network cooperates and the user defects (A1); that is, the user decides to leave the network or network accepts users' demand. D1 is the user payoff, when the user cooperates but the network defects. B1 is the user payoff when both user and network cooperate. B2 is the network payoff when both user and network cooperate. D2 is the payoff of the network, when network cooperates and user defects. A2 is the network payoff, when the user cooperates, but the network defects. C1 is the user payoff, when both user and network defect, and C2 is the payoff of the network when both user and network defect.

4.4 Results and Discussions

These four ranking methods, namely SAW, MEW, GRA, and TOPSIS, are used to evaluate the score value for each RANs such as WLAN and WiMAX using MATLAB, for five quality levels with corresponding user weight for quality utility. The quality levels for multimedia service such as video considered are QL1 = 120 kbps, QL2 = 240 kbps, QL3 = 480 kbps, QL4 = 960 kbps, and QL5 = 1920 kbps. The user weights for each quality levels are 0.2, 0.4, 0.6, 0.8, and 1.0, respectively. Using each ranking method, the score value for WLAN and WiMAX is calculated, in which network has the highest score value for a particular video quality and user demands are considered as target network.

(1) MEW method

The score is calculated using Eq. (4.7) mentioned in the MEW method. As shown in Figs. 4.3 and 4.4, the score is calculated for each quality level and with corresponding quality weight for each network. Quality level in which network has the highest score is selected as the target network.

(2) SAW method

As in MEW method and in SAW method also, we calculate the score for each network using Eq. (4.6) already mentioned for the corresponding quality level of

Fig. 4.3 MEW analysis of WLAN

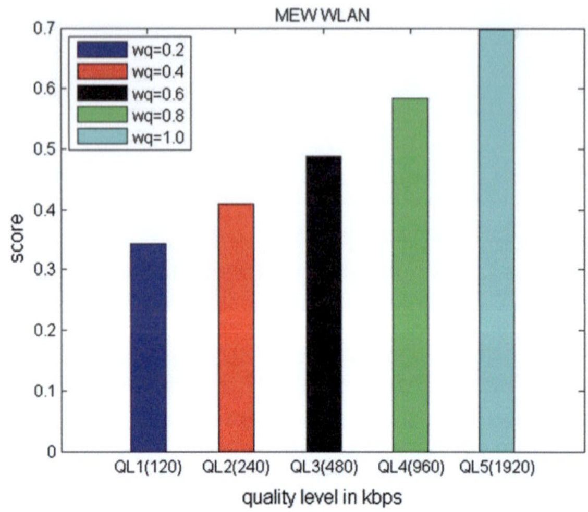

Fig. 4.4 MEW analysis of WiMAX

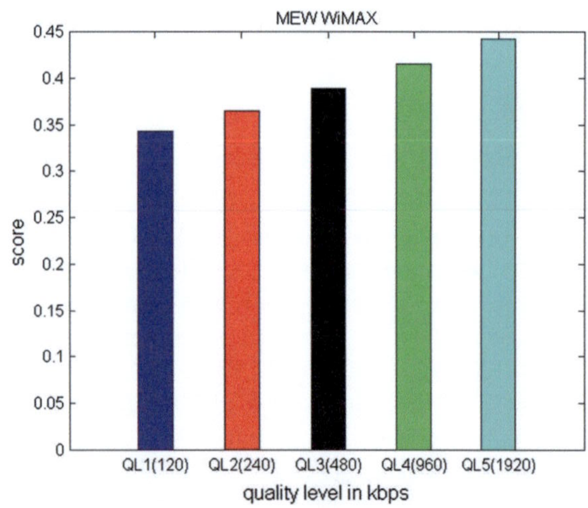

the multimedia service and the targeted network is selected. From Figs. 4.5 and 4.6, we can conclude that for QL1, the score is less compared to QL5, and for both the network, it shows that user is provided at a high value for the quality weight for high quality. Figures 4.5 and 4.6 show the SAW method analysis for WLAN and WiMAX networks, respectively.

(3) GRA method

Gray relational analysis takes a reference value for a best network. Here, the highest values for all the weights of the best reference are considered as 1. GRA

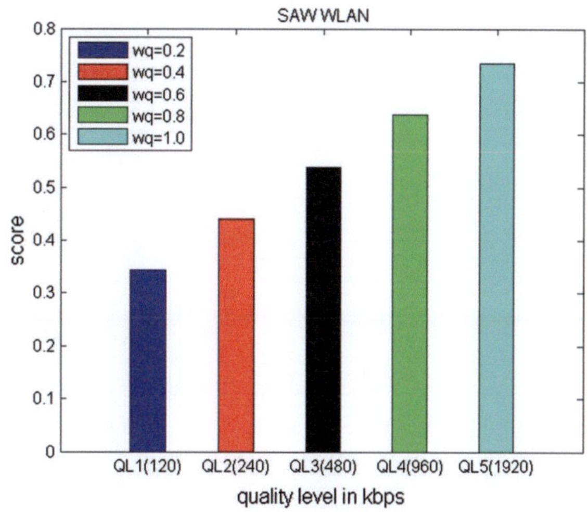

Fig. 4.5 SAW analysis of WLAN

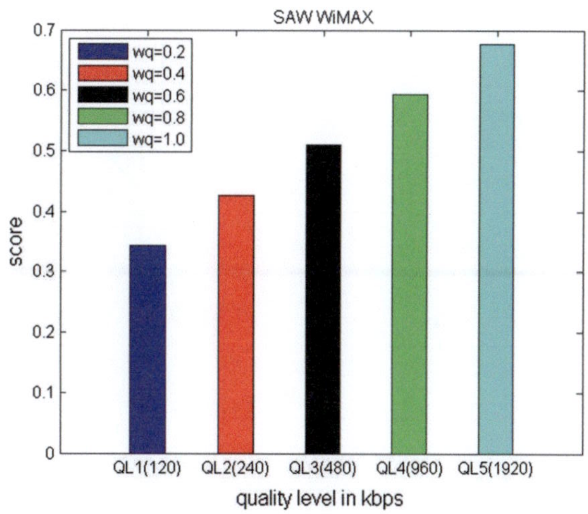

Fig. 4.6 SAW analysis of WiMAX

methods compare both the networks with the best reference network. The results for WLAN and WiMAX are shown in Figs. 4.7 and 4.8.

(4) TOPSIS method

In TOPSIS method, the score is calculated using the values taken from the Euclidean distance from both the best and worst reference networks. Here, the weights for the best reference networks are taken as 1, and the weights for the worst reference networks are taken as 0. The TOPSIS method analysis of WLAN and WiMAX networks is shown in Figs. 4.9 and 4.10.

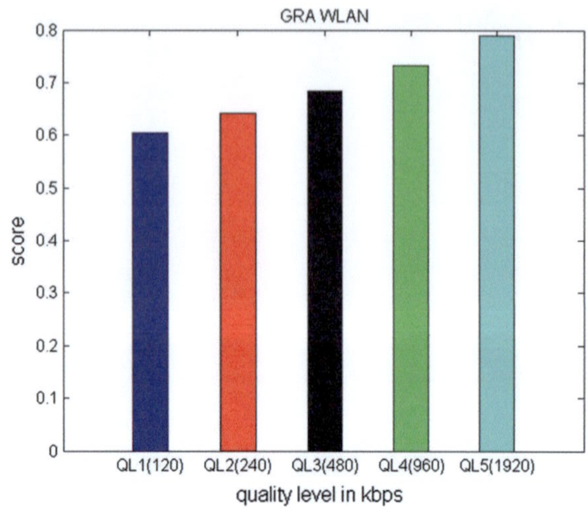

Fig. 4.7 GRA analysis of WLAN

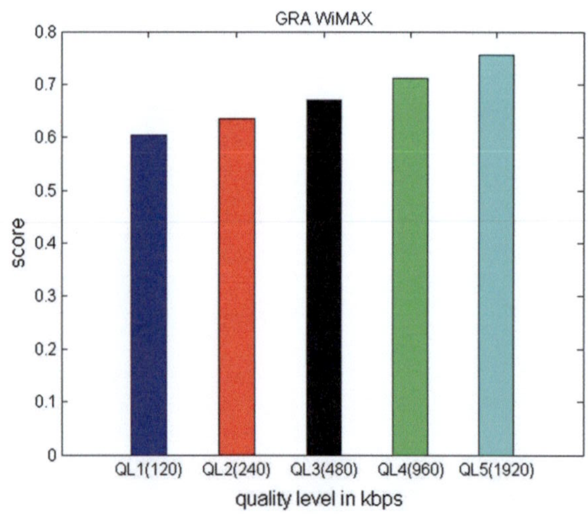

Fig. 4.8 GRA analysis of WiMAX

Table 4.4 shows that when the user and network cooperate, both of them get an equal payoff of 3 each. When user cooperates and network defects means user will get a payoff of only 1 and network will get a payoff of 4. If the user defects and the network cooperates then the user will get a payoff of 4 and the network will get a payoff of 1. When both the network and user defect means they will get a less payoff of 2 each. The process is continued till the Nash equilibrium is reached.

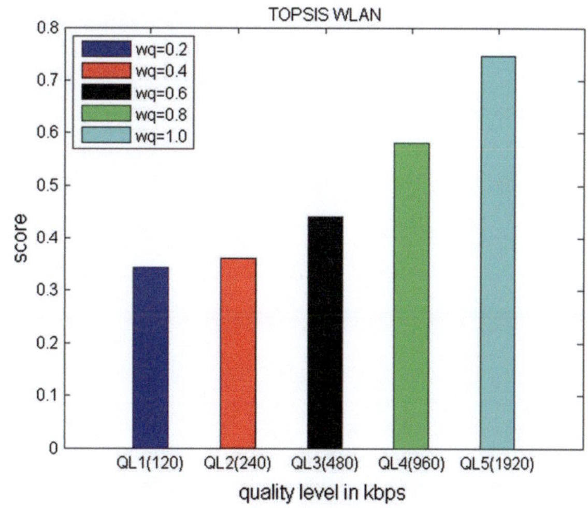

Fig. 4.9 TOPSIS analysis of WLAN

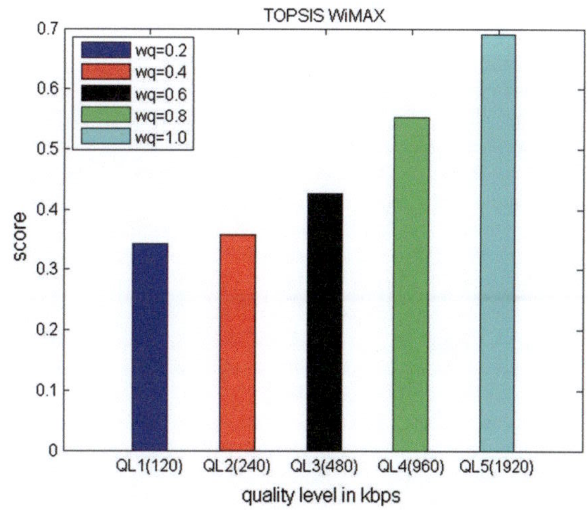

Fig. 4.10 TOPSIS analysis of WiMAX

Table 4.4 User/network payoff

User/networks	Cooperate	Defect
Cooperate	(3,3)	(1,4)
Defect	(4,1)	(2,2)

4.5 Conclusion and Future Work

In this paper, a network selection algorithm is used to calculate the score for the available radio access networks like WLAN and WiMAX. Four ranking methods of MADM are considered: MEW, SAW, GRA, and TOPSIS. Among the four methods, MEW is the most appropriate method for network selection, and it shows the clear score value for WLAN and WiMAX. The network selection is modeled as a Prisoner's dilemma game, and the best network is selected when it reaches Nash equilibrium.

In future, a repeated Prisoner's dilemma game theory method is used to choose a better network for user demand.

References

1. Wang HJ, Katz RH, Giese J (1999) Policy-enabled handoffs across heterogeneous wireless networks. In: Proceedings of second IEEE workshop on mobile computing systems and applications, (WMCSA)
2. Bakmaz B, Bojkovic Z, Bakmaz M (2007) Network selection algorithm for heterogeneous wireless network environment. In: IEEE international symposium on personal, indoor and mobileradio communications, (PIMRC)
3. Song Q, Jamalipour A (2005) A network selection mechanism for next generation networks. In: IEEE international conference on communications, (ICC)
4. Liu B, Tian H, Wang B, Fan B (2014) AHP and game theory based approach for network selection in heterogeneous wireless networks. In: The 11th annual IEEE CCNC-wireless communications track, pp 501–506
5. Machado R, Tekinay S (2008) A survey of game-theoretic approaches in wireless sensor networks. Comput Netw 52(16):3047–3061
6. Wang B, Wu Y, Ray Liu KJ (2010) Game theory for cognitive radio networks: an overview". Comput Netw 54(14):2537–2561
7. Sengupta S, Anand S, Chatterjee M, Chandramouli R (2009) Dynamic pricing for service provisioning and network selection in heterogeneous networks. Phys Commun 2(1):138–150
8. Malanchini I, Cesana M, Gatti N (2013) Network selection and resource allocation games for wireless access networks. IEEE Trans Mob Comput 12:2427–2440
9. Xu Y, Wang J, Wu Q, Anpalagan A, Yao Y (2012) Opportunistic spectrum access in cognitive radio networks: global optimization using local interaction games. IEEE J Sel Top Signal Process 6(2):180–194
10. Alpcan T, Başar T, Dey S (2006) A power control game based on outage probabilities for multicell wireless data networks. IEEE Trans Wirel Commun 5:890–899
11. Trestian R, Ormond O, Muntean G-M (2011) Reputation-based network selection mechanism using game theory. Phys Commun 4:156–171
12. Charilas DE, Panagopoulos AD (2010) A survey on game theory applications in wireless networks. Comput Netw 54(18):3421–3430
13. Khan MA, Touseef U, Marx S, Goerg C (2010) Game-theory based user centric network selection with media independent handover services and flow management. In: Eighth annual research conference on communication networks and services (CNSR)
14. Charilas DE, Panagopoulos AD, Vlacheas P, Markaki OI, Constantinou P (2009) Congestion avoidance control through non-cooperative games between customers and service providers. In: International Conference on mobile lightweight wireless systems (mobilight)
15. Antoniou J, Papadopoulo V, Vasiliou V, Pitsillides A (2010) Cooperative user-network interactions in next generation communication networks. Comput Netw 54(13):2239–2255
16. Mahmud K, Inoue M, Murakami H, Hasegawa M, Morikawa H (2004) Measurement and usage of power consumption parameters of wireless interfaces in energy-aware multi-service mobile terminals. In: IEEE international symposium on personal, indoor and mobile radio communications, (PIMRC)

17. Nguyen-Vuong Q-T, Ghamri-Doudane Y, Agoulmine N (2008) On utility models for access network selection in wireless heterogeneous networks. In: IEEE network operations and management symposium, (NOMS)
18. Trestian R, Moldovan AN, Muntean CH, Ormond O, Muntean GM (2012) Quality utility modelling for multimedia applications for Android mobiledevices. In: Proceedings of IEEE international symposium on BMSB

Chapter 5
Leakage Reduction by Test Pattern Reordering

Shrabanti Chakraborty, Trupa Sarkar and Sambhu Nath Pradhan

5.1 Introduction

With the improvement in semiconductor manufacturing technology, hundreds of millions of transistor are now placed in a single VLSI device. As the scaling continues to rise, the complication of testing each transistor increases. So testing has become the most time-consuming and important step in the development process of VLSI circuits. In addition, the diminishing feature size and speedy growth of the portable electronic products have made power minimization an important issue during testing to extend battery life and to increase reliability.

However, power consumption during test mode is generally higher than during normal functional mode of the circuit [1]. The main reason behind this is the negligible correlation between consecutive test patterns applied to the circuit under test (CUT) which increases the switching activity of internal nodes [2]. Parallel testing generally employed for SOCs to reduce the test application time also increases test power. The activation of additional DFT circuitry during test mode also enhances power dissipation. As the heat produced in a circuit during operational mode is almost proportional to the test power dissipation [3], it can cause irreversible structural damage to the circuit [4]. Moreover, excessive power supply noise and voltage drop can inaccurately change some logic states of internal nodes

S. Chakraborty · T. Sarkar (✉) · S.N. Pradhan
Department of ECE, National Institute of Technology, Agartala, India
e-mail: trupa.sarkar@gmail.com

S. Chakraborty
e-mail: shrabantic6@gmail.com

S.N. Pradhan
e-mail: sambhu.pradhan@gmail.com

V. Nath (ed.), *Proceedings of the International Conference on Nano-electronics, Circuits & Communication Systems*, Lecture Notes in Electrical Engineering 403,
DOI 10.1007/978-981-10-2999-8_5

which result in the failure of some good dies, thus leading to unnecessary loss of yield [5]. Increased test power requires expensive packaging and cooling techniques to dissipate the excessive power which crosses the final product cost and area constraint of circuit [6].

Power dissipation in the CMOS circuits is mainly divided into two components—dynamic and static. Dynamic power dissipation is caused by switching of logic levels at a particular node. Static power dissipation occurs when circuits are in steady state that means when the inputs are not changing. Leakage power is the main component of static power. Although dynamic power is the main source of power dissipation, however, as the technology progresses towards the deep submicron regime, leakage power dominates over dynamic power. So minimizing leakage power is the primary concern in most circuit designs and testing.

So far, it is considered that leakage power depends only on the current input pattern applied to the circuit. But runtime leakage power [7] is a function of both the previous and present input patterns applied to the circuit [8]. With the change of test application time, this leakage power dependence on previous and present input patterns also changes. So, leakage power is dependent on the order of test patterns to be applied in the circuit. In this paper, a method has been proposed to reorder the applied test vectors so that the transition density between the consecutive test vectors is minimized which further decreases the leakage power. Here, only the sequence of test patterns are changed which will not affect the fault coverage. A genetic algorithm (GA)-based approach has been used to reorder the test patterns.

This paper is organized as follows. Section 5.2 deals with the previous work. Section 5.3 details proposed algorithm. Section 5.4 gives an insight to the calculation of runtime leakage power. Section 5.5 enumerates the experimental results.

5.2 Related Work

Badereddine et al. [9] proposed a technique in which the don't care bits of test cubes are assigned in such a manner so that the number of bit transitions in consecutive test patterns is minimized which also lowers scan cell transitions. An approach based on GA is used to fill the test cubes to minimize both dynamic power and leakage power. In addition to this, the newly generated test patterns are reordered again to decrease the dynamic power.

Sharifi et al. [10] presented a new method which uses multiplexers at scan cell outputs to suppress some transitions in scan chains without affecting the normal mode circuit operation. Multiplexers are generally used to apply any desired values to pseudo-inputs while scanning the test patterns in the scan chain of the circuit. To reduce both dynamic power and leakage power, a suitable order of test patterns is applied to the controlled inputs (primary inputs and multiplexed inputs) during scan mode. This ordered test pattern also blocks the transitions which are sourced from non-multiplexed pseudo-inputs. Test time, fault coverage and operating frequency of the circuit remain unaffected by this method. No additional control signal and routing overhead are needed.

Kumar et al. [11] studied that if the specific bits which are generated by ATPG tool and which are a part of the deterministic set of test vectors are identified first, they will detect the targeted fault models. The remaining bits are used as don't care (X) bits to transform the original vector to a power-aware vector. The knowledge of fault propagation path and fault activation path is also utilized here. PSO (particle swarm optimization)-based approach is used for vector reordering which uses travelling salesman problem (TSP) concept. This proposed technique reduces both runtime leakage power and dynamic power without any change in fault coverage.

5.3 Proposed Algorithm

In this GA-based approach, input test patterns are referred to as chromosomes. To evaluate the fitness of chromosomes, first the cumulative leakage power is calculated which is generated when a sequence of specific ordered test patterns are applied to the circuit input. The proposed algorithm contains three operators which are briefly described below:

5.3.1 Direct Copy

To avoid an inferior solution, 20% of the best test patterns are copied to the next generation without any modification.

5.3.2 Crossover

Generally, fittest chromosomes are crossed to generate new offspring. For the selection of the fittest test patterns which are to be crossed, a number between 0 and 1 is taken. The number is generated randomly. Now, if the number is greater than or equal to 0.5, then the test patterns are randomly chosen from the 20% best test patterns. If it is less than 0.5, they are selected from rest of the 80% test patterns. Then, one crossover point is selected randomly at each of the test patterns. The first offspring test pattern will take the characteristics of first parent up to the selected crossover point, and it will take the characteristics of second parent after crossover point. The process will be reversed for second offspring test pattern. This step is repeated for 60% of the test patterns.

5.3.3 Mutation

Mutation is used to introduce distinctiveness of chromosomes between generations. Here, first, a parent test pattern is selected randomly. This random selection procedure is same as that of crossover. Each bit of the selected test pattern is switched, that is, 1 is replaced by 0 and 0 is replaced by 1. Then, it is examined whether this newly created test pattern already exists or not. Mutation is performed for 20% of test patterns.

The flow chart of proposed algorithm is given below (Fig. 5.1):

Here, the best solution is ordered test patterns which will generate minimum cumulative leakage power when applied to a particular circuit. The proposed GA is given below:

Step 1. Initialization
Generate an initial population of n test patterns.
generation_no=0;
Step 2. Evaluation
Calculate the fitness function of initial population produced in initialization step.
Step 3. Crossover
A pair of parent test patterns is randomly selected from current population. Crossover is done at a randomly chosen point to form two new test patterns. Repeat this process for 60% of n test patterns.
Step 4. Mutation
Mutate a test pattern at each bit position and generate a new test pattern. Repeat this process for 20% of n test patterns.
Step 5. New population generation
Place the 80% newly generated test patterns and 20% original test patterns in a new population.
generation_no=generation_no+1;
Calculate the fitness function of new population.
Step 6. loop1
while generation_no<max_gen
for each generation repeat step 3, 4, 5 and 6
end for
end while
Step 7. loop2
while generation_no<max_gen
for each generation
if(fitness function of present generation<fitness function of previous generation)
select the present generation as best solution
else select the previous generation as best solution
end if else
end for
end while

Fig. 5.1 Flow chart of proposed algorithm

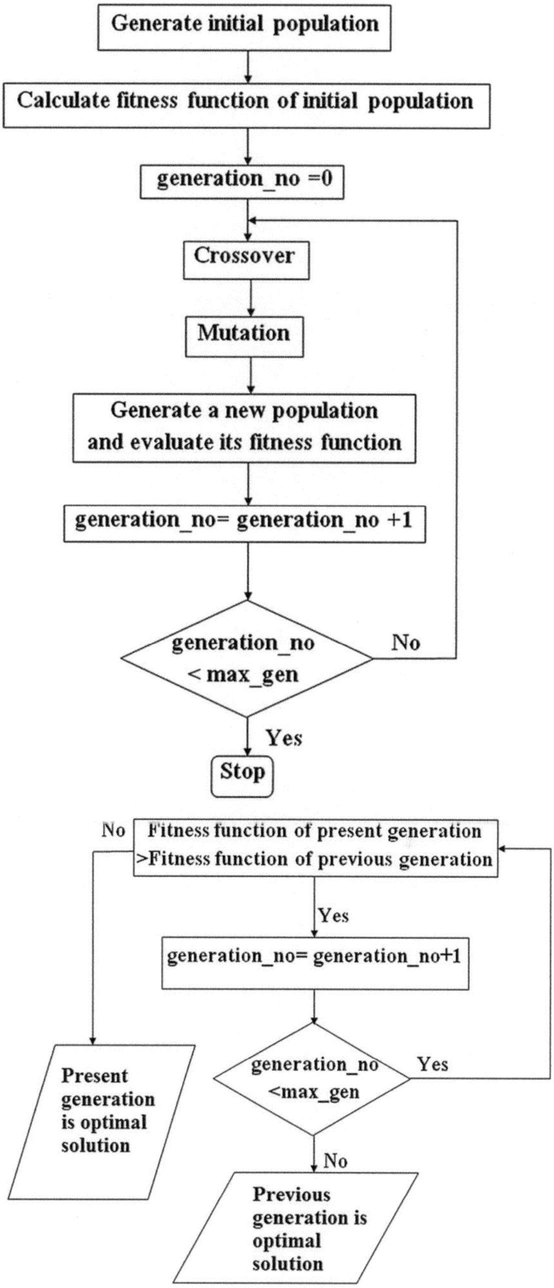

5.4 Calculation of Runtime Leakage Power

Kennedy et al. [12] suggested a technique to calculate runtime leakage power. The switching power depends on input patterns, and the short-circuit power is a function of both rise time and fall time. These two facts are used to calculate the runtime leakage power. It is considered that two vectors, 00 followed by 01, are applied to a 2-input NAND gate (Fig. 5.2) and the total energy (E) consumed by the NAND gate is measured. Then, the same two patterns are again applied, but the time period is doubled. The total energy (E_{new}) is calculated again. The sequence of input patterns and rise time are kept constant in both the cases. So, switching power and short-circuit power do not change. The only variable quantity is the leakage power. The energy difference ($E_{new} - E$) is divided by the difference in time period, and the runtime leakage power consumed by the NAND gate when 00 is applied followed by 01 is obtained. Similarly, the leakage power for all possible combinations of input pattern is calculated, and a lookup table (LUT) is prepared. The block level representation of the proposed method is shown in Fig 5.3.

The runtime leakage power of 2-input NAND gate is shown in Tables 5.1 and 5.2 that represents the runtime leakage power for 2-input NOR gate for all the possible combinations of input pattern. The tables have been prepared for a supply voltage (V_{DD}) of 1 V. The leakage power also changes with time period of simulation. So, leakage power is also calculated for different time periods. Here, leakage power for three different time periods (10, 20 and 40 ns) is considered.

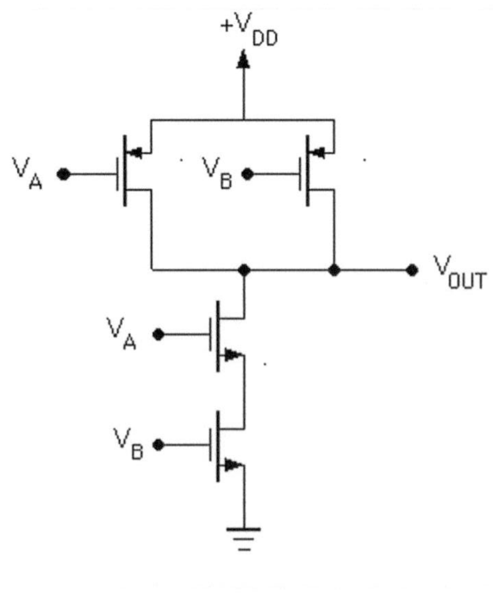

Fig. 5.2 2-input NAND gate

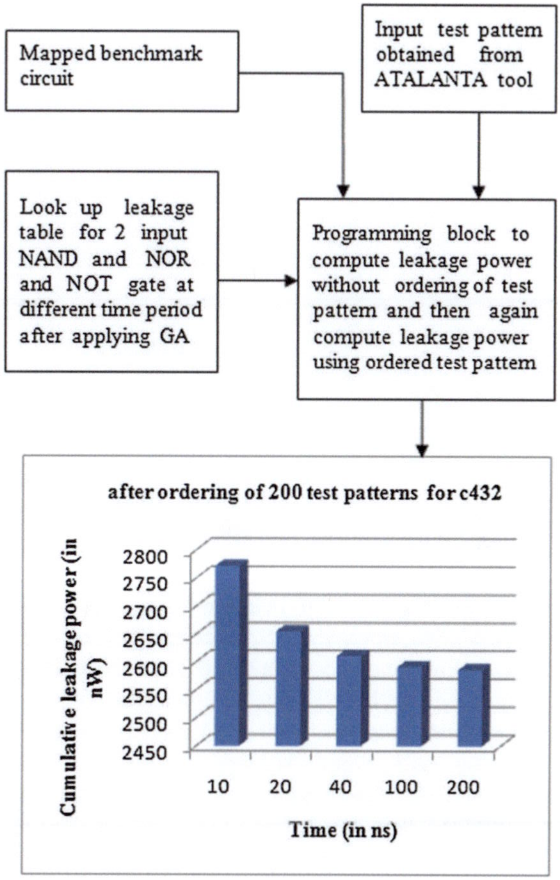

Fig. 5.3 Block diagram of proposed method

Table 5.1 Runtime leakage power for 2-input NAND gate

Previous pattern	Current pattern	10 ns (in nW)	20 ns (in nW)	40 ns (in nW)
00	01	0.54	0.002636	0.1296
00	10	2.059	1.102	0.5846
00	11	65.26	63.57	63.29
01	00	0.006477	0.00541	0.004932
01	10	54.32	51.52	50.10
01	11	81.84	79.09	78.16
10	00	0.003301	0.003087	0.003033
10	01	45.27	46.63	47.51
10	11	73.23	71.01	70.45
11	00	68.59	65.72	64.55
11	01	80.40	78.33	77.78
11	10	78.37	73.78	71.91

Table 5.2 Runtime leakage power for 2-input NOR gate

Previous pattern	Current pattern	10 ns (in nW)	20 ns (in nW)	40 ns (in nW)
00	01	80.94	78.40	77.60
00	10	80.39	75.50	73.38
00	11	70.80	68.02	66.84
01	00	79.58	77.69	77.15
01	10	49.87	46.69	45.07
01	11	0.006478	0.00522	0.004673
10	00	72.77	71.31	71.18
10	01	39.15	40.88	42
10	11	0.002972	0.002818	0.002779
11	00	65.57	64.86	65.68
11	01	0.6356	0.004057	0.2007
11	10	1.773	0.009718	0.5243

5.5 Experimental Result

A test pattern set with maximum fault coverage is generated using the tool ATALANTA. Here, ISCAS'85 benchmark circuits are used for experimentation purpose. First, the benchmark circuits are technology mapped into a library which contains 2-input NAND gate, 2-input NOR gate and NOT gate. Runtime leakage power is calculated in 'Cadence Virtuoso Analog Design Environment' in 45 nm technology using supply voltage (V_{DD}) of 1 V. The proposed method is simulated in C language. The cumulative leakage power is calculated first for unordered test patterns and then for ordered test patterns after applying GA for different time periods. To address the input dependency problem more clearly, the cumulative leakage power for 100 test patterns and then for 200 test patterns is calculated separately.

Figure 5.4 shows the graphical representation of cumulative leakage power (in nW) for c432 circuit after reordering of 100 test patterns at different time periods.

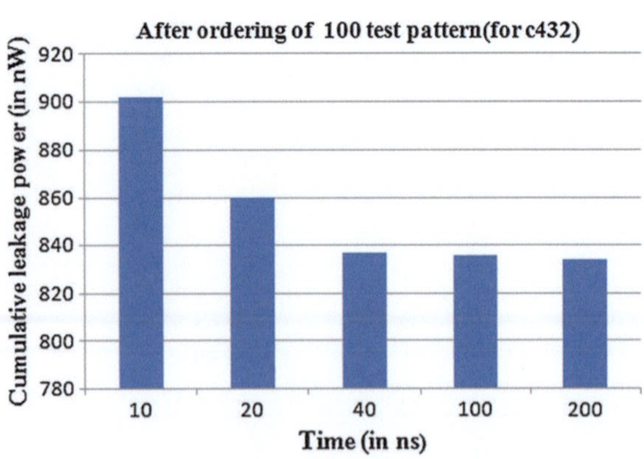

Fig. 5.4 Graphical representation of cumulative leakage power for c432 after ordering of 100 test patterns at different time periods

From this, it can be said that as time period approaches 200 ns, the decrease in cumulative leakage power is not that much compared to the leakage power at 40 ns.

In Table 5.3, there are 5 columns for five different time periods—10, 20, 40, 100 and 200 ns. Except the first column, the remaining columns have two sub-columns. The first sub-column shows the cumulative leakage power without ordering of test patterns, and the second one shows the same for ordering of 100 test patterns after application of GA.

Similarly, in Table 5.5, the first sub-column details the cumulative leakage power without ordering of test patterns and the second one shows the same for ordering of 200 test patterns after application of GA.

Tables 5.4 and 5.6 show cumulative leakage power savings at different time periods for ordered 100 test patterns and for ordered 200 test patterns, respectively. The last column shows the average leakage power savings for different circuits where the average is taken over five different time periods (10, 20, 40, 100 and 200 ns).

Table 5.7 shows the cumulative leakage power savings in percentage after ordering of 100 test patterns where five different columns are for five different time periods. The cumulative leakage power savings is maximum for c432 at 10 and 40 ns which is 79.07%. The leakage power savings is almost 43.6% for c880, 35.6% for c1355 and 22.7% for c1908 circuit.

Table 5.8 shows the cumulative leakage power savings in percentage after ordering of 200 test patterns where five different columns show the data for five time periods. The cumulative leakage power savings is maximum for c432 at 10 ns which is 64.18%. The leakage power savings are almost 45.6% for c880, 28.5% for c1355 and 22% for c1908 circuit.

Following are the observations that can be made from the calculated results:

1. From the Tables 5.3 and 5.5, it is evident that the cumulative leakage power is maximum for unordered test patterns and it decreases with optimized order of test patterns.
2. As leakage power changes with simulation time period, test application time is also considered as a variable in calculation of runtime leakage power.
3. For all the benchmark circuits, it can be seen that the leakage power is maximum at 10 ns and it decreases with increase in time period.
4. It is clear from the Tables 5.4 and 5.6 that for a large time period, the cumulative leakage power for ordered test sequence becomes almost constant.
5. From the Tables 5.7 and 5.8, it can be commented that the improvement in leakage power in c432 is maximum which is almost 79% for 100 test patterns and almost 64% for 200 test patterns.

Table 5.3 Cumulative leakage power without ordering and after ordering of 100 test patterns at different time period

	10 ns		20 ns		40 ns		100 ns		200 ns	
Circuit	Without ordering of test patterns (in nW)	After ordering of test patterns (in nW)	Without ordering of test patterns (in nW)	After ordering of test patterns (in nW)	Without ordering of test patterns (in nW)	After ordering of test patterns (in nW)	Without ordering of test patterns (in nW)	After ordering of test patterns (in nW)	Without ordering of test patterns (in nW)	After ordering of test patterns (in nW)
C432	4309.74	901.99	4088.21	859.88	3998.79	836.86	3958.36	835.82	3948.11	833.94
C880	94583.21	53298.93	90766.82	51124.29	89372.94	50166.86	88648.85	49937.41	88461.47	49832.13
C1908	19888.94	15339.24	19376.84	14976.72	19224.14	14881.60	19151.97	14838.72	19134.33	14758.96
C1355	136208.26	87351.26	132440.53	85155.49	131328.84	84444.89	131024.65	84257.08	131027.25	84252.76

Table 5.4 Cumulative leakage power and average leakage power savings after ordering of 100 test patterns at different time periods

	Cumulative leakage power (in nW) savings in different time period					Average leakage power savings (in nW)
Circuit	10 ns (in nW)	20 ns (in nW)	40 ns (in nW)	100 ns (in nW)	200 ns (in nW)	
C432	3407.75	3228.33	3161.94	3122.54	3114.16	3207.04
C1908	4549.71	4400.12	4342.54	4313.25	4275.36	4376.19
C880	41284.26	39642.52	39206.07	38711.44	38629.35	39494.73
C1355	48857.26	47285.04	46883.95	46767.57	46774.48	47313.66

Table 5.5 Cumulative leakage power without ordering and after ordering of 200 test patterns at different time periods

	10 ns		20 ns		40 ns		100 ns		200 ns	
Circuit	Without ordering of test patterns (in nW)	After ordering of test patterns (in nW)	Without ordering of test patterns (in nW)	After ordering of test patterns (in nW)	Without ordering of test patterns (in nW)	After ordering of test patterns (in nW)	Without ordering of test patterns (in nW)	After ordering of test patterns (in nW)	Without ordering of test patterns (in nW)	After ordering of test patterns (in nW)
C432	7735.68	2770.76	7353.7	2654.89	7201.32	2610.27	7133.15	2590.77	7116.07	2585.97
C880	193749.98	105312.09	185979.4	101101.41	183156.29	99601.86	181674.34	98808.43	181292.71	98603.55
C1908	44229.34	34447.11	43148.38	33630.65	42837.07	33411.45	42695.68	33313.68	42663.59	33293.29
C1355	276229.65	198527.34	268712.6	192109.62	266518.41	190534.67	265971.46	190187.79	265947.25	190146.75

Table 5.6 Cumulative leakage power and average leakage power savings after ordering of 200 test patterns at different time periods

	Cumulative leakage power (in nW) savings in different time period					Average leakage power savings (in nW)
Circuit	10 ns (in nW)	20 ns (in nW)	40 ns (in nW)	100 ns (in nW)	200 ns (in nW)	
C432	4964.92	4698.81	4591.04	4542.38	4530.09	4665.45
C1908	9782.24	9517.73	9425.62	9382	9370.31	9495.58
C880	88437.89	84878.06	83554.43	82865.91	82689.15	84485.09
C1355	77702.31	76603.03	75983.73	75883.67	75800.50	76394.65

Table 5.7 Cumulative leakage power savings (in %) after ordering of 100 test patterns at different time periods

Circuit	10 ns (in %)	20 ns (in %)	40 ns (in %)	100 ns (in %)	200 ns (in %)
C432	79.07	78.97	79.07	78.88	78.87
C880	43.65	43.67	43.87	43.66	43.69
C1355	35.87	35.71	35.69	35.69	35.69
C1908	22.87	22.71	22.59	22.52	22.86

Table 5.8 Cumulative leakage power savings (in %) after ordering of 200 test patterns at different time periods

Circuit	10 ns (in %)	20 ns (in %)	40 ns (in %)	100 ns (in %)	200 ns (in %)
C432	64.18	63.89	63.75	63.68	63.66
C880	45.64	45.64	45.62	45.61	45.61
C1355	28.48	28.51	28.51	28.49	28.5
C1908	22.12	22.06	22	21.97	21.96

5.6 Conclusion

In this paper, an approach based on GA has been proposed for reordering of test patterns to reduce runtime leakage power dissipation. We have seen that for two consecutive test patterns, if the correlation of the corresponding bits is maximum, then runtime leakage power reduces. Runtime leakage power also changes with different time period of simulation. With increasing time period, the leakage power decreases, and for a large time period, it becomes almost constant.

References

1. Zorian Y (1999) Testing the monster chip. IEEE Spectr 36(7):54–60
2. Wang S, Gupta SK (1997) DS-LFSR: a new BIST TPG for low heat dissipation. In: Proceedings of international test conference, pp 848–857

3. Altet J, Rubio A (2002) Thermal testing of integrated circuits. Springer Science, New York
4. Pouya B, Crouch A (2000) Optimization trade-offs for vector volume and test power. In: Proceedings of international test conference, pp 873–881
5. Chang YS, Gupta SK, Breuer MA (1997) Analysis of ground bounce in deep sub-micron circuits. In: Proceedings of VLSI test symposium, pp 110–116
6. Nicolaidis M, Zorian Y (1998) On-line testing for VLSI: a compendium of approaches. JETTA J Electron Test—Theor Appl 12(1–2):7–20
7. Lin YS, Sylvester D (2007) Runtime leakage power estimation technique for combinational circuits. In: Asia and South Pacific design automation conference
8. Kumar SK, Kaundinya S, Kundu S, Chattopadhyay S (2010) Particle swarm optimization based vector reordering for low power testing. In: International conference on computing, communication and networking technologies (ICCNT), pp 1–5
9. Badereddine N, Girard P, Pravossoudovitch S, Landrault C, Virazel A (2006) Minimizing peak power consumption during scan testing: test pattern modification with X filling heuristics. In: International conference on design and test of integrated systems in nanoscale technology, pp 359–364
10. Sharifi S, Jaffari J, Hosseinabady M, Afzali-Kusha A, Navabi Z (2005) Simultaneous reduction of dynamic and static power in scan structures. In: IEEE Proceedings of the design, automation and test in Europe conference and exhibition, pp 1530–1591
11. Kumar SK, Kaundinya S, Kundu S, Chattopadhyay S (2010) Customizing pattern set for test power reduction via improved X-identification and reordering. In: ACM/IEEE international symposium on low-power electronics and design (ISLPED), pp 177–182
12. Kennedy I, Eberhart RC (1995) Particle swarm optimization. In: IEEE proceedings of international conference on neural networks, pp 1942–1948. Piscataway, NJ

Chapter 6
Review on Band-Notching Techniques for Ultrawideband Antenna

Ankur Saxena and R.P.S. Gangwar

6.1 Introduction

The ultrawideband (UWB) technology utilizes low power over a short distance besides achieving high bandwidth (3.1–10.6 GHz) in wireless communication [1, 2]. UWB has been finding applications in diverse fields from communication to medical research. The UWB system transmits a bulk of data over a wide bandwidth but for a short distance. The UWB technology can carry signals across the various obstacles such as doors or walls which tend to reflect signals for limited bandwidths. UWB coexists with other users and achieves large data for distance of 230 feet.

UWB technology has mainly two types of applications.

(i) In radar applications, the signal detects the objects which may be behind the wall. The signal penetrates the near surfaces but reflects from far surfaces.
(ii) The second type of application involves voice and data transmission. The digital signals having low power may carry information at a high data rate in a defined range.

A system having a bandwidth larger than 500 MHz is categorized under ultra-wideband. UWB antenna comprises of wider bandwidth, omnidirectional radiation

A. Saxena (✉) · R.P.S. Gangwar
Department of Electronics and Communication Engineering, College of Technology, GBPUAT, Pantnagar 263145, Uttarakhand, India
e-mail: ank.saxena@gmail.com

R.P.S. Gangwar
e-mail: profrpsgangwar@yahoo.co.in

V. Nath (ed.), *Proceedings of the International Conference on Nano-electronics, Circuits & Communication Systems*, Lecture Notes in Electrical Engineering 403, DOI 10.1007/978-981-10-2999-8_6

pattern, and good impulse response having very less distortion. The Federal Communications Commission (FCC) [3] designated the range from 3.1 to 10.6 GHz frequency band for UWB systems. The challenges to design a UWB antenna include compact size, constant gain, and constant group delay over the operating range. The wide operating bandwidth provided by a UWB antenna creates a possibility for problem of interference with other bands such as WiMAX (3.3–3.88 GHz), WLAN (4.96–6.23 GHz), and ITU (7.9–8.7 GHz). A remedy for this problem would be to design UWB antennas using various band-notching techniques with objective to minimize the interference.

6.2 Design of Band Notch

The values of the width and length of the microstrip patch antenna have been given by Balanis [4].

For operating frequency, the width of microstrip patch is calculated by

$$W = \frac{c}{2f_r}\sqrt{\frac{2}{\varepsilon_r + 1}} \tag{6.1}$$

and the actual length of the patch

$$L = \frac{1}{2f_r\sqrt{\varepsilon_{\text{reff}}}\sqrt{u_0\varepsilon_0}} - 2\Delta L \tag{6.2}$$

where c denotes the velocity of light, f_r denotes the resonant frequency, and ε_r is effective dielectric constant of the substrate.

The notch frequency [5] for the band-notched region may be calculated for the length, L, of the patch

$$f_{\text{notch}} = \frac{c}{4L \cdot \sqrt{\varepsilon_r}}. \tag{6.3}$$

6.3 Band-Notching Techniques

In the past few years, researchers have been implementing various techniques to create notches in the ultrawide frequency bandwidth. This section reviews the techniques and tabulates the achieved results of various antenna parameters. Broadly, the techniques have been categorized as follows:

- Defected ground structure (DGS),
- Slot-loading effects,
- Split ring resonator (SRR).

6.3.1 Defected Ground Structure (DGS)

Defected ground structure may be seen as a structure made of unit cells which may be repeated in one, two, or three dimensions in the ground. These unit cells may be periodic or non-periodic in nature. These structures interrupt the flow of current and change the distribution of current in the conducting ground plane which leads to changes in line capacitance and inductance of a transmission line. Any defect removed from the ground surface of microstrip may change and increase the effective capacitance and inductance.

DGS may be preferred [6] due to the following reasons:

(i) The circuit area is effectively reduced without periodic structures. The reason is that only a few DGS structures have the same distinctive features as that of periodic structure like the stop-band characteristic.
(ii) Various designs of DGS have been investigated. The common among them are concentric ring, circle, ellipse, dumbbells, and U- and V-shaped slots. Each shape may be represented as an equivalent circuit consisting of inductance and capacitance. The equivalent circuit for DGS is easy to calculate, and its fabrication is also simple.
(iii) DGS requires small circuit sizes for performing slow-wave effect. On comparing DGS with electronic band gap (EBG) structure, the design and implementation of DGS has become easier in addition to higher accuracy.

An inverted fork-shaped cut in the ground plane [7] leads to increase in the current path. This excites another resonance. An inverted U-ring strip has been used in the conducting patch to create notches in the impedance bandwidth. Two I-shaped structures [8] are designed on both the sides of the microstrip feed line to achieve wide impedance bandwidth. A G-shaped slot defected ground structure and a pair of Γ-shaped structures in the conducting patch provides dual band notch characteristics. A pair of hook-shaped slots [9] has been introduced to achieve the notch at 5.5 GHz. A slot that is rectangular in shape [10] has been removed in the ground plane. This affects the lower frequency band and improves the impedance matching. To enhance the impedance bandwidth [11], the ground plane has been modified to a symmetrical sawtooth shape by cutting trapezoid-shaped slots on its top edge. A novel fractal triangular monopole microstrip antenna with a truncated notched ground [12] has been proposed. With a truncated notched ground, a better input impedance matching over a wide frequency range has been achieved. A comparison of band-notched antennas designed using defective ground technique is shown in Table 6.1.

Table 6.1 Comparison of designed antenna using defective ground technique

Reference antenna	Size (mm^2)	BW (GHz)	Frequency bands-notched (GHz)	Antenna gain (dBi)	Simulation tool
[7]	12 × 18	3.02–13.86	3.38–4.31 5.1–5.95	3–5	HFSS
[8]	20 × 18	2.8-11.8	33–3.8 5.1–6	2.5–5.5	HFSS
[9]	24 × 32	2.8–11	3.3–3.7 5–6	1.5–4	HFSS
[10]	40 × 30	3.1–9.65	3.2–4	3.5	CST Microwave studio
[11]	30 × 25	3.1–17	3.26–3.68 5.1–5.89	0.5–4	HFSS
[12]	32.4 × 27	2.5–11	3.3–3.7 5.15–5.825	2–5	HFSS

6.3.2 *Slot-Loading Effects*

The loading of radiating patch with narrow slot affects [13] the resonance, input impedance, and bandwidth. The slots may be made in two ways:

(a) On microstrip patch antenna

The slot corresponding to desired wave length along the radiating surface of microstrip patch antenna is created that increases the length of current path which in turn decreases the efficiency and bandwidth.

(b) In ground plane

The slot length may be varied around the axis of the resonating length of rectangular patch. The fundamental resonant length shifts to lower frequency side with the increase in slot length. This indicates the size reduction of antenna.

In [14], by creating and varying the slots within the microstrip feed line, a band-notched UWB antenna has been achieved. The antenna has been designed to cover 3.1–10.6 GHz and provide 3 band notches, thus reducing the interference problem. In [15], a dual band notch characteristic has been achieved. Partial annular slot provides one notch, and semicircular slot in circle-shaped ring patch gives second slot. In [16], an inverted U-ring strip in the conducting patch has been used to create dual band notch characteristic. An inverted fork-shaped slot in ground plane produces an additional resonance. In an open slot antenna [17], two symmetrical slits have been used to obtain band notching. Ultrawideband has been achieved by using stepped strip. A comparison of band-notched antennas designed using slot-loading technique is presented in Table 6.2.

Table 6.2 Comparison of designed antenna using slot-loading technique

Reference antenna	Size (mm^2)	BW (GHz)	Frequency bands-notched (GHz)	Antenna gain (dBi)	Simulation tool
[14]	27 × 30.5	3.1–10.6	3.36–3.88 4.96–6.23 7.9–8.7	3–5	Not mentioned
[15]	24 × 26	2.79–11	3.3–3.8 5.1–6	2.9	IE3D
[16]	12 × 18	3.02–13.86	3.38–4.31 5.1–5.95	3–5	HFSS
[17]	26.5 × 12	2.95–12	5.1–6	2–5	CST Microwave Studio
[18]	30 × 30	3–11	5.3–5.7	2–5.9	HFSS
[19]	50 × 50	2.72–12	4.9–6	4–6	HFSS
[20]	30 × 10.5	3.1–11	4.95–5.98	2–8	HFSS
[21]	25 × 26	2.9–14.5	5–5.86	2.1–4.3	CST Microwave Studio

The antenna [18] having circular shape and defective ground structure has been implemented. Three C-shaped slots are cut in the radiating structure. The antenna [19] has a stub which is elliptical in shape. The band notch is created by cutting H-shaped slot in the stub. An octagonal patch [20] is implemented with two L-shaped slots in the ground plane. A triangle-shaped patch with staircase [21] has been proposed to achieve broad bandwidth. A U-shaped slot provides the frequency notch in the WLAN range.

6.3.3 *Split Ring Resonators (SRRs)*

A split ring resonator consists of two concentric rings. The rings are metallic in nature and are cut on the substrate. SRRs are capable of producing electromagnetic oscillations besides having small electrical length. Composite material is used for preparing split ring resonators. The response of each SRR to EM field is unique. Three types of UWB antenna [22] that consists of planar circular patch and multiple etched slots on the patch are proposed. SRRs have been coupled to feed line. By inserting [23] three resonating structures, namely curl slot, open stub, and C-shaped structure, three band notch characteristics have been achieved. The proposed [24] antenna has been implemented using an ellipse which has an opening. Two resonating structures are cut to create band notches. One is located on the conducting patch and other on the back of the substrate. The two resonating structures are

Table 6.3 Comparison of designed antennas using split ring resonators

Reference antenna	Size (mm^2)	BW (GHz)	Frequency bands-notched (GHz)	Antenna gain (dBi)	Simulation tool
[22]	54 × 47	2.1–105	2.23–2.45 3.26–3.48 5.54–5.88	2–5.9	CST Microwave Studio
[22]	78 × 47	1.89–9.8	2.24–2.52 3.78–4.03 5.94–6.4	2–5	CST Microwave Studio
[22]	68 × 47	1.85–10.4	2.25–2.52 3.53–3.77 5.96–6.3	2–6	CST Microwave Studio
[23]	48 × 60	1.7–11	23–2.8 3.1–3.9 5.6–6.1	2–5	HFSS
[24]	21 × 24	3–11	3.4–4 5–5.9	1–4	HFSS
[25]	43.5 × 39	3.05–11.7	3.3–3.7 5.2–5.4 5.7–6	2.1–5.9	EM simulation software
[26]	35 × 40	2.5–10.9	33–3.6 5.15–5.825	2.1–8	HFSS
[27]	34 × 32	3.1–10.6	33–3.7 5.15–5.82 8.02–8.4	1–5	EM simulation software
[28]	30 × 29	2.9–12	5.2–5.95	3–4	IE3D

connected using metallic holes. A comparison of band-notched antennas designed using split ring resonators' technique is given in Table 6.3.

Two spiral-shaped loop resonators [25] have been combined with the conducting patch to achieve the first notched band for the WiMAX system. Also, microstrip resonators in the ground produce the second and third notched bands. The antenna [26] has been implemented using a circle-shaped monopole patch and a U-shaped aperture on the ground surface. A rainbow-shaped resonant structure is joined on the aperture plane. A complementary split ring resonator (CSRR) is removed from the monopole. This provides two notched bands. The CSRR- and SRR-loaded UWB antenna [27] with triband notching has been demonstrated. Triband capability has been achieved with only SRR and CSRRs introduced at different locations with respect to patch and on the ground surface of the antenna. With this, additional space has not been required. A frequency-notched UWB microstrip slot antenna [28] with a ring resonator of square shape that is fixed firmly in the tuning stub has been proposed. By uniting square ring resonator and slot antenna's tuning stub, a high and sharp band rejection characteristic of more than −20 dB has been achieved.

6.4 Simulations of Band-Notching Technique

The proposed antenna [1] has been designed, and the simulation results have been reproduced using HFSS [29]. The antenna's conducting patch has an arc shape and has a defected ground structure. The modification in ground structure consists of two bevel slots on upper edge and two semicircle slots on the bottom edge. Figure 6.1 shows the geometry of antenna of the reference design. The arc-shaped edge of the radiation patch has a half-ellipse. The major axis of ellipse is 8 mm and has an axial ratio of 1.5. The S-parameter and group delay for the antenna under reference are shown in Figs. 6.2 and 6.3, respectively. The radiation pattern along with gain is shown in Fig. 6.4. A pair of T-shaped stub has been designed inside the elliptic slot to achieve the frequency notch corresponding to WLAN band.

The coplanar-fed monopole antenna [2] is reproduced using HFSS [29]. The ground plane has been slotted to provide additional current path. Figure 6.5 shows the geometry of antenna of the reference design. The S-parameter and group delay for the antenna under reference are shown in Figs. 6.6 and 6.7, respectively. The radiation pattern plus gain is shown in Fig. 6.8. The slots in ground plane change the inductance and capacitance of input impedance to produce WLAN frequency notch.

In Table 6.4, merits and demerits of band-notching techniques have been summarized [30–33].

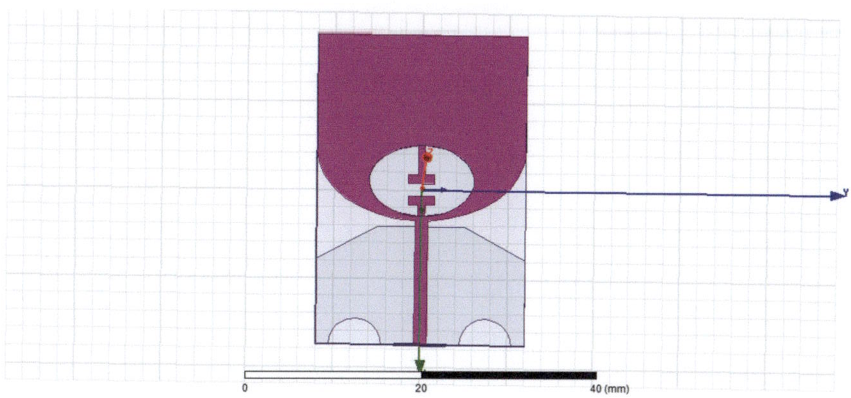

Fig. 6.1 Geometry of antenna

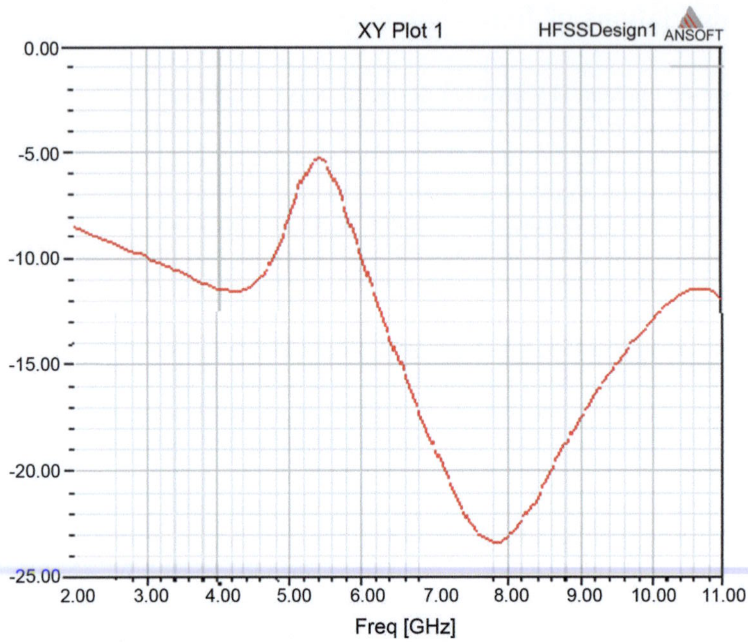

Fig. 6.2 Simulated S-parameters

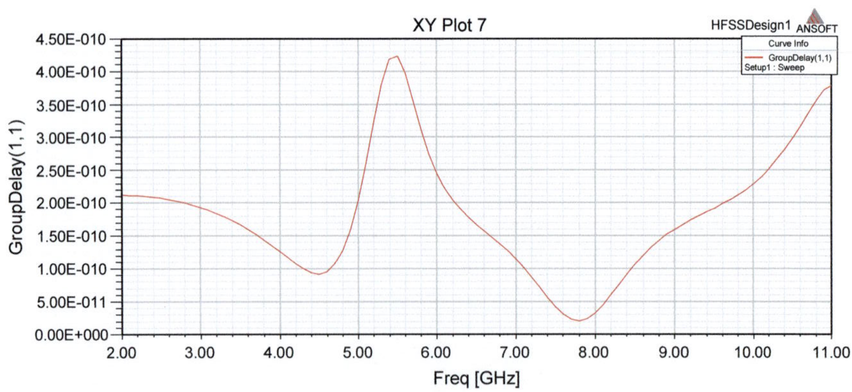

Fig. 6.3 Group delay

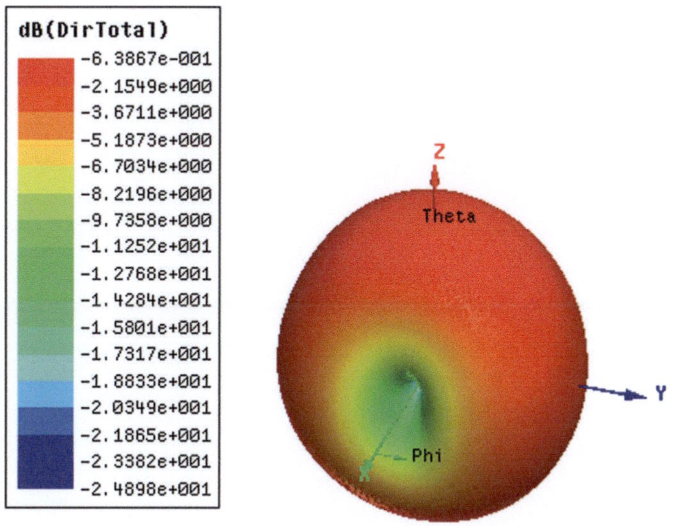

Fig. 6.4 Radiation pattern

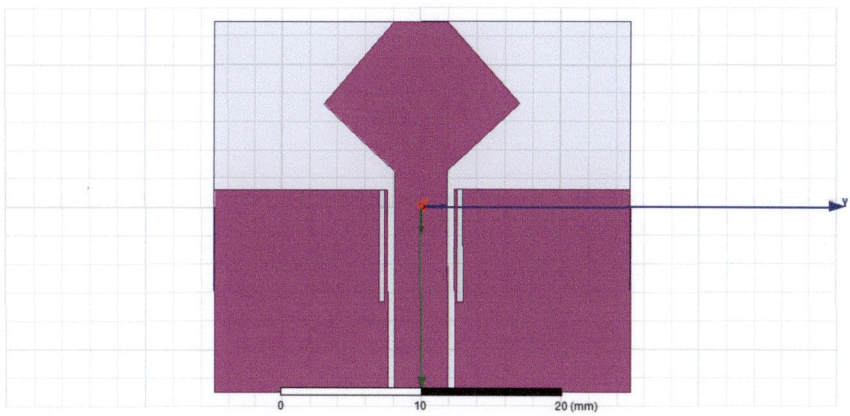

Fig. 6.5 Geometry of antenna

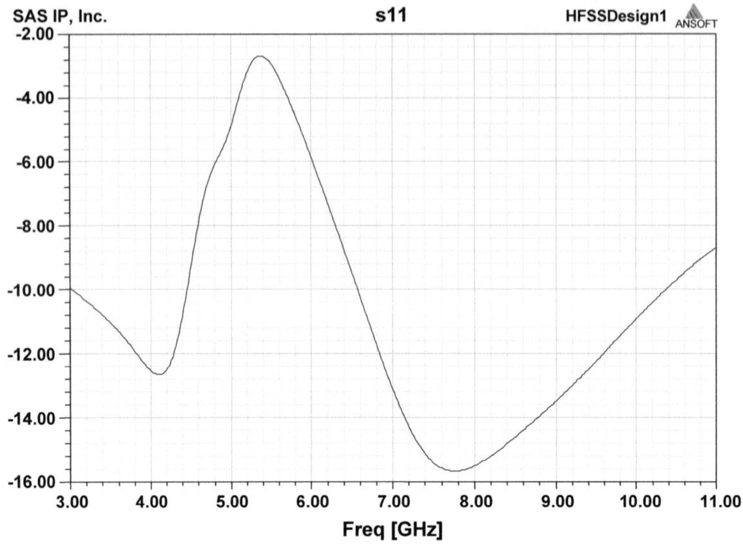

Fig. 6.6 Simulated S-parameters

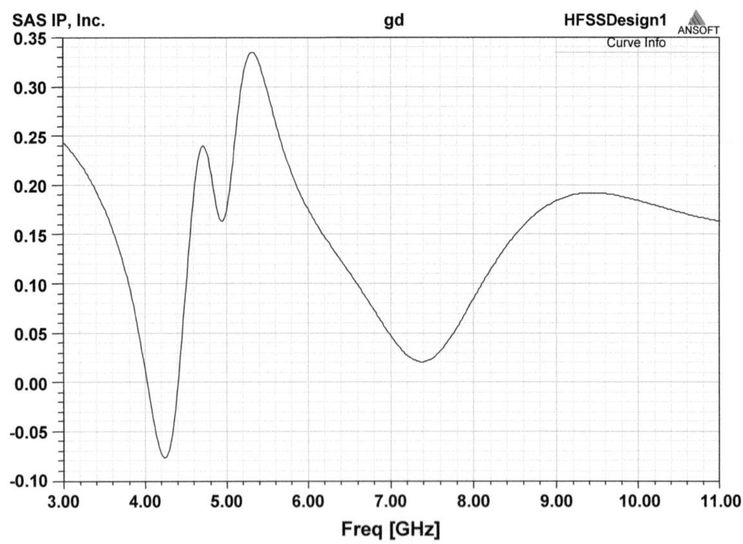

Fig. 6.7 Group delay

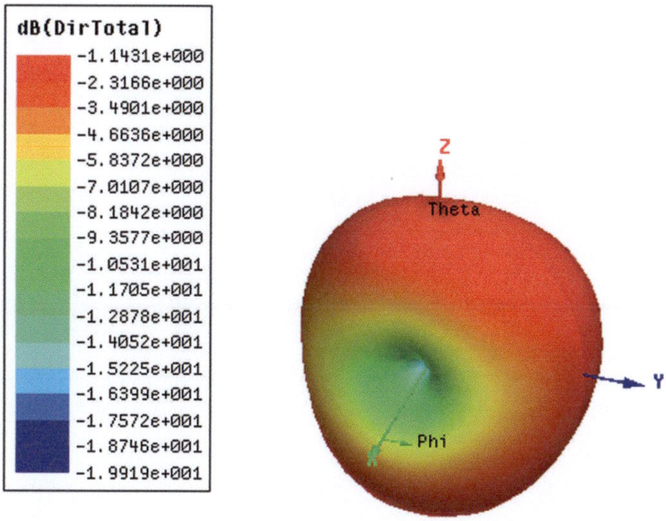

Fig. 6.8 Radiation pattern

Table 6.4 Merits and demerits of band-notching techniques

	Merits	Demerits
Defective ground structure	1. Size reduction 2. Easy to design and fabricate 3. The stop band is wide and deeper	1. Increase radiation 2. Reduction in gain
Slot-loading effects	1. Easy to design 2. Better control of notch band	1. One slot provides one notch band
Split ring resonators	1. Multiple notch bands possible 2. Mutual coupling between cells	1. Q factor may need to be compromised for high band notch characteristic

6.5 Conclusion

In this paper, different band-notching techniques for ultrawideband antenna have been reviewed briefly. The bandwidth of the notched band directly depends on the longitudinal length of the slot, and the shape of the slot has no effect. Each resonator design corresponds to one frequency notch in an antenna. Moreover, a designer may incorporate more than one technique in a single design to achieve band notches. Two antennas under reference [1, 2] have been implemented.

Acknowledgements The authors are thankful to anonymous reviewers for valuable suggestions. The authors also acknowledge the College of Technology, Pantnagar, for providing financial support under TEQIP-II.

References

1. Hong C-Y, Ling C-W, Tarn IY, Chung S-J (2007) Design of a planar ultrawideband antenna with a new band-notch structure. IEEE Trans Antennas Propag 55(12):3391–3397
2. Rajput A, Verma D, Kishan J, Singh T (2012) Design a compact CPW-fed UWB antenna with WLAN band notch characteristic. Int J Sci Eng Res 3(7):1–3
3. FCC (2002) Ultra-wideband operation FCC report and order, Technical Report US 47 CFR Part 15
4. Balanis CA (2013) Antenna theory—analysis and design. Wiley, London
5. Cai Y-Z, Yang H-C, Cai L-Y (2014) Wideband monopole antenna with three band notched characteristics. IEEE Antennas Wirel Propag Lett 13:607–610
6. Weng LH, Guo YC, Shi XW, Chen XQ (2008) An overview on defected ground structure. Prog Electromagnet Res B 7:173–189
7. Ojaroudi M, Ojaroudi N, Ghadimi (2013) Dual band-notched small monopole antenna with novel coupled inverted U-ring strip and novel fork-shaped slit for UWB applications. IEEE Antennas Wirel Propag Lett 12:182–185
8. Abdollahvand M, Dadashzadeh G, Mostafa D (2010) Compact dual band-notched printed monopole antenna for UWB application. IEEE Antennas Wirel Propag Lett 9:1148–1151
9. Ren L-S, Li F, Zhao J-J, Zhao G, Jiao Y-C (2010) A novel compact UWB antenna with dual band-notched characteristics. J Electromagn Waves Appl 24:1521–1529
10. Saada AAR, Alib MMM, Khaled EEM (2013) An integrated 3G/Bluetooth and UWB antenna with a band-notched feature. J Electromagn Waves Appl 27(18):2430–2441
11. Li T, Zhai H-Q, Zhu C, Li L, Liang C-H, Han Y-F (2013) Design and analysis of compact printed dual band-notched ultrawideband (UWB) antenna. J Electromagn Waves Appl 27 (5):560–571
12. Zhang C-W, Yi Y-Z, Liu P-A, Xie J-J (2013) Compact dual band-notched UWB antenna with hexagonal slotted ground plane. J Electromagn Waves Appl 27(2):215–223
13. Shivnarayan S Sharma, Vishvakarma BR (2005) Analysis of slot-loaded rectangular microstrip patch antenna. Indian J Radio Space Phys 34:424–430
14. Cai You-Z, Yang Hong-C, Cai Ling-Y (2014) Wideband monopole antenna with three band-notched characteristics. IEEE Antennas Wirel Propag Lett 13:607–610
15. Azim R, Islam MT, Mandeep JS, Mobashsher AT (2012) A planar circular ring ultra-wideband antenna with dual band-notched characteristics. J Electromagn Waves Appl 26(14–15):2022–2032
16. Ojaroudi M, Ojaroudi N, Ghadimi (2013) Dual band-notched small monopole antenna with novel coupled inverted U-ring strip and novel fork-shaped slit for UWB applications. IEEE Antennas Wirel Propag Lett 12:182–185
17. Su M, Liu Y-A, Li S-L, Yu C-P (2010) A compact open slot antenna for UWB applications with band-notched characteristic. J Electromagn Waves Appl 24:2001–2010
18. Yang G, Chu Q-X, Tu Z-H (2011) A compact band-notched UWB antenna with controllable notched bandwidths by using coupled slots. J Electromagn Waves Appl 25:2148–2157
19. Neyestanak AAL, Kalteh AA (2008) Band-notched elliptical slot UWB microstrip antenna with elliptical stub filled by the H-shaped slot. J Electromagn Waves Appl 22:1993–2002
20. Sim C-Y-D, Chung W-T, Lee C-H (2008) Printed UWB monopole antenna embedded with L-shaped slots in the ground plane for 5 GHz band-notch function. J Electromagn Waves Appl 22:1965–1974
21. Cho YJ, Kim KH, Choi DH, Lee SS, Park S-O (2006) A miniature UWB planar monopole antenna with 5-GHz band-rejection filter and the time-domain characteristics. IEEE Trans Antennas Propag 54(5):1453–1460
22. Zhang Y, Hong W, Chen Y, Kuai Z-Q, Don Y-D, Zhou J-Y (2008) Planar ultrawideband antennas with multiple notched bands based on etched slots on the patch and/or split ring resonators on the feed line. IEEE Trans Antennas Propag 6(9):3063–3068

23. Gong J-G, Li Q, Zhao G, Song Y, Jiao Y-C (2009) Design and analysis of a printed UWB antenna with multiple band-notched characteristics. J Electromagn Waves Appl 23: 1745–1754
24. Ni T, Li W-M, Jiao Y-C, Ren L-S, Han L (2011) Novel compact UWB antenna with 3.5/5.5 GHz band-notched characteristics. J Electromagn Waves Appl 25:2212–2221
25. Natarajamani S, Kumar Behera S, Kumar Patra S (2013) A triple band-notched planar antenna for UWB applications. J Electromagn Waves Appl 27(9):1178–1186
26. Deng JY, Yin YZ, Ren XS, Liu QZ (2009) Study on a dual-band notched aperture UWB antenna using resonant strip and CSRR. J Electromagn Waves Appl 23:627–634
27. Yoo Minyeong, Lim Sungjoon (2013) SRR- and CSRR-loaded ultra-wideband (UWB) antenna with tri-band notch capability. J Electromagn Waves Appl 27(17):2190–2197
28. Lui Wen-jun, Cheng Chong-hu, Zhu Hong-bo (2007) Improved frequency notched ultrawideband slot antenna using square ring resonator. IEEE Trans Antennas Propag 55(9): 2445–2450
29. HFSS, Ansys Corporation
30. Lim J-S, Park J-S, Lee Y-T, Ahn D, Nam S (2002) Application of defected ground structure in reducing the size of amplifiers. IEEE Microwave Wirel Compon Lett 12(7):261–263
31. Ahn D, Park J-S, Kim C-S, Kim J, Qian Y, Itoh T (2001) A design of the low-pass filter using the novel microstrip defected ground structure. IEEE Trans Microw Theory Tech 49(1):86–93
32. Sung YJ, Kim Y-S (2005) An improved design of microstrip patch antennas using photonic bandgap structure. IEEE Trans Antennas Propag 53(5):1799–1804
33. Kamma A, Shrikanth Reddy G, Mishra SK, Mukherjee J (2012) Printed band-notched u shaped ultra wide band antenna with modified split ring resonator. IEEE Asia-Pacific Conference on Antennas and Propagation

Chapter 7
Real-Time Simulation Design for Continuous Process Industries

S. Sridevi and P. Sakthivel

7.1 Introduction

Simulation software is a key for both analysis and training. The end result depends on the capacity of the process engineer to integrate the tools' capabilities with his personal domain skills. Today, steady-state simulation is commonly available. Our objective is to enhance customized modelling and dynamic simulation features. Also, conducting training in real-time environment has many challenges related to variable production capacities, constant technology upgrades, and complex equipment designs. The objective of this paper was to build generic process simulation software, enabling trainees to perform all operations in a risk-free environment in configurable pseudo-real-time simulation workbench.

7.2 Literature Survey and Related Work

In recent years, different frameworks and tools have been developed and used to facilitate process modelling and simulation objectives, such as MATLAB, Modelica, modelling languages, and Aspen. Table 7.1 briefly compares the same [1–3].

These tools have strengths in individual areas more like a tool box, data-driven analytical systems. They are not complete in all aspects to depict intelligent

S. Sridevi (✉)
R&D Centre, Bharathiar University, Coimbatore, India
e-mail: sridevi@shasuncollege.edu.in

P. Sakthivel
Department of ECE, Anna University, Chennai, India
e-mail: psv@annauniv.edu

V. Nath (ed.), *Proceedings of the International Conference on Nano-electronics, Circuits & Communication Systems*, Lecture Notes in Electrical Engineering 403,
DOI 10.1007/978-981-10-2999-8_7

Table 7.1 Tool, advantages, and constraints

Tool	Advantages	Constraints
MATLAB	Tools/scripts for simulation	Expert KB systems, qualitative modelling
Modelica	Discrete equations based cross-domain modelling	More of a language; lacking mature UI features/frameworks
UML/SRML systems	Standardized and widely used and generic	Lacking integrated environment for simulation systems
LISP/CLIPS	Good in engineering modelling of systems	Not suitable for optimization constraints resolution
Commercial: Aspen+/ Unisim, etc.	Scalable/configurable/generic features/libraries	Cost involved/customization/ Convergence issues, non-integrated platform, weak UI

problems with knowledge-based approach and lack mature development environment to design end-to-end simulation systems.

The need is insisted on the process data to be analyzed and presented to reflect the critical underlying trends and process events supporting the operators to get trained on process dynamics [4]. This paper aims at integrating the positive characteristics of such systems and designing the generic and comprehensive process simulation environment [5].

7.3 Simulation Software Architecture

The proposed system consists of trainer and trainee processes (Fig. 7.1).

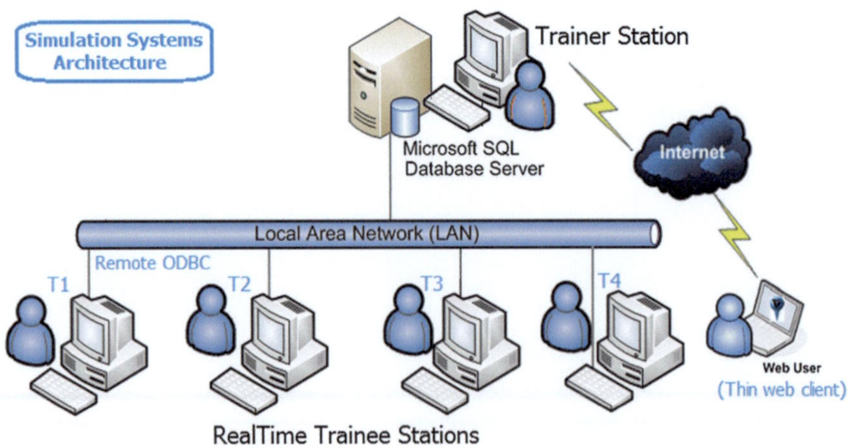

Fig. 7.1 Architecture of simulator

7.3.1 Trainer Console

Typical simulation softwares to train the production plant operators consists of a trainer station where server applications run and are connected to a set of trainee stations in a closed network or through internet. IDE such as visual studio (VC++) suits the system-level programming. The trainer system has database server installed capable of storing pseudo-real-time transactions (storing snapshot of thousands of variable values every second and their retrieval). RDBMS like SQL server/Oracle and other real-time DBs found to be capable of addressing the needs of data capturing and history data storage/retrieval for trends and replay.

The trainee stations are registered first in the trainer for dynamically accessing the processes running in the trainees from server through *remote named pipes* IPC method. Trainer also has list of plant sections/profiles (plant scenarios) that can be loaded to a specific trainee related to the training session.

Trainer station also has features to introduce disturbances during a particular point of training session to check how it is handled by the operator. It can be a simple change to a digital signal (stop of induction fan), or a continuously ramping disturbance of a process variable.

Trainer system also has recording/replay features of how an operator executed the control of the specific processes and grading systems to evaluate the trainees based on a predefined set of rules like output produced, emissions control, equipment trips, critical alarms encountered, and consistency trends.

7.3.2 Trainee Stations

Each of the trainee stations has a set of components as shown in Fig. 7.2 and connected to the trainer data server through remote data connections. The trainee can also be from a remote location, as a web client to trainer's web server. This usually requires high bandwidth internet connection for seamless operations.

All these processes can be remotely started/stopped by the trainer through named pipe communication, and specific profiles can be loaded to a trainee. A trainer also can save a snapshot of a user scenario and load the same at a later point of time to same or different users.

The system also should have a set of predefined/prebuilt scenarios like start-up/steady state of a section of a grinding mill or a burning kiln. It can also be profiles of how new or old plants of different production capacity will behave for specific loads. All of these can be achieved by storing a specific process rules profile dump with all the process variables stored as a snapshot in the database, which can be loaded to the data server.

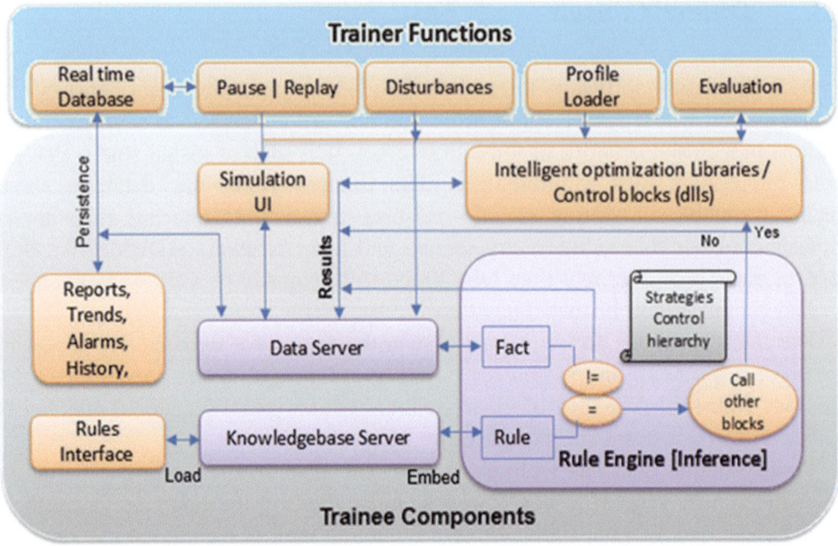

Fig. 7.2 Components of a process simulator

7.4 Simulation System Components

A phased approach is to be planned for design and construction of the below comprehensive generic process simulator components. These components enable users to configure controls/rules at design time and fine tune/simulate at runtime.

Figure 7.2 shows the typical modules in a trainee station. These enables the subject matter expert to model the process in design mode, which once tuned can be simulated at runmode.

Process Rules Editors: An interface for the domain expert to configure rules required for simulating various scenarios of a production system. The rules are built using prebuilt library functions ranging from simple “if-then” statements to PID controller, math functions, and fuzzy/neural blocks.

Knowledge Base [KB] Server: To hold different profiles of rule bases configured. This can be simple conditional rules to complex fuzzy, neural, and MPC blocks whose input/outputs are integrated by the simulation rules/strategies. KB profiles are stored as text file serialization of defined structures which can be updated/retrieved by different components.

The process knowledge base comprises computational blocks that can be independently built as required by a process. The knowledge base is typically made up of a number of rules in a control engine. Also the easy-to-use editors envisioned in following sections are to be built for the configuration of different types of blocks.

User can configure all of these blocks by specific editors, and these are grouped to form a profile, which is saved in flat files. All such combination of rules and blocks for a specific section of the plant (say burning operations of a 900 TPD modern plant) can be grouped and serve as a prebuilt knowledge base for plugging into required scenarios.

The knowledge base server is the running process that loads the specific profiles (rules/fuzzy/neural blocks saved in files with related data loaded from data server) at runtime and acts like a server to other components: read (engine/UI) and write (editor/UI) information at runtime (refer Fig. 7.2).

Process logic group's hierarchy is shown above, where the rules in turn call the special control blocks, during execution. Each of these sections can be sequenced, delayed, or disabled at runtime by specific functions provided by knowledge base server.

Rule Engine: To interpret and execute semantics and process, the rules strategy is written to produce the required output ensuring continuous simulation at preset speed. Once the process rules are built and loaded by the knowledgebase process at runtime, the rule engine executes them by interpreting each line of the rule. As the model of process logic is based on *if and then* with the help of control blocks described below, the engine requires no complex compilation capabilities apart from fetching the current values of variables from data server and assigning them to the logic in knowledge base server.

The whole logic of the currently loaded profile in the defined sequence/delays (refer Fig. 7.3) is executed every second by a separate thread in engine, which is the heart of the simulation process for the logics to get executed and enable the user see the changes in process variables in the user interface.

In real world, the process variables managed by data server has to be refreshed each second by a communication driver connected to PLC/SCADA (programmable logic controllers/supervisory control and data acquisition system) systems, so that these rules get actual values. But in simulated environment, the pattern of the variables value changes to be generated by the rules we write with dependency,

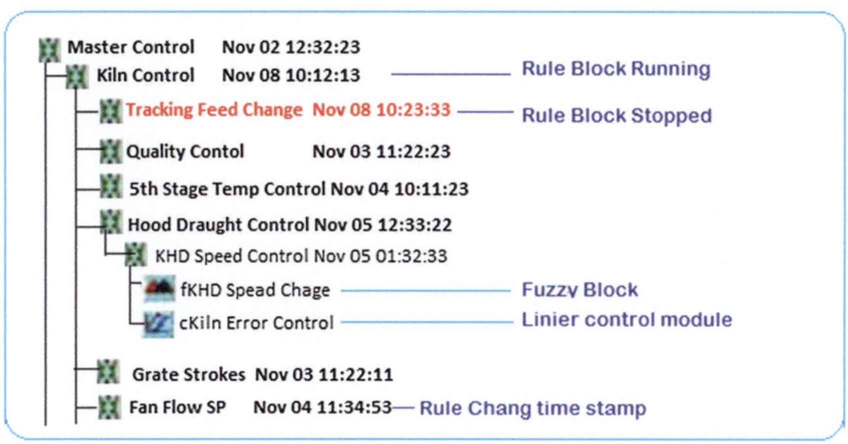

Fig. 7.3 Rules hierarchy in a knowledgebase server

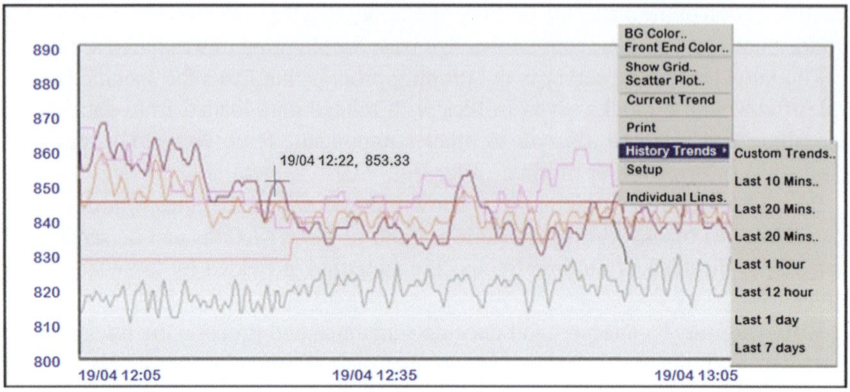

Fig. 7.4 Replay function and control restoration

sequencing, delays, ramps/PID functions, to model scenarios and react to user control actions and set points.

User interface [**UI**]: The UI should act in design and execution mode differently. During design, we configure operational diagrams using controls from tool box and write simulation logic. At runtime, the UI mimics a SCADA-like screens (Fig. 7.5), controls for trainee to observe process variable changes/alarms/trends/process interlocks and enter set points.

Data Server: To hold number of plant parameters/variables to simulate real-time environment and act like a soft PLC, to hold real-time data of different types. This data structure for each profile is persisted on files and loaded accordingly. All other processes access the current value of variables through IPC.

Database: To store the history of high-volume simulated data and render trends/alarms/replay of steps, etc. for each profile.

Knowledge Base Serialization and Execution: The primary task of any simulate package is to have a configurable process knowledge/scenarios builder (Fig. 7.7), which manipulates the process variables to replicate different process conditions like plant start-up, and reaching steady state and shutdown. Our aim was to build a framework/architecture which can be used to model process knowledge with the help of building blocks such as process rules block, fuzzy/neural, MPC blocks, and correlation blocks.

Replay features: As every process/internal variable is recorded in database per second, the trainer can explain the trainee by choosing a specific period of operation of the trainee and replaying it by fetching the past process variable values and loading it to data server module. User can stop this replay at a particular event/stage and take manual control to bring the process back to stability (Fig. 7.4).

7.4.1 Communication Methods

All of these components listed above are running as separate processes in each of the trainee systems. These processes interface with each of these processes or exchange data with them using a concept called IPC (interprocess communication)

Different IPC mechanisms like shared memory, pipes, message queues, and memory mapped files are commonly used and we propose "named pipes" here. It is a two-way data exchange system, where the data is read or written back by one data element or structure. The pipes communication can also be implemented between 2 running processes and threads either in same or different computers, and the latter is called "remote named pipes." The server application in trainer interfaces/exchanges data with each of the trainee processes using this mechanism.

Typically pipes are created in C/C++ like CreateNamedPipe (\\<server_-name_or_ip\\pipe\\ <Pipename>). A client process connects to a server named pipe by using CreateFile function, with read or write attributes.

As all the components in Fig. 7.2 run as separate processes (.exe), there should be a master spawning process that initiates these components and also watches their running status, so that it reinitiates the instance/connections in case of failover at runtime.

7.4.2 Simulation Interface

User views the different sections of a plant in simulation in an interface which mimics a control room system.

Design mode: An interface to be provided for the process engineer to build these diagrams by drag and drop of different ActiveX controls which can be chosen from tool box.

An ActiveX control helps to show animations and based on component object model (COM) and supports different OLE functions. There will be a control box with different ActiveX control elements like placing an image file, a motor with changing colors based on a status, a variable display text box for data monitoring, or to display a simple flow diagram as a configurable control as in Fig. 7.5, which when used in different combinations depict different scenarios. The control attributes (position, states, mapped variable, etc.) of each control configured are serialized in standard structures. In design mode, visual simulation to be provided for tuning individual blocks for specific range of inputs is developed.

Runmode/simulation: Controls placed in a MFC ActiveX container shall be executing their functions as set in attributes.

The mimics display is implemented by ActiveX container which gives an environment to host and run ActiveX controls. The container framework provides support for event handling, in-proc server (dll) implementation of embedded objects and multiple events handling during execution of the controls.

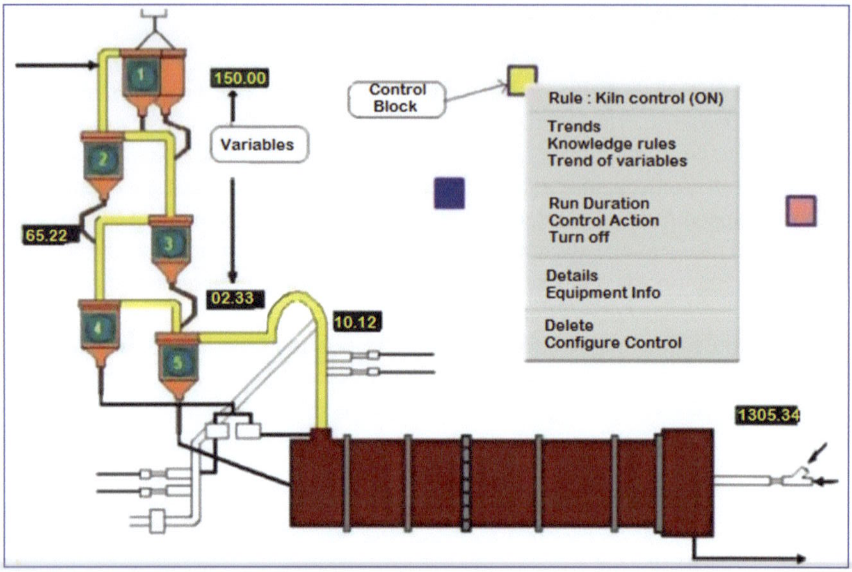

Fig. 7.5 UI building phase by drag and drop of controls

S.No	ID	Variable	Type	Attribute	Length	Trending	Description
1	35	ElecEarArray	Internal	Integer array	4	No	Array to store electronic Ear values
2	44	N-MotarKWArray	Internal	Float array	120	No	
3	56	RetrunFeedAvg	Internal	Float	1	Yes	
4	57	VRMIDFanRun	Internal	Integer	1	No	

Fig. 7.6 Sample list of configured variables and attributes

The ActiveX controls are packaged in an.ocx file, which has to be registered in windows system registry (using regsvr32). COM creates a ClasssID which is a unique GUID for each control. Different versions of a same control can be maintained. The container loads the control using LoadLibrary (dll name) and uses a GetProcAddress function to get the Dll entry point.

7.5 Process Model Design and Execution

7.5.1 Data Server and Variables

A simulation UI is based out of thousands of variables, which have mapping to process variables and those used for internal calculations. User defines/updates these variables (system maintains unique ids). The system supports different types like int, float, arrays, and digital (Boolean) variables (Fig. 7.6).

Arrays are often used to collect the process variable trend over a period of time and to take the average for smoothening, change in error, and PID functions. These configured variables are created at runtime and maintained by data server block which is used as real-time server for the data sharing across modules.

7.5.2 *Process Rule Configuration*

A rule editor is an interface given to process engineer to sequence various rules based on the components available.

Executable statements (rules) are put together in logically related groups to form process rule blocks. These act as a supervisory layer over fuzzy blocks, correlation graphs, or other sub-logics. Control objectives are defined in different rules, and these can be executed collectively or on priorities as required by the process.

A rule block is typically made up of a number of steps with no restriction on the number of steps within a rule. Each step groups a set of actions to be taken, conditionally or otherwise. If the step is conditional, then it contains a decision part and an action part.

Each of these parts is made up of blocks. In the conditional part, the blocks are separated by connectors "And"/"Or." Non-conditional steps contain only actions. Thus, all steps contain actions. A step can contain multiple actions. The rule is built through the addition of steps to it. Steps are built block by block.

Each block represents certain operations (with associated function is referred in Math library) like conditional checks, calculations, block executions, and so on. The functions in a step are written by selecting the parameters required by it. These parameters can be variables or other computational blocks (Fig. 7.7).

```
Step No : 2
 If
  PreCalCoalChange IsGreaterThanorEqualT (0.10)
Then
 Transmit(W_PreCalCoalFeed, PreCalCoalChng, PCLL,PCHL,10)(PCCoalOut)
DelaySteps (2,2,2)
Time : 1
Step Status : Running
Step No : 3
 If
  ArrayValueAtIndex(RFeed_array,10) IsLessThan AvgFeedTPH
 Then
 FeedTemp = Max(FidAvg(R_CalcArray),20)
  Execute(F_CControBlk)
```

Fig. 7.7 Sample rules block with different parameters

These rule blocks linked in a hierarchy as shown in Fig. 7.3 act as a decision tree and decide the functioning of the framework through arithmetic and logical functions. Different tasks like time delay between steps, ordering, enabling a step or deactivating it are defined in the rules setup. Process logic block provides different primitive functions, which carry out the suggested action. Prebuilt functions such as "Run" enables a strategy to execute a fuzzy block, correlation table, or other logical blocks.

To communicate the actions to an external module, a function like "Send" function is provided. Process logic block support a number of decisions making, reasoning, and arithmetic functions using predefined libraries (dll). The rule engine parses the semantics by classifying tokens/positions in defined syntax, interprets natural text patterns, and initiates appropriate action [6].

7.5.3 Computational Block Library

The math block caters to all the typical mathematical calculation needs. This math block provides additional functions like derivative and sliding average. A library of standard functions is provided for building process rules, which the user can use to write simple *if..then..else* statements to complex event sequencing logics.

Different functions are available to the user in if and then cases like comparison (equal, >, <, etc.), mathematical (add, multiply, etc.), logarithmic, PID-based, etc.; each function takes predefined set of inputs and gives the output which can be assigned in a variable. These functions are packaged in a C/C++ dll, which can be loaded at runtime by rule engine. The process rule editor has features to plug-in these math functions as building blocks and is loaded by rule engine dynamically at runtime.

7.5.4 Predictive Control Blocks

Apart from the math library, the simulator logic can be built by advanced control blocks with specific UI and tuned model, which can be loaded in process rules as one block by providing input/outputs. The below section describes few of them in detail (Fig. 7.8).

7.5.5 Fuzzy Blocks Configuration

Complex process loops where multiple input parameters have an effect on one or more output parameters are best modeled using fuzzy logic. The fuzzy editor facilitates the building of fuzzy logic blocks for control. Any number of fuzzy

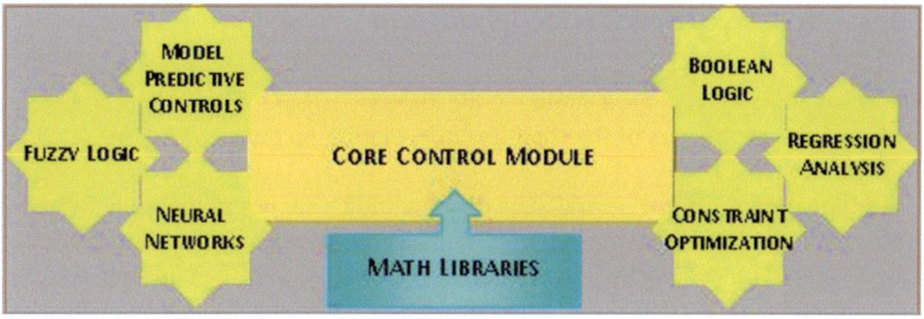

Fig: 7.8 Special control blocks

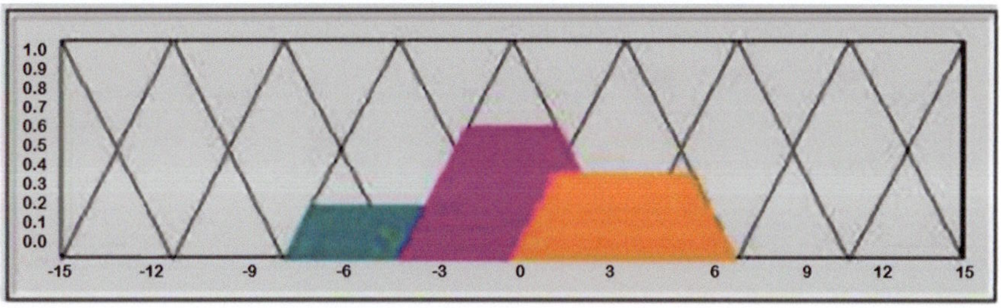

Fig. 7.9 Sample fuzzy block simulation for inputs

blocks can be built, and the first step in configuring a fuzzy block is to define the input and output parameters. For each parameter, user can define membership sets, range, shape, and count of members (Fig. 7.9).

Rules for each fuzzy logic block are framed, and the entire block is built as per the knowledge elicitation of the process experts. The process rules validate all inputs and outputs of a block before controlling the process.

Fuzzy logic configurator provides the power to perform heuristic modeling. The input and output fuzzification/de-fuzzification can be achieved through a group of membership functions. The associated fuzzy rules help to simulate the linguistic terms like "slightly higher than."

Linear Correlation modelling: Represents the statistical relationship between 2 random variables. Here, a simple single input, single output from the process system is required; the user does not have to go in for fuzzy calculations. Instead, the lookup kind of matrix block can be used. This block is configured using a graphical editor where the relationships between an independent and a dependent variable are drawn.

7.5.6 Neural Network (NN) prediction

NN adds adaptability and self-tuning to the knowledge base. This block can be used to change the parameters of the shell itself depending on the environment in which it operates.

Training phase of the neural network (training can be built on simple back propagation algorithm/supervised learning) results in adapting to the relationship between input/output parameters through repeated process of adjusting weights. Once the network is trained and fine tuned for sample inputs-output set, this prediction block, when looped to the simulation model, predicts the output every second based on the current input parameter variations. Based on the error that we receive from the input parameters are adjusted to achieve the desired output [7]. The algorithm is written in a dll, which can be called in process rule block. NN configurator used to set all training attributes for each problem.

7.5.7 Model Predictive Control (MPC) Blocks

MPC works well to achieve nonlinear MIMO objectives. MPC are useful to control where standard PID delays and inverse responses fail to take control. This works on receding horizon principle to work on historic data of control actions and predict set of MV in target horizon [8]. The dynamic model of the system is used to find the impact of future actions of the MV on the output. The moves of the MV in future horizon are decided by how the predicted error is minimized within the operating range.

7.5.8 Real-Time Tuning and Validation of Simulation Block

Simulation deign has a system identification and continuous learning phases to fit the model to process dynamics.

The tuned model when looped as a predictor block responds to input changes, and the predicted output is fed to the optimizer module to decide steps required to transmit the control action back to closed process loop, as depicted in Fig. 7.10.

To tune these control blocks, to a specific control objective, they have to be plugged into a real-time system for continuous model learning. To model the process behavior, the past history data of inputs and outputs are collected, validated, and used to train the neural network or nonlinear MPC blocks offline. Fig. 7.11 summarizes the proposed model for simulation blocks tuning.

Once tuned with multiple parameters, we leverage optimal prediction output of neural/MPC/PLS (partial least square) modules with required data validation and suitable training/prediction algorithms like backpropagation for neural and receding

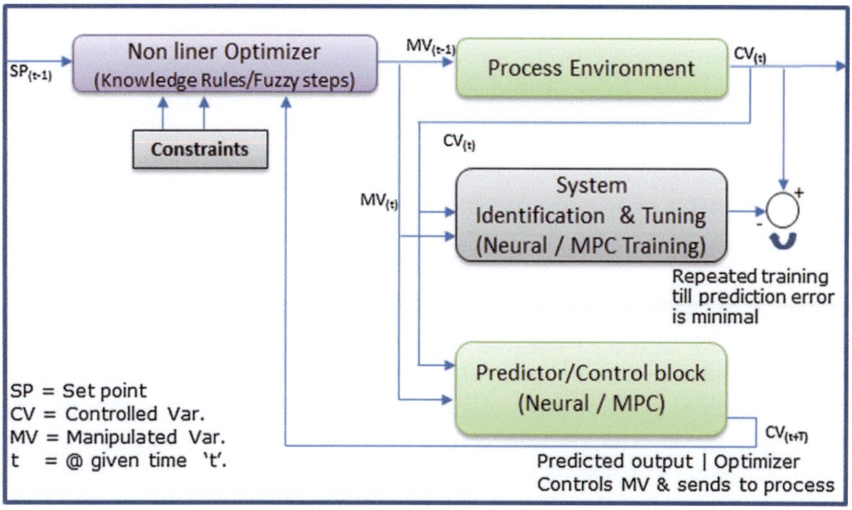

Fig. 7.10 Simulation data generation: neuro-MPC with fuzzy/KB optimizer

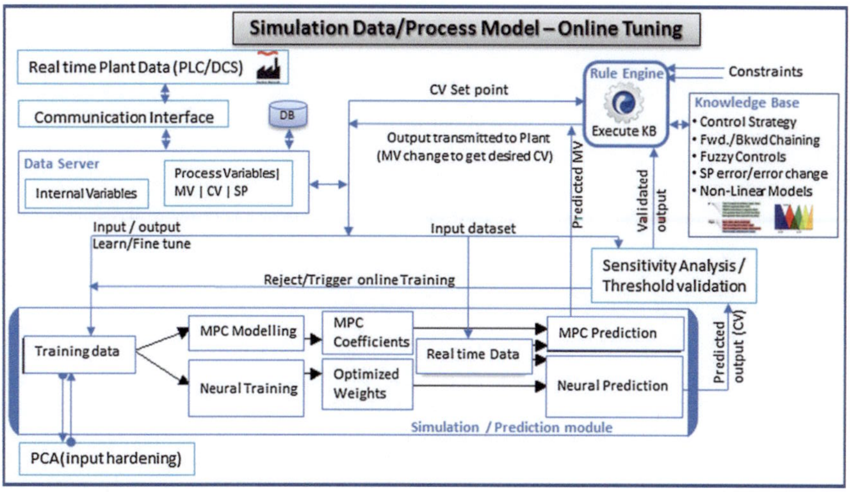

Fig. 7.11 Simulation block tuning with online training

horizon method for MPC. These algorithms are parameterized to customize for tuning the modelling and control aspects of the model to suit specific needs [9].

To maintain desired set point of the output variable (CV), the error and the direction change of error are given to a typical fuzzy membership functions for determining the desired process variable changes. A control strategy component which integrates the calling of these model blocks with if-then-else statements is built over process variables and constraints. The communication interface fetches

Table 7.2 Sample training file format/data collected from a grinding unit

Manipulated variable				Control variable
Input-1	Input-2	Input-3	Input-4	Output-1
Feed (TPH)	Pressure	Airflow	Inlet-temperature (C)	Fineness (cm^2/g)
144.52	301.83	95.92	59.69	3214
145.72	302.79	102.44	61.58	3090
145.97	254.48	101.67	60.58	3351

the process values to these strategies in real time, and the rule engine executes these strategies every second to integrate the system's input/outputs and transmit the suggested control actions by the model to plant automation systems.

The fine-tuned simulation block is capable of predicting the manipulated variable (MV) through MPC and the control variable (CV) in NN and MPC. These blocks generate the range of simulated I/O variables to represent a closed loop nonlinear dynamics of the system under control. The objective is to predict a CV outcome each second that is otherwise measureable at higher intervals physically and taking continuous proactive changes on inputs to get desired output.

For example, predicting fineness of a material ground (output) is based on input parameters like grinding pressure, moisture, temperature, and load on the grinding table.

The impact of each input parameter on the output is analyzed by principal component analysis (PCA), and the most influential input parameters are fed to the training model [10]. Sample training file format/data collected from a grinding unit is given in Table 7.2.

7.5.9 Training the Prediction Network

Soft computing models such as neural/MPC enhances nonlinear statistical, probabilistic data and act as adaptive optimization tools for learning, predicting, and classifying new patterns based on historical data [11]. The input and output set is given to the neural network to enable supervised learning in back propagation algorithm. Once the model is trained, the final weights are saved in weights file. Similarly, the MPC training stores the resultant coefficients.

7.5.10 Result Analysis

The simulation model is tested for controlling a material grinding unit with all the UI designed with variables positioned appropriately and required rules/fuzzy/NN blocks configured. Getting the simulated value of single output variable (fineness) is described here. For NN training, we get optimal results for the parameters such as 5 hidden layers with 6 nodes each and momentum of 0.2 in a 1000 epoch cycles with a sigmoidal activation function resulted in optimal outputs for the problem chosen. The trained model (optimized weights and coefficients) could predict the output with the simulated inputs and without the guiding output. By linking the trained model [weights/coefficients optimized for near-zero errors] to a control engine program, prediction process can be triggered for each second.

To make the control more proactive and to predict like real time, the random functions generate the MV (input) in preset ranges, every second, and feed to the model. The output from neural/MPC predictions are compared and averaged. The output is fed to *sensitivity analysis* with major influencing input variables to compare whether the predicted output has any major variance, to ensure the results are in the range. The validated output flows through the optimizer which is a control strategy/fuzzy block.

The difference between the predicted output and the set point are derived (error). To minimize the error (objective function), rule strategies/fuzzy blocks are configured to suggest input parameter changes (MV) and transmitted to field. Based on this, the inputs like mill feed rates, grinding pressure, airflow, and separator settings are adjusted to maintain constant specific target/set point (CV).

As a closed loop control system, the output error (set point-predicted) drives the input changes direction and magnitude for reaching process stability. If the prediction range is moving away from control objectives constantly, offline model training is triggered based on more recent plant data, to get tuned to correct process dynamics. With the simulated fineness values changing on UI, trainee takes appropriate control on MV and brings the CV near to set point. The trainer could record/replay control action taken by different trainees and introduce disturbances and share evaluation reports of critical parameters controlled with trend.

7.6 Conclusion and Future Research

The concept and design considerations of building comprehensive, indigenous simulator suitable for process training, is elaborated in this paper. As less research has gone into the area, this cost-effective solution shall be helpful for process industries.

The industry or domain agnostic simulation is aimed through designing process profile set with expandable function libraries and controller blocks. Also the system is scalable to configure and execute multiple profiles and scenarios, which are uniquely combination of knowledge, data, and controls, which addresses the training needs of different operational sections/conditions.

This prototype helps to achieve enhanced productivity, safety, and quality resulting in financial benefits and reduce equipment trips/shutdowns/material wastage, etc. These simulators have potential usage in educational institutions to display the functioning of sections of plant and reactions.

Testing simulation model for grinding section of cement plant with sufficient data collection is done to test the process rules/knowledge with the help of the domain expert.

The overview of generic simulator components and their interactions discussed has good research potential in each of the sections elaborated, to be developed into a fully operational and cost-effective process simulator, having varied usage.

References

1. Chiang LH, Russell EL, Braatz RD (2001) Fault detection & diagnosis in industrial systems. Springer, London
2. Sanz V (2012) Modeling of hybrid control systems using the Modelica library. Control Eng Pract 10:24–34
3. Li N, Wenqing W (2012) Development & application of intelligent system modeling. Simul Model Pract Theory 29:149–162
4. Venkatasubramanian V, Rengaswamy R (1992) A framework for process monitoring, diagnosis, & control using knowledge-based systems & NN. IFAC, Delaware, pp 49–54
5. Uraikul V, Chan CW (2007) Artificial intelligence for monitoring & supervisory control of process systems. Eng Appl Artif Intell 20(2):115–131
6. Vlas R (2011) Rule-based natural language techniques. In: 44th Hawaii international conference on system sciences
7. Li N, Deng (2010) A novel optimization algorithm: particle back propagation NN optimization, CACS
8. Holkar KS, Waghmare LM (2010) Overview of MPC. Int J Control Autom 3
9. Konakom K, Kittisupakorn P (2010) Neural network-based model predictive control (NNMPC) strategy. In: World congress on engineering and computer science 2010, vol II
10. Perros H (2009) Computer simulation techniques: definitive introduction. NC State University
11. Chandwani V, Agarwal V (2014) Applications of ANN in strength of concrete: a state of the art review. Int J Curr Eng Technol 4(4):2949–2956

Chapter 8
Performance Enhancement of LTE HetNet Using EVM-Based Constellation Combiner (ECC) in WARP

R.K. Mugelan, M. Anusiya and M.A. Bhagyaveni

8.1 Introduction

Long-term evolution (LTE) allows users to operate in a new and wider spectrum and complements 3G networks with a flat IP-based architecture, lower latency, and higher data rates [1]. For the service implementation and network design of LTE, the heterogeneous network or "HetNet" architecture is the preferred architecture. In basic terms, HetNets have three major components: the Macro, or "traditional," base station; a dense network of small cells (pico and femto) that delivers bandwidth, particularly in dense traffic areas. LTE-based heterogeneous networks are about improving the spectral efficiency per unit area. The network energy efficiency (EE) considers the energy consumed by the base station (BS) and also the capacity and coverage of the network. LTE-femtocells that are known as 'home evolved node base station (HeNB)' have low-cost, low-power cellular network access points that use cable broadband connections, residential DSL, or optical fibers to connect standard mobile devices to a mobile operator's network [2].

In [3], to get better data coverage and indoor voice, the data access points which are less expensive alternatives are installed by home users. Femtorelays has been discussed in [4], for improving the transmission reliability of femtocell users and possibly, also macro-cell users. In [5], a larger indoor environment is setup using femtorelays instead of femtocells, which performs well at both low interference and high interference. In [6], the two most popular relaying techniques are

R.K. Mugelan (✉) · M. Anusiya · M.A. Bhagyaveni
Department of Electronics and Communication Engineering, College of Engineering Guindy, Anna University, Chennai, Tamil Nadu, India
e-mail: mugelan.r.k@gmail.com

M. Anusiya
e-mail: anud120@gmail.com

M.A. Bhagyaveni
e-mail: bhagya@annauniv.edu

V. Nath (ed.), *Proceedings of the International Conference on Nano-electronics, Circuits & Communication Systems*, Lecture Notes in Electrical Engineering 403,
DOI 10.1007/978-981-10-2999-8_8

amplify-and-forward (AF) relaying and decode-and-forward (DF) relaying which are discussed with linear combining techniques. The authors in [7] proposed and analyzed selection combining (SC) at the destination for differential amplify-and-forward (D-AF) relaying over slow Rayleigh-fading channels. In [8], the EVM is being calculated from receiver constellation diagram which can be used to assess the quality of digitally modulated signal. Since there exists direct relationship between BER and EVM, it is better to calculate EVM, because EVM is an equivalent performance metric of BER which can be calculated by symbol level instead of bit level, which saves considerable time and complexity in the receiver.

In this paper, EVM-based constellation combiner (ECC) is used to evaluate the performance of the LTE HetNet. The proposed work has been done in wireless research test bed called wireless open-access research platform (WARP) [9]. WARP provides a scalable and configurable platform to implement wireless communication algorithms for educational and research purposes in a real-time scenario.

The rest of the paper is organized as follows. In Section 8.2, the LTE HetNet system and transceiver design are discussed. In Sect. 8.3, experimental setup of the LTE HetNet in WARP is explained. Performance evaluation of EVM-based constellation combiner (ECC) technique is presented in Sect. 8.4, and conclusion and future work are given in Sects. 8.5 and 8.6.

8.2 LTE HetNet Network

8.2.1 System Model

Consider a LTE HetNet system model as shown in Fig. 8.1. The transmission of data from eNodeB (EB) to user equipment (UE) is organized in two time slots. In the first time slot, the eNodeB transmits the data to both the HeNodeB (HEB) and the user equipment (UE) (bold lines in Fig. 8.1). In the second time slot, the HEB retransmits a version of received source signal by amplifying them with a factor called amplification factor to the UE (dotted lines in Fig. 8.1).

The received signals $r_{b,u}$ and $r_{b,h}$ at the destination and the relay, respectively, can be written as follows [9]:

$$r_{b,u} = T_B H_{b,u} x + n_{b,u} \tag{8.1}$$

$$r_{b,h} = T_B H_{b,h} x + n_{b,h} \tag{8.2}$$

where T_B is the transmitted power at the EB, x is the transmitted information symbol, and $n_{b,u}$ and $n_{b,h}$ are the additive noises. In Eqs.(8.1) and (8.2), $H_{b,u}$ and $H_{b,h}$ are the channel coefficients from the EB to UE and HEB links, respectively.

In the second time slot, the HEB amplifies the received signal from EB with an amplification factor α and retransmits the revised version of the received signal to

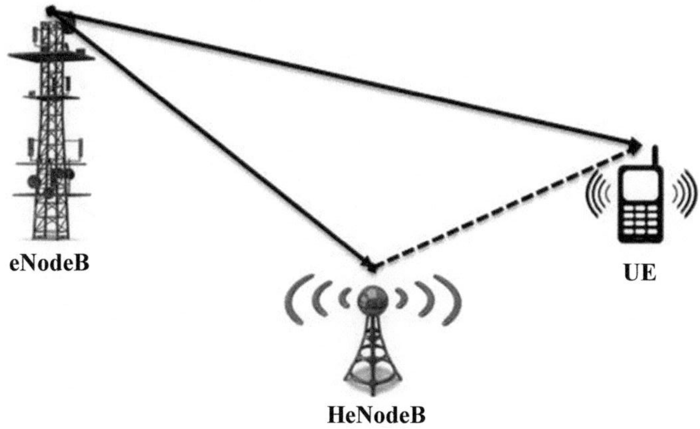

Fig. 8.1 LTE HetNet

the UE with the transmitted power T_H. The received signal at the UE in the second time slot is specified in (8.3) as follows:

$$y_{r,m} = \frac{\sqrt{T_B T_H}}{\sqrt{T_B |h_{b,h}|^2 + N_0}} H_{h,u} H_{b,h} x + n'_{r,u} \tag{8.3}$$

$$\alpha = \frac{\sqrt{T_B T_H}}{\sqrt{T_B |h_{b,h}|^2 + N_0}} \tag{8.4}$$

$$n'_{r,m} = \frac{\sqrt{T_H}}{\sqrt{T_B |h_{b,h}|^2 + N_0}} H_{r,u} n_{b,h} + n_{r,m} \tag{8.5}$$

After receiving the transmitted data from both the direct link and the relay link, the UE will undergo the proposed combining technique called EVM-based constellation combiner (ECC) to get a better QoS using both the links, which is explained below.

8.2.2 *Transceiver Chain*

In this section, the detailed flow of the transceiver chain used in EB, HEB, and UE integrated with WARP test bed for single HEB network has been explained with respect to the flowchart as shown in Fig. 8.2.

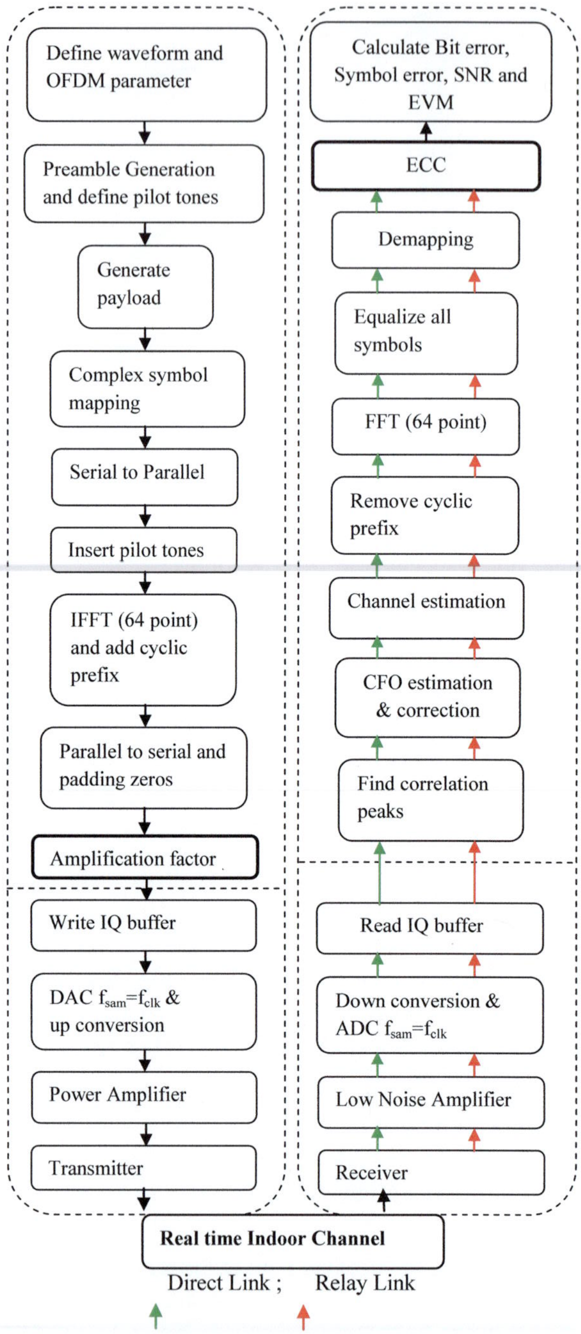

Fig. 8.2 Flowchart of transceiver chain

8.2.2.1 Transmitter

After setting the experimental parameters as given in Table 8.1, the preamble sequences are generated. The short training sequence is chosen as follows: (STS-f) = 64 samples and (STS-t) = 16 samples, which is repeated 30 times giving 480 samples followed by the long training sequence (LTS) = 64 samples repeating 2.5 times giving 160 samples. Altogether, we get the entire preamble sequence of length 640 samples (LTS + STS). The transmitting data sequence is being generated randomly with respect to N_{OFDM} number of symbols and N_{DATA} number of data symbols.

The data bits inserted from the source are mapped using 16-QAM modulation technique and converted from serial to parallel through S-P convertor. The pilot tones are defined as [1 1 −1 1], and the pilot subcarriers are [8 22 44 58]. Then, these pilot tones are inserted and repeated for every N_{OFDM}.

The 64-point IFFT is applied to the code word $C = [c_0\, c_1 \ldots \ldots c_{63}]$ generated to convert it to time domain. IFFT output at nth time interval ($n = 0, 1, \ldots 63$) is given by Eqs. (8.6), (8.7), and (8.8):

$$x_n = \frac{1}{8}\sum_{k=0}^{63} c_k e^{\frac{j2\pi kn}{64}} \tag{8.6}$$

In matrix form,

$$[x_0 x_1 \ldots\ldots x_{63}]^T = D^H [c_0 c_1 \ldots\ldots c_{63}]^T \tag{8.7}$$

$$\left[x_0^T x_1^T \ldots\ldots x_{63}^T\right]^T = (D^H \otimes I_2)\left[c_0^T c_1^T \ldots\ldots c_{63}^T\right]^T \tag{8.8}$$

The 64 × 64 matrix D^H realizes the IFFT operation. Hence, D is a FFT matrix.

After converting the code word into time domain to avoid inter-symbol interference, a guard interval vector (cyclic prefix) of length 16 denoted as

Table 8.1 Experimental parameters

Parameters	Values
Number of OFDM Symbols	180
Modulation order	16 (16-QAM)
Number of sub-carriers	64
Cyclic prefix length	16
FFT offset	4
LTS correlation threshold	0.8
Transmitter RF gain mode [12]	High
Transmitter power (dBm) [12]	−18.5 to −9
Automatic gain control	Enabled
Number of EB, HEB, and UE	One

$X_g = [x_{-16}\, x_{-15} \ldots\ldots x_1]$ is added in front of the code word $X = [x_0\, x_1 \ldots\ldots x_{63}]$ so that OFDM symbol $X' = [X_g\; X]$.

After IFFT processing and cyclic prefix addition, parallel streams are converted into serial time-domain samples using multiplexer and transmitted through air using multiple carrier frequencies. The data to be transmitted have to be padded with zeros in order to match the buffer size of the WARP kit. The buffer size of each radio board is 32768 bits.

The data to be transmitted are given as an input to WARP kit which will be time-domain samples with real and imaginary parts in the first time slot. In the case of the second time slot, the data to be transmitted are multiplied with an amplification factor α. These data to be transmitted are stored separately on I & Q buffer. The stored data are converted into analog waveforms using a DAC. Two analog waveforms are generated separately from real-valued samples obtained from I & Q. Here, $f_{\mathrm{sam}} = f_{\mathrm{clk}}$. The baseband signal is upconverted to RF carrier frequency. Then, finally, the modulated signal is amplified using RF amplifier and transmitted using 2.4/5 GHz Omni Antenna.

8.2.2.2 Receiver

In the receiver end, the RF signal is received by an Omni Antenna, which will be amplified using power amplifier to increase its signal strength and then downconverted into baseband frequency. The incoming signal is correlated with inphase and quadrature-phase carrier separately to recover inphase and quadrature-phase baseband components separately. Further, the analog waveform is converted into digital samples using ADC. The sampling frequency of ADC is equal to buffer clock frequency. Inphase and quadrature-phase samples are obtained from separate ADC. After conversion, the samples are stored in a buffer. As the samples are not passed through the FFT block, the samples are in time domain. So received time-domain LTS (long training sequence) samples are correlated with received vector samples.

$$\text{LTS} = 64 \text{ Samples.}$$

The time-domain OFDM waveform involves repeated block of [LTS STS (short training sequence) payload]. As LTS is transmitted twice while performing correlation, it gives two peaks and autocorrelation gives a peak in the middle. The difference between the two peaks gives timing information about one OFDM symbol because the number of samples in LTS and OFDM symbol is equal.

In a communication system, CFO refers to the difference in carrier frequency at transmitter and receiver. This offset is termed as CFO. The input carrier frequency at the receiver can vary due to Doppler effect caused by relative motion between Tx and Rx. The frequency offset value is determined using Eq. (8.9):

$$r(t) = x(t)e^{j2\pi f_\Delta t} \tag{8.9}$$

Given that short preamble is periodic with $\delta_t = 0.8\,\mu\text{s}$, Eq. (8.9) is written as (8.10)

$$r(t - \delta t) = x(t)e^{j2\pi f_\Delta (t-\delta t)} \tag{8.10}$$

At the receiver, both $x(t)$ and $r(t)$ are known.

Calculating the angle at both sides of Eq. (8.10), we get the frequency offset as in (8.11):

$$\Delta_f = -\frac{\forall y(t-\delta t)y^*(t)}{2\pi\delta t} \tag{8.11}$$

The LTS samples that are not CFO-corrected are extracted. After estimating the carrier frequency offset value, each sample is corrected with the offset value.

To reduce the computation complexity, zero-forcing technique is used for estimating the channel. For a perfect CSI, the ZF detection weight matrix W is given by the pseudo-inverse of the channel matrix H, as in (8.12):

$$W = \left(H^H H\right)^{-1} H^H \tag{8.12}$$

At the receiver, the guard interval is removed and the 64 output samples are gathered from the received signal which gives the expression (8.13)

$$\left[r_0^T r_1^T \ldots\ldots r_{63}^T\right]^T = H_g\left[x_{-16}^T x_{-15}^T \ldots x_{-1}^T x_0^T \ldots x_{63}^T\right]^T + \left[n_0^T n_1^T \ldots\ldots n_{63}^T\right]^T \tag{8.13}$$

where H_g is 128 × 158 ($64n_r \times n_t(64 + 16 - 1)$) matrix representing the channel seen by the OFDM symbol.

Eq. (8.13) is rewritten as (8.14):

$$\left[r_0^T r_1^T \ldots\ldots r_{63}^T\right]^T = H_{cp}\left[x_0^T \ldots x_{63}^T\right]^T + \left[n_0^T n_1^T \ldots\ldots n_{63}^T\right]^T \tag{8.14}$$

where H_{cp} is a block-wise circulant matrix of size 128 × 128. As a result, its SVD decomposition $H_{\text{cp}} = (D^H \otimes I_2)D_{\text{cp}}(D \otimes I_2)$ is such that D_{cp} is a block diagonal matrix whose blocks are obtained by a block-wise FFT. The full information about the channel is accounted for in D_{cp}, and the eigenvectors of H_{cp} are independent of the channel matrix H. So Eq. (8.14) is depicted as (8.15):

$$\left[r_0^T r_1^T \ldots\ldots r_{63}^T\right]^T = (D^H \otimes I_2)D_{\text{cp}}\left[c_0^T \ldots c_{63}^T\right]^T + \left[n_0^T n_1^T \ldots\ldots n_{63}^T\right]^T \tag{8.15}$$

Applying a 64-FFT operation to the received vector, we finally obtain Eq. (8.16) as follows:

$$R = r_k = \sqrt{E_s} H_{(k)} c_k + n_k = XH + N \tag{8.16}$$

The ZF symbol vector estimate, $\widehat{X}$, is given in (8.17):

$$\begin{aligned}\widehat{X} &= WR = \left(H^H H\right)^{-1} H^H X + \left(H^H H\right)^{-1} H^H N \\ &= X + \left(H^H H\right)^{-1} H^H N\end{aligned} \tag{8.17}$$

That is, ZF detection attempts to eliminate the inter-stream interference.

Given the estimate $\widehat{H}$ of H, ZF detection estimates the symbol transmitted through the kth antenna by mapping the kth element of the two-dimensional vector into the closest modulation constellation symbol by using (8.18):

$$\widehat{W}R = \left[\widehat{H}^H \widehat{H}\right]^{-1} + \widehat{H}^H X + \left[\widehat{H}^H \widehat{H}\right]^{-1} \widehat{H}^H N \tag{8.18}$$

8.2.3 *EVM-based Constellation Combiner (ECC)*

After receiving the symbols, error vector magnitude is found to assess the quality of received signal. Error vector magnitude is the vector difference between transmitted signal (ideal symbol) and received signal (measured symbol). EVM helps in identifying the sources of signal degradation, i.e., phase noise, I-Q imbalance, amplitude nonlinearity, and filter distortion. Error vector magnitude [10] is mathematically expressed as Eqs. (8.19) and (8.20).

$$\text{EVM} = \sqrt{\frac{\frac{1}{N}\sum_{K=1}^{N}(e_K)}{\frac{1}{N}\sum_{K=1}^{N}(I_K^2 + Q_K^2)}} \tag{8.19}$$

$$e_K = (I_K - \widetilde{I}_k)^2 + (Q_K - \widetilde{Q}_k)^2 \tag{8.20}$$

where

I_k and Q_k	are ideal values (transmitted symbols)
$\widetilde{I}_k$ and $\widetilde{Q}_k$	are received symbols
I_k	is the inphase measurement of the kth symbol in the burst
Q_k	is the quadrature-phase measurement of the kth symbol in the burst
N	is the input vector length

The EVM is calculated for the symbols that are received in direct link (EB to UE) and in relay link (HEB to UE) with data-aided estimation. The minimum of the Euclidean distance is calculated for each symbol received in both the links, and a

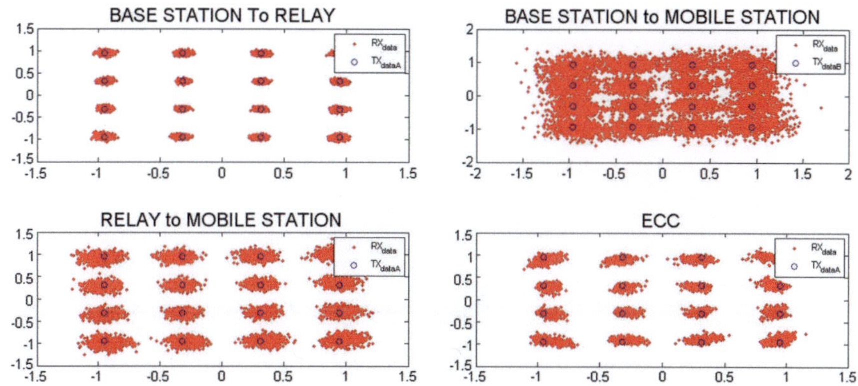

Fig. 8.3 Effect of ECC

new version of the constellation is obtained with an optimal EVM as shown in Fig. 8.3.

ECC can be mathematically depicted as in Eqs. (8.21) and (8.22):

$$\mathrm{EVM}_{\mathrm{ECC}} = \sqrt{\frac{\frac{1}{N}\sum_{K=1}^{N}(e_{\mathrm{ECC}})}{\frac{1}{N}\sum_{K=1}^{N}(I_K^2 + Q_K^2)}} \tag{8.21}$$

$$e_{\mathrm{ECC}} = \min\left[(I_K - I_K^r)^2 + (Q_K - Q_K^r)^2, (I_K - I_K^b)^2 + (Q_K - Q_K^b)^2\right] \tag{8.22}$$

where I_K^r and Q_K^r are the inphase- and quadrature-phase-received symbols from the HEB to UE (relay) link. I_K^b and Q_K^b are the inphase- and quadrature-phase-received symbols from the EB to UE (direct) link.

8.3 Experimental Setup

8.3.1 Wireless Open-Access Research Platform

The WARP platform shown in Fig. 8.4 was designed at Rice University and is used by a number of academic and industrial research laboratory for clean state protocol implementation of the MAC and PHY [9]. A WARPV3 kit usually contains two RF radio boards (RF-A and RF-B) that are suitable for SISO transmission, and each RF board can be configured as either transmitter or receiver. Two more RF boards (RF-C and RF-D) can be added via FMC (FPGA Mezzanine Connector) for MIMO transmission. MIMO-capable radio is capable of operating in 2.4/5 GHz frequency ranges and can support wideband applications. Multiple WARP boards can be connected to the same computer via Ethernet switch.

Fig. 8.4 WARP SDR kit

8.4 Performance Evaluation of ECC

After depicting the combined received symbols from both the direct link (EB to UE) and the relay link (HEB to UE) using ECC, the revised symbols are demapped. The performance of LTE HetNet with ECC and without ECC is evaluated using the BER, EVM, and SNR parameters.

The relationship between EVM, SNR, and BER is given by Eqs. (8.23), (8.24), and (8.25):

$$\mathrm{EVM} \approx \left[\frac{N_0}{E_s}\right]^{\frac{1}{2}} \approx \left[\frac{1}{\mathrm{SNR}}\right]^{\frac{1}{2}} \tag{8.23}$$

$$\mathrm{SNR} \approx \frac{1}{\mathrm{EVM}^2} \tag{8.24}$$

$$\mathrm{BER} \approx \frac{2\left(1 - \frac{1}{L}\right)}{\log_2 L} Q\left[\sqrt{\left[\frac{3\log_2 L}{L^2 - 1}\right] \frac{2}{\mathrm{EVM}_{\mathrm{RMS}}^2 \log_2 M}}\right] \tag{8.25}$$

where N_0 denotes the noise power, E_s is the energy per symbol, and L is the number of levels in each dimension of the M-ary modulation system [10].

The LTE HetNet system is tested for various transmission powers. Here, both the EB and the HEB have same transmitting power. There is a relation between BER and P_T [11] which is given by Eq. (8.26):

$$P_T = \frac{(2/3)N_0 N_f B(M-1)(G_{\mathrm{eh}}h_{\mathrm{eh}} + G_{\mathrm{eu}}h_{\mathrm{eu}})ln\left(2\frac{2^{(b/2)-1}}{b2^{b/2}(\mathrm{BER})}\right)}{(G_{\mathrm{eh}}h_{\mathrm{eh}} + G_{\mathrm{hu}}h_{\mathrm{hu}})G_{\mathrm{eu}}h_{\mathrm{eu}} + G_{\mathrm{eh}}h_{\mathrm{eh}}^2 G_{\mathrm{hu}}h_{\mathrm{hu}}} \tag{8.26}$$

where N_0, N_f, and B are the receiver noise power, the thermal noise power spectral density, and the signal bandwidth. h_{XY} and G_{XY} are the channel fading power gain and the path loss for the X–Y link, $X, Y \in \{e, h, u\}$, respectively.

Table 8.1 gives the experimental setup parameters for the proposed incremental ratio combining.

Figure 8.5 shows the BER variation with respect to the increase in the transmission power of both the EB and the HEB.

While transmitting with −18.5 to −15.5 dBm, (EB-HEB) and (HEB-UE) links give an acceptable QoS, while (EB-UE) link could not produce an acceptable BER. By using ECC, the operator can provide a better QoS to user. When operating above −15.5 dBm, it can be seen that there is a gradual decrease in the (HEB-UE) link performance and gradual increase in the (EB-UE) link performance. Considering at −12.5 dBm of transmission power, the (EB-UE) link gives a better performance compared to (HEB-UE) link. But with the ECC, the operator can able to achieve an incredible QoS compared to both the links. Transmitting above −11.5 dBm, there is degradation of (EB-HEB) link.

Figure 8.6a, b shows the relation between EVM and SNR with respect to the increase in the transmission power of both the EB and the HEB, respectively.

The above result portrays that the EVM decreases with an increase in the transmission power for the direct (EB-MS) link. From Figs. 8.5 and 8.6, it is evident that by using ECC technique, the operator can offer a better QoS to the user with low SNR or high EVM while operating in the range of −18.5 to −15.5 dBm. With the experimental setup as mentioned in Table 8.1, operating at −12.5 dBm, the operator can provide an acceptable QoS with very low SNR by using ECC.

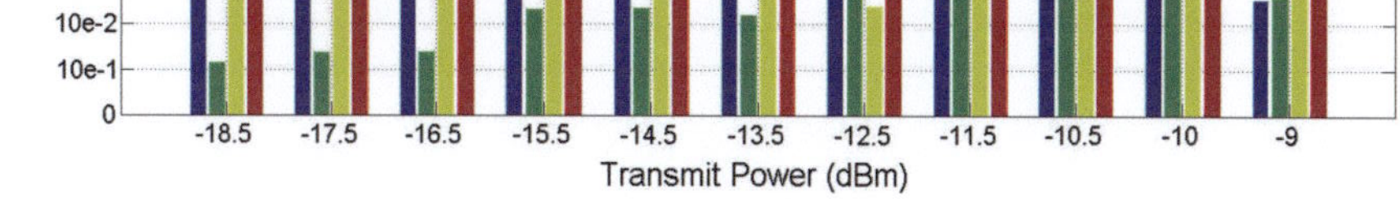

Fig. 8.5 BER performance metrics for various transmission powers

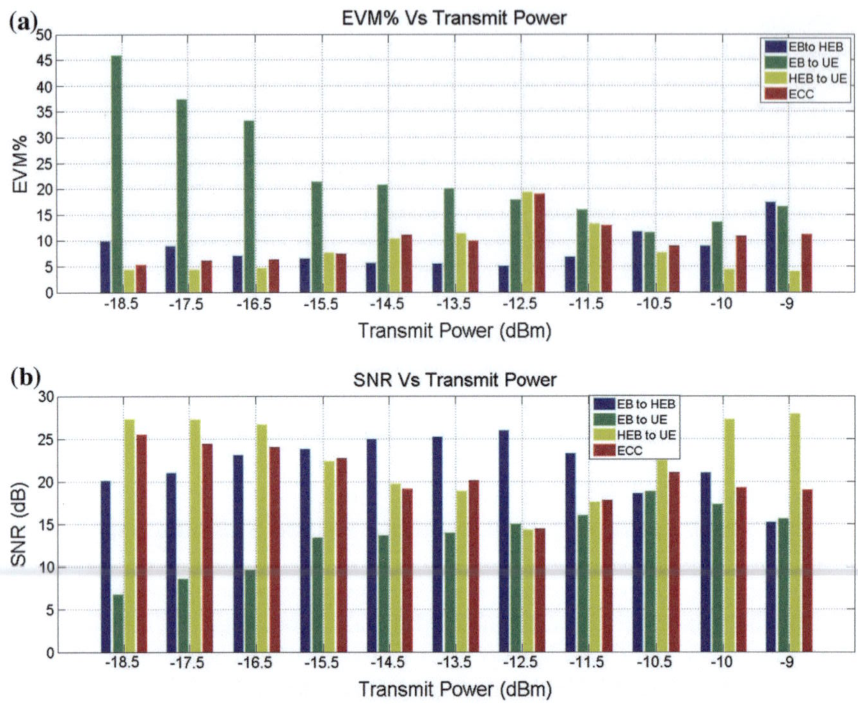

Fig. 8.6 EVM and SNR performance metrics for various transmission powers

8.5 Conclusion

This paper aims to evaluate the performance of LTE HetNets using ECC technique. The real-time results with one eNodeB, one HeNodeB, and one user equipment are obtained under with and without ECCs. The results prove that LTE HetNet with ECC provides 49, 58, and 35% improvement in BER, EVM, and capacity, respectively.

8.6 Future Work

The future work will be extending the proposed technique in a multiple relaying system with different retransmission modes and analyzing its performance.

Acknowledgments We would like to thank Department of Science and Technology, Government of India, for supporting the work under the scheme of "FAST TRACK SCHEME FOR YOUNG SCIENTIST" under the grant number SR/FTP/ETA-92/2011.

References

1. Khandekar A, Bhushan N, Tingfang J, Vanghi V (2010) LTE-advanced: heterogeneous networks. In: European wireless conference, pp 978–982, 12–15 April 2010
2. Arif M, Yameen IM, Matin MA (2013) Femtocell suburban deployment in LTE networks. Int J Inf Electron Eng 3(2):208–212
3. Chandrasekhar V, Andrews JG (2008) Femtocell networks: a survey. The University of Texas at Austin Alan Gatherer, Texas Instruments, pp 59–67
4. Elkourdi T, Simeone O (2011) Femtocell as a relay: an outage analysis. IEEE Trans Wireless Commun 10(12):4204–4213
5. Akyildiz IF, Chavarria-Reyes E, Gutierrez-Estevez DM, Balakrishnan R, Krier JR (2013) Enabling next generation small cells through femto relays. Broadband Wireless Networking Laboratory, School of Electrical and Computer Engineering, Georgia Institute of Technology, Atlanta, GA 30332, United States, pp 1–15, 23 April 2013
6. Joshi G (2010) On relay-assisted cellular networks. Thesis-Indian Institute of technology-Bombay
7. Avendi MR, Nguyen HH (2013) Selection combining for differential amplify-and-forward relaying over rayleigh-fading channels. Signal Process Lett IEEE 20(3):277–280
8. Mugelan RK, Thilaga P, Bhagyaveni MA (2015) EVM based performance analysis of LTE Downlink & it's real time evaluation using WARP. Int J Appl Eng Res 10(2):2401–2418
9. Murphy P, Sabharwal A, Aazhang B (2006) Design of WARP: wireless open-access research platform. In: European Signal Processing Conference, June 2006
10. Shafik RA, Rahman MS, Islam AR (2006) On the extended relationships among EVM, BER and SNR as performance metrics. In: 4th International Conference on Electrical and Computer Engineering ICECE 2006, pp 19–21, Dhaka, Bangladesh
11. Hussain E, Bhargava VK, Fettweis GP (2012) Green radio communication networks. Cambridge University Press
12. Data sheet—MAX2828/MAX2829 Single/Dual-Band 802.11a/b/g World-Band Transceiver ICs, pp 1–39, 2014

Chapter 9
Performance Analysis of 3-Phase Induction Motor: Through Simulation in SIMULINK and TMS320C6713

Nutan Lata Nath, Pallavi Dutta, Shilpi Kumari and Urjaswit Lal

Symbols

V_{as}, V_{bs} and V_{cs}	Three-phase stator voltages
Ω	Supply frequency
V_{ds}, i_{ds} and V_{qs}, i_{qs}	Direct and quadrature axis components of the stator voltage/current vector V_s and i_s
V_{dr}, i_{dr} and V_{qr}, i_{qr}	Direct and quadrature axis components of rotor voltage/current vector V_r and i_r
L_{ls}	Stator leakage inductance
L_{lr}	Rotor leakage inductance
L_m	Magnetizing inductance
ω_r	Rotor speed
J	Moment of Inertia of rotor and load
T_L	Load Torque

9.1 Introduction

In current scenario, it is evident from literature that the demand for a more comprehensive and uniform model for analysis of induction motor is prevalent. The modelling of induction motor gives us proper insights into the behavior of a healthy

N.L. Nath · P. Dutta (✉) · S. Kumari · U. Lal
Department of E.E.E., B.I.T. Mesra, Ranchi, India
e-mail: pallavidutta.bit@gmail.com

N.L. Nath
e-mail: nutanlatanath@bitmesra.ac.in

S. Kumari
e-mail: shilpi6484@gmail.com

U. Lal
e-mail: urjaswit.lal@gmail.com

V. Nath (ed.), *Proceedings of the International Conference on Nano-electronics, Circuits & Communication Systems*, Lecture Notes in Electrical Engineering 403,
DOI 10.1007/978-981-10-2999-8_9

motor. This paper focuses on the implementation of the induction motor model in platforms like MATLAB/SIMULINK and TMS320C6713 DSP kit. The results from these platforms are compiled and compared with each other.

Behavior of the healthy induction motor in transient state can be studied very well using the TMS320C6713 DSP Kit which allows a very fast analysis of signals generated using the virtual model.

For steady state analysis of Induction motor characteristics, a MATLAB/SIMULINK model was developed using the same mathematical equations. Again, various characteristics of the Induction motor were obtained.

9.2 Literature Review

The simulation of induction motor requires in-depth analysis of the mathematical model of the motor [1]. Among various softwares and platforms used for the modelling of an induction motor, MATLAB/SIMULINK presents itself as a strong tool where transient state and steady state behavior can be studied [2]. There have been several approaches to the development of mathematical model for the motor [1, 3, 4]. Modelling using two axis Voltage transformation is one of them [4]. The two-axis model for induction motor has been derived using state space equations [5]. The modelling of three-phase induction motor in MATLAB/SIMULINK environment using this mathematical model for the dynamic *d–q* voltage transform has also been developed in [6]. TMS320C6713 DSP starter kit is a very powerful tool for signal analysis [7]. Simulation of 3-phase induction motor using TMS320C6713 DSK has not been done previously and is a novel methodology to study the transient behavior of healthy induction motor.

9.3 Mathematical Model of Induction Motor

The three phase equations of the induction motor are converted to the two axis format. The *d–q* axis model of induction motor is the preferred mode of study of the dynamic performance of the motor as its reliability and accuracy has been well proven [5]. The conversion is obtained with the given equation:

$$\begin{bmatrix} V_{ds} \\ V_{qs} \end{bmatrix} = (2/3)\begin{bmatrix} 1 & -\frac{1}{2} & -\frac{1}{2} \\ 0 & \frac{\sqrt{3}}{2} & -\frac{\sqrt{3}}{2} \end{bmatrix}\begin{bmatrix} V_{as} \\ V_{bs} \\ V_{cs} \end{bmatrix} \tag{9.1}$$

$$\begin{aligned} V_{as} &= |V|\sin(\Omega t) \\ V_{bs} &= |V|\sin(\Omega t - 2\pi/3) \\ V_{cs} &= |V|\sin(\Omega t + 2\pi/3) \end{aligned} \tag{9.2}$$

where $|V|$ is the amplitude of terminal voltage, V_{as}, V_{bs}, V_{cs} denote the three phase stator voltages and Ω is the supply frequency.

Applying KVL to the motor circuit, we get:

$$L\frac{\mathrm{d}i}{\mathrm{d}t} = V - iR \tag{9.3}$$

Writing as a state space equation,

$$\dot{i} = Ai + Bv \tag{9.4}$$

where, $A = -L^{-1}R$ and $B = L^{-1}$.

The stator inductance L_s is the sum of the stator leakage inductance and magnetizing inductance. Similarly, the rotor inductance L_r is the sum of the rotor leakage inductance and magnetizing inductance.

$$L_s = L_{ls} + L_m \tag{9.5a}$$

$$L_r = L_{lr} + L_m \tag{9.5b}$$

The *d–q* components of the stator and rotor current and voltage are given in the following vector:

$$i = \begin{bmatrix} i_{ds} & i_{qs} & i_{dr} & i_{qr} \end{bmatrix}^{\mathrm{T}} \tag{9.6a}$$

$$v = \begin{bmatrix} v_{ds} & v_{qs} & v_{dr} & v_{qr} \end{bmatrix}^{\mathrm{T}} \tag{9.6b}$$

Equation of the induction motor in the *d–q* axis frame is given by [4]:

$$\begin{bmatrix} i_{ds} \\ i_{qs} \\ i_{dr} \\ i_{qr} \end{bmatrix} = \int_{t=0}^{t} \left\{ \begin{bmatrix} L_s & 0 & L_m & 0 \\ 0 & L_s & 0 & L_m \\ L_m & 0 & L_r & 0 \\ 0 & L_m & 0 & L_r \end{bmatrix}^{-1} \times \left(\begin{bmatrix} v_{ds} \\ v_{qs} \\ v_{dr} \\ v_{qr} \end{bmatrix} - \begin{bmatrix} R_s & 0 & 0 & 0 \\ 0 & R_s & 0 & 0 \\ 0 & \frac{P}{2}\omega_r L_m & R_r & \frac{P}{2}\omega_r L_r \\ -\frac{P}{2}\omega_r L_m & 0 & -\frac{P}{2}\omega_r L_r & R_r \end{bmatrix} \begin{bmatrix} i_{ds} \\ i_{qs} \\ i_{dr} \\ i_{qr} \end{bmatrix} \right) \right\} \mathrm{d}t \tag{9.7}$$

Of the various reference frames, the synchronously rotating reference frame is preferred since it gives constant steady state variables that do not exhibit sinusoidal variation with time. The *d–q* axes in this frame of reference rotate at the synchronous speed [5].

To simulate the induction motor and obtain transient as well as steady state nature of the currents, speed and torque the rotor d–q axis voltages were assigned a value of zero considering the rotor terminals to be short circuited.

I.e.

$$v_{dr} = 0$$
$$v_{qr} = 0$$

The initial value of rotor speed ω_{r} is assigned zero taking into consideration the fact that the motor is at standstill initially. This is assumed in order to obtain the transient behavior of the induction motor.

Figures 9.1 and 9.2 show the equivalent q-axis and d-axis dynamic model of induction motor respectively. ω is the synchronous speed of motor and ω_{r} is the rotor speed. Hence, $(\omega - \omega_{\mathrm{r}})$ is the slip. λ_{ds} and λ_{dr} are the direct-axis stator and rotor flux and λ_{qs} and λ_{qr} are the quadrature-axis stator and rotor flux respectively.

The torque equation of the Induction motor in the d–q axis frame is given by [4]:

$$T = \left(\frac{3}{2}\right)\frac{PL_{\mathrm{m}}}{2}\left(i_{ds}i_{qs} - i_{dr}i_{qr}\right) \tag{9.8}$$

The rotor speed ω_0 is obtained by the following equation:

$$\omega_{\mathrm{r}} = \int\limits_{\tau=0}^{t} \frac{T - T_{\mathrm{L}}}{J}\mathrm{d}\tau \tag{9.9}$$

where

J Moment of Inertia of rotor and load
T_{L} Load Torque.

The viscous friction is neglected to obtain the above equation [4].

The induction motor model was developed using Eqs. (9.1–9.9). The SIMULINK model developed is shown in Fig. 9.3.

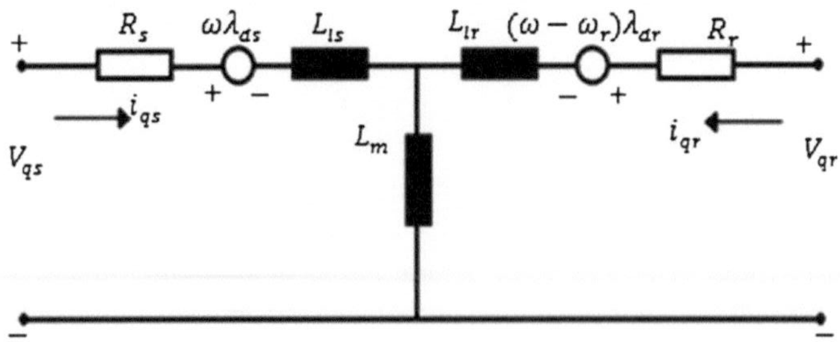

Fig. 9.1 q-axis model of induction motor

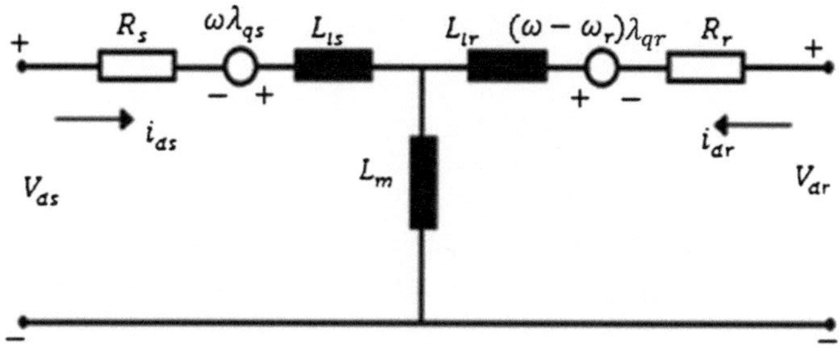

Fig. 9.2 *d*-axis model of induction motor

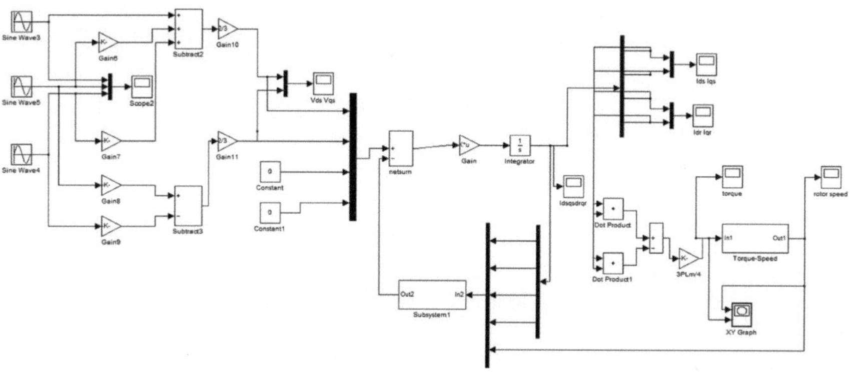

Fig. 9.3 Proposed SIMULINK model of induction motor

9.4 Results

9.4.1 Simulation Results

The specifications of wye connected Squirrel cage type induction motor chosen for simulation are:

Number of phases: 3,
Number of poles: 2,
Power rating: 3 HP,
Supply voltage: 380 V,
Supply frequency: 50 Hz.

The parameters of the motor that have been taken are same as those in [4]:

Stator resistance = 3.5 Ω
Rotor resistance = 3.16 Ω
Stator leakage inductance = 6.90732 mH
Magnetizing inductance = 0.26674 H

Rotor leakage inductance = 6.81183 mH
Moment of Inertia = 0.4 kg m^2
Load Torque = 11.9 N m.

The Induction motor was simulated on MATLAB/SIMULINK as well as TMS320C6713 DSK.

The results obtained from the SIMULINK model are shown in Figs. 9.4 and 9.5.

Figures 9.4 and 9.5 show the *d–q* components of stator current. The direct and quadrature axes components of stator current are displaced by 90°. Figure 9.4 shows the transient as well as the steady state nature of stator current. Figure 9.5 shows the steady state portion of stator current.

Figures 9.6 and 9.7 show the *d–q* components of rotor current. The direct and quadrature axes components of rotor current are displaced by 90°. Figure 9.6 shows the transient as well as the steady state nature of rotor current. Figure 9.7 shows the steady state portion of rotor current. The magnitude of rotor current is less than stator current.

The peak values of current for transient state were observed as follows:

$$I_{ds} = I_{qs} = 50.1\text{ A}$$
$$I_{dr} = I_{qr} = 47.6\text{ A}$$

Similarly the peak values of current for steady state were observed as follows:

$$I_{ds} = I_{qs} = 5.55\text{ A}$$
$$I_{dr} = I_{qr} = 3.49\text{ A}$$

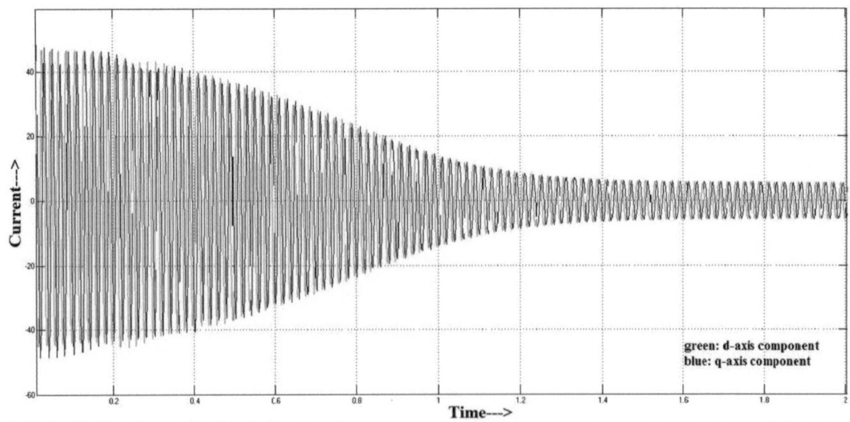

Fig. 9.4 I_{ds} and I_{qs}: *d–q* components of stator current

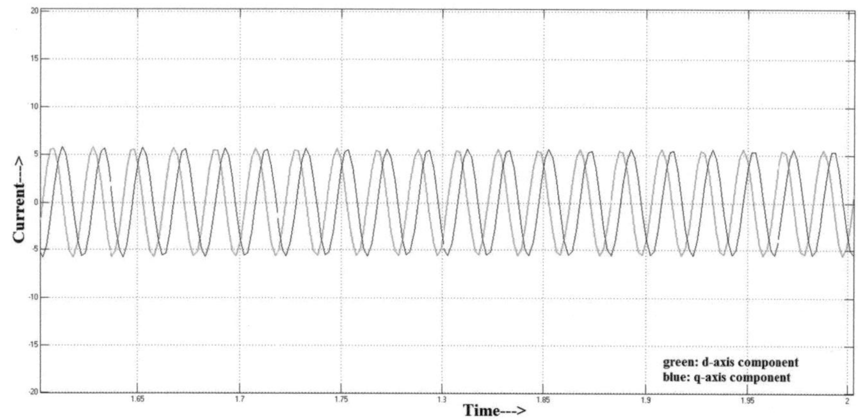

Fig. 9.5 Steady state portion of *d–q* components of stator current

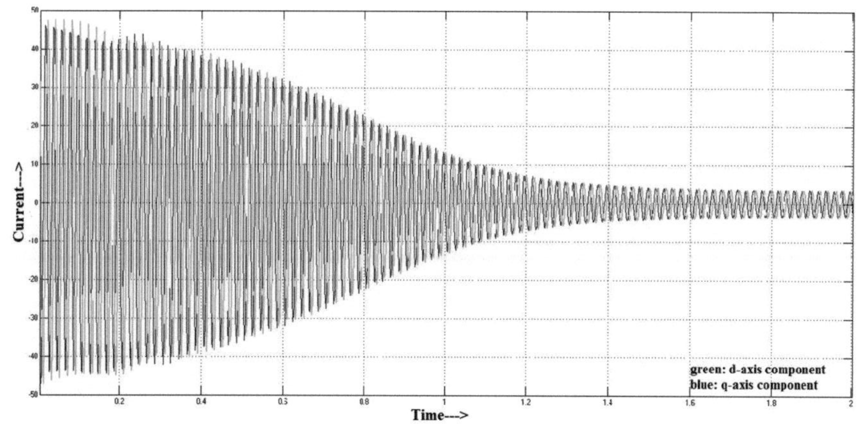

Fig. 9.6 I_{dr} and I_{qr}: *d–q* components of rotor current

Figure 9.8 shows that the speed of healthy induction motor increases almost linearly in the transient state and attains a constant value at steady state.

Figure 9.9 shows that the torque of healthy induction motor decreases almost linearly in the transient state and attains a constant value at steady state.

Figure 9.10 shows the torque-speed characteristics of healthy induction motor obtained from the SIMULINK model.

The nature of torque initially was observed to be oscillatory and the peak value was 142 N m and the steady state value was equal to load torque value of 11.9 N m. The steady state speed was observed to be 152.3 rad/s.

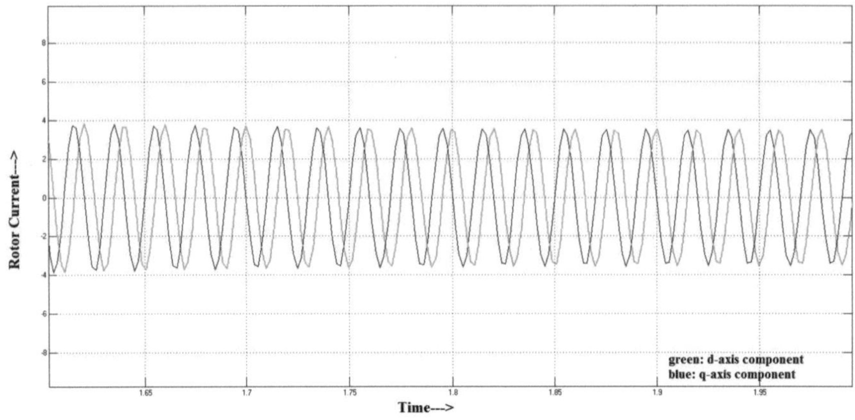

Fig. 9.7 Steady state portion of *d–q* components of rotor current

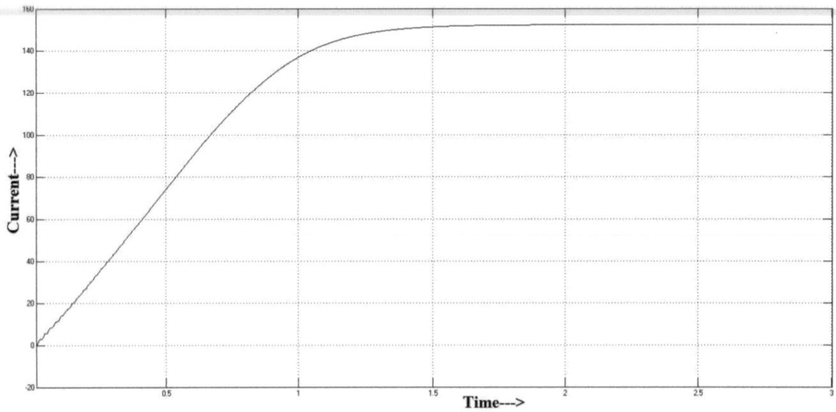

Fig. 9.8 Speed curve of induction motor

9.4.2 *Graphical Results Obtained from TMS320C6713 DSK*

Figure 9.11 shows the *d–q* components of rotor current. The direct and quadrature axes components of rotor current are displaced by 90°. This shows the transient state of rotor current.

Figure 9.12 shows the *d–q* components of stator current. The direct and quadrature axes components of stator current are displaced by 90°. This shows the transient state of stator current.

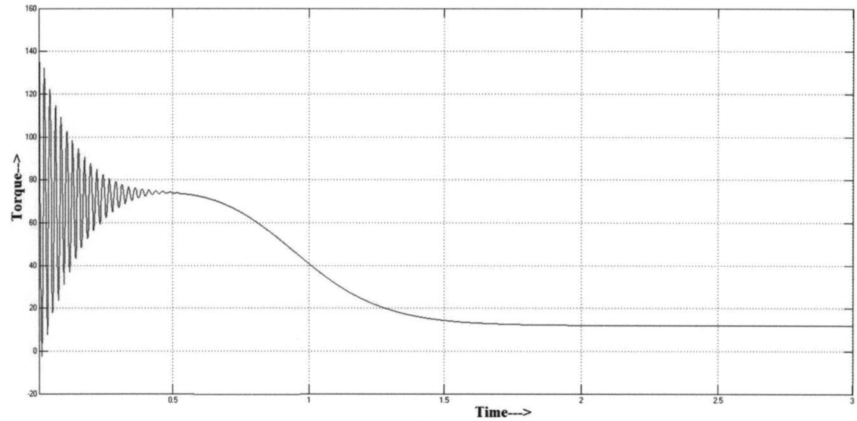

Fig. 9.9 Torque curve of induction motor

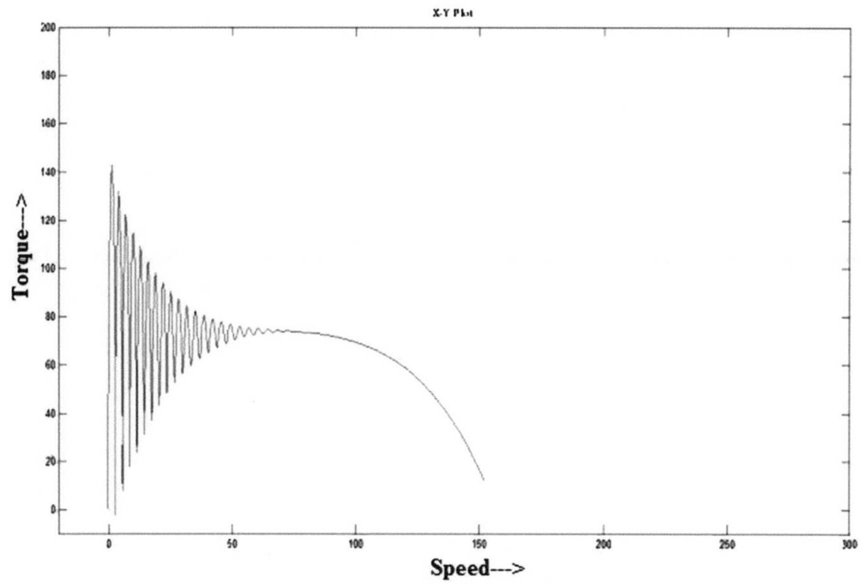

Fig. 9.10 Torque-speed characteristic of induction motor

Figure 9.13 shows the *d*–*q* components of supply voltage. The direct and quadrature axes components of the supply voltage are displaced by 90° to each other.

The torque curve of induction motor in the transient state is displayed in Fig. 9.14.

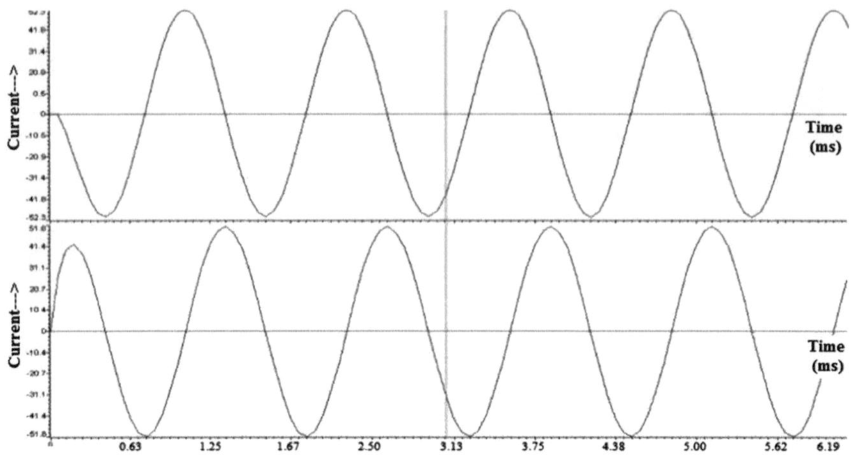

Fig. 9.11 *d–q* components of rotor current

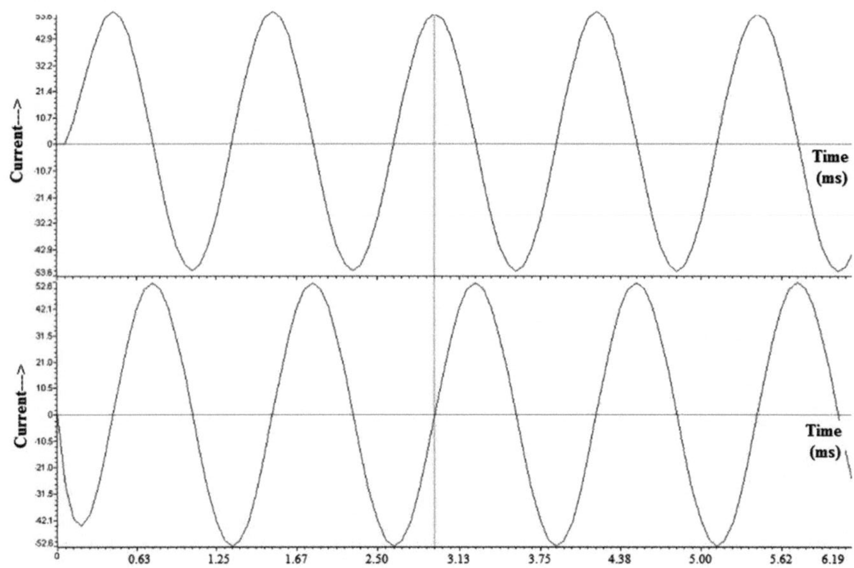

Fig. 9.12 *d–q* components of stator current

The value of maximum torque obtained is 82.4 N m.

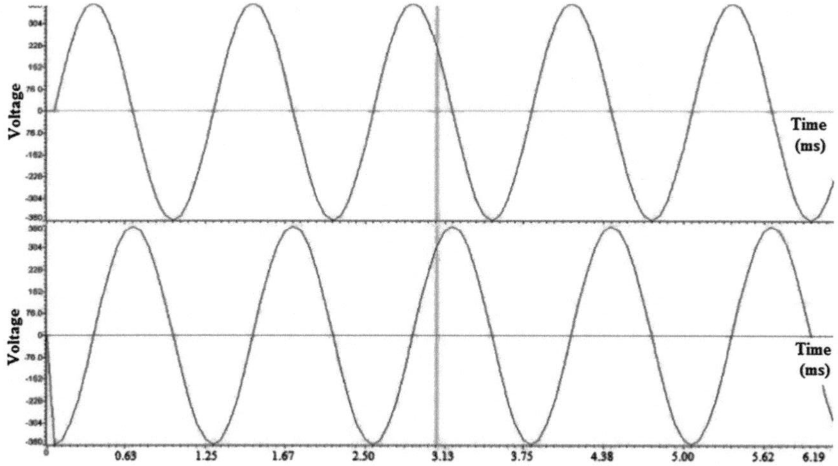

Fig. 9.13 *d–q* components of stator voltage

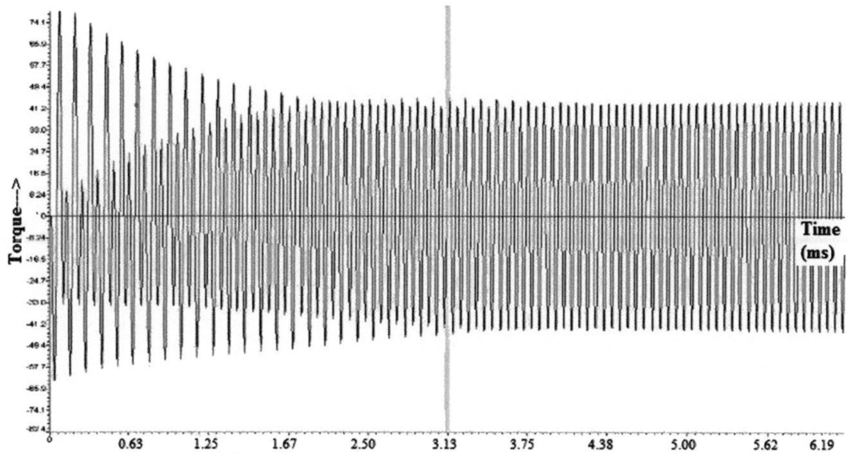

Fig. 9.14 Torque curve of healthy induction motor

9.5 Discussion

The results of stator current, rotor current, speed and torque obtained from the SIMULINK model of the induction motor that we developed were found to be similar to the expected waveforms as seen from the papers [1, 4, 6, 8]. The stator current, rotor current and torque developed by the motor take 1.6 s to reach steady state values prior to which they exhibit transient behavior.

The same model was simulated in TMS320C6713 DSP starter kit [7] in order to study the transient state of the various signals and the waveforms recorded were in coherence with the previously obtained results. It was observed that the transient torque of induction motor is oscillatory in nature that dies down to the value of load torque during steady state.

9.6 Conclusion

The healthy induction motor was modelled in SIMULINK and implemented using TMS320C6713. The stator current, rotor, current, torque and speed curves were obtained as a function of time. The results obtained from the two processes were similar, thus, verifying the results of the other. The output from SIMULINK model was that of the transient as well as steady state.

The signals obtained from the simulation of induction motor can be used to analyze its performance using various signal processing techniques like FFT, STFT, Wavelet etc. Variations made in the parameters can help in the study of abnormalities in the motor.

References

1. Shi KL, Chan TF, Wong YK, Ho SL (1999) Modelling and simulation of the three phase induction motor using Simulink. Int J Electr Eng Educ 36:163–172. Manchester U.P.
2. Leedy AW (2013) Simulink/MATLAB dynamic induction motor model for use in undergraduate electric machines and power electronics courses, Southeastcon. In: Proceedings of IEEE, pp 1–6
3. Wieczorek M, Rosolowski E (2010) Modelling of induction motor for simulation of internal faults. In: MEPS, proceedings of international symposium, pp 1–6
4. Palpankar PM, Ghanmare RU, Makade N (2014) A generalized dynamic model of induction motor using Simulink. ITSI-TEEE 1(5):118–122
5. Okoro OI (2003) Steady state and transient analysis of induction motor driving a pump load. Niger J Technol 22(1)
6. Shakuntla B, Agarwal SK, Sandhu KS (2013) Dynamic D-Q axis modeling of three-phase asynchronous machine using Matlab. Int J Adv Res Electr Electron Instrum Eng (An ISO 3297: 2007 Certified Organization) 2(8)
7. TMS320C6713 Technical reference. Spectrum Digital Inc., Nov 2003
8. Jagadanand G, Gopi L, George S, Jacob J (2012) Inter-turn fault detection in induction motor using stator current wavelet decomposition. IJEET 3(2):103–122

Chapter 10
A Distributed Fault-Tolerant Multi-objective Clustering Algorithm for Wireless Sensor Networks

Nabajyoti Mazumdar and Hari Om

10.1 Introduction

Wireless Sensor Networks (WSNs) have gained considerable research interest in last few years. They are applicable for a wide range of applications such as environment monitoring, surveillance, and vehicular tracking [1]. A WSN consists of a large number of sensor nodes, each having capability of sensing, computing, and communicating. Sensor nodes are however equipped with limited powered batteries and also have limited processor power and low bandwidth communication channels. Among all these constraints, the energy of sensors accounts for maximum constricting since it disturbs the network lifetime. Therefore, energy-efficient communication techniques are particularly desired for WSNs. In the recent years, clustering has become a very effective energy preservation mechanism in which sensor nodes are grouped to form clusters, each cluster has a controller, called Cluster head (CH), and a set of cluster members (CMs). Thus, the CMs of a cluster sense the data in the monitoring environment and send to their respective CHs. CHs have more load as they have to aggregate the packets received from their CMs and send the aggregated packets to the base station (BS), which is situated far from the network. Since the CHs spend more energy as compared to other nodes, i.e., non-CHs, they die earlier. In [2–5], some special nodes, called gateway nodes, which are equipped with extra energy and larger communication range, have been used. The gateway nodes perform the functionality of CHs; however, these gateway nodes are also equipped with non-replaceable limited power batteries. So, energy of gateway nodes must also be conserved for extensive runs of network. In rest of the

N. Mazumdar (✉) · H. Om
Department of Computer Science & Engineering,
Institute of Technology (Indian School of Mines), Dhanbad, India
e-mail: nabajyoti@cse.ism.ac.in

H. Om
e-mail: hariom4india@gmail.com

V. Nath (ed.), *Proceedings of the International Conference on Nano-electronics, Circuits & Communication Systems*, Lecture Notes in Electrical Engineering 403,
DOI 10.1007/978-981-10-2999-8_10

paper, we will use gateway nodes and CHs interchangeably. The gateway and sensor nodes are deployed randomly in the target area. After deployment of sensor nodes, cluster formation begins in which each sensor node decides which CH in its communication range to join among the several candidate CHs. A sensor node selects a CH from a set of possible CHs based on some metric. If the selection of CH is based on the single objective optimization technique, then it may lead to a bad choice for the overall network. For example, if the nearest CH is selected but that CH is further away from the base station as compared to other candidate CHs, then that CH may not be the best choice due to the fact that it has to communicate at a greater distance. Consider Fig. 10.1 in which the sensor node S_i receives CH-ADV-MSG messages from three gateways G_1, G_2, and G_3. If the node S_i selects CH based on single parameter like residual energy, then the G_1 will be selected as CH; however, it is farthest from the base station as compared to G_2 and G_3. Similarly, if selection is done based on the shortest distance from the base station, then G_2 will be selected, but G_2 has minimum residual energy. Therefore, each node must consider multiple parameters while selecting its CH. This evidence forms the base of our work in this paper. Moreover, the sensor nodes are prone to failure due to energy depletion, environment hazards, etc. The death of nodes will certainly affect the performance of WSNs. Furthermore, if a gateway node G_i dies, the corresponding CMs become inactive, although they have sufficient residual

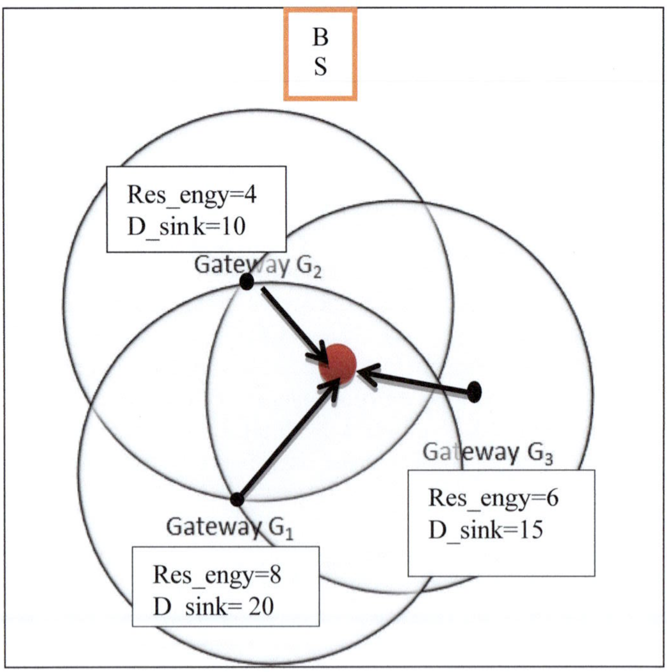

Fig. 10.1 Illustration of CH selection based single parameter

energies with them. So, some recovery mechanism should be incorporated to overcome such faults. Hence, the fault tolerance is also very important for long run of WSNs.

In this paper, we propose a *d*istributed *f*ault-tolerant *m*ulti-objective clustering *a*lgorithm (DFMCA) for WSNs. The motivation behind this algorithm is to maximize the network lifetime by considering various objectives, such as distance between node and CH, distance of CH from base station, and the residual energy of the CH. We also propose a distributed fault-tolerant algorithm in case of sudden death of the CHs.

The remaining part of this paper is structured as follows. In Sect. 10.2 the related works are reviewed. In Sect. 10.3, we describe the network model, and in Sect. 10.4, we present our proposed algorithm. In Sect. 10.5, the simulation results of our algorithm are given, followed by conclusion in Sect. 10.6.

10.2 Related Work

In last decade and so, several works have been done on WSNs, especially related to clustering [3–5]. The *L*ow-*E*nergy *A*daptive *C*lustering *H*ierarchy (LEACH) protocol [6] is a well-known distributed clustering algorithm in which nodes elect themselves as CH based on some probability. Though the LEACH provides significant energy saving and prolongs the network lifetime, it does not ensure the selected CHs to be evenly positioned in the network. The *P*ower-*E*fficient *G*athering in *S*ensor *I*nformation *S*ystems (PEGASIS) [7], *H*ybrid *E*nergy-*E*fficient *D*istributed (HEED) protocol [8], *T*hreshold-sensitive *E*nergy-*E*fficient sensor *N*etwork (TEEN) [9], etc. are improved algorithms of LEACH. In these algorithms, CH selection is done based on different parameters, such as residual energy of the node [8] and distance [10]. All these algorithms elect the CHs from the homogeneous sensor nodes; as a result, the CH nodes will deplete quickly their energy due to higher energy consumption, which affects network lifetime.

In order to solve the problem of homogeneous WSNs, many efficient clustering algorithms have been introduced for heterogeneous WSNs [2–5]. In these algorithms, some higher energy sensor nodes are used along with traditional sensor nodes, where the higher energy sensor nodes act as cluster heads and low-energy sensor nodes as CMs. The main challenge for clustering algorithms in heterogeneous WSNs is to balance the network load by assigning proper CH to a sensor node. Most of the algorithms use single parameter such as energy of a CH, distance from the CH to optimize the selection of CH. Pratyay and Jana [5] have proposed a distributed clustering algorithm in which the nodes join their CH based on the residual energy. However, selection of CHs based on a single parameter may not give optimal results for the overall network. Furthermore, none of these algorithms address fault tolerance in WSNs. In [11], Gupta et al. discuss a fault-tolerant technique in which the sensor nodes can recover from the faulty CHs in runtime. It considers two CHs in each cluster: one primary and one secondary. All the CMs are connected with the primary CH. In case of failure of the primary CH, the CMs will join the secondary CH. However, as the gateway nodes are deployed randomly, such scenario is not practically feasible.

10.3 Network Model: Basic Assumptions and Terminologies

We make the following assumptions for the network:

(i) The network consists of two types of nodes: sensor nodes and gateway nodes.
(ii) The nodes are deployed randomly in network and are stationary after deployment.
(iii) The gateway nodes are less energy constraint.
(iv) A sensor node can join only one gateway node/CH within the communicating range of sensor node.
(v) The base station is situated outside the target area and has no energy constraint.
(vi) The wireless link is symmetric and bidirectional.

Following are the terminologies that help in presenting and understanding our proposed algorithm:

- $S = \{S_1, S_2 \ldots S_n\}$, S_i is a sensor node: Set of sensor nodes
- $G = \{G_1, G_2 \ldots G_m\}$, G_i is a *ga*teway node: Set of gateway nodes
- $Dist(i, j)$: distance between the nodes S_i and S_j
- $D_{sink}(i, \text{sink})$: distance of node S_i from the base station/Sink
- $\text{Energy}_{res}(i)$: residual energy of node S_i
- R_s: sensing range of node S_i
- $Neighbor(S_i) = \{S_j{:}Dist(S_i, S_j) \le R_s\}$: set of neighbors of node S_i
- $Cand\_CH(S_i) = \{G_j{:}\ Dist(S_i, G_j) \le R_s\}$: set of candidate CHs within sensing radius of sensor node S_i

Based on the connectivity of the sensor nodes to gateway nodes, the sensor nodes can be divided into the following sets: connected nodes and their sets, unconnected nodes and their sets, and agent nodes and their sets.

(a) *Connected nodes and connected sets*: A connected node is that node which has at least one gateway node within its sensing radius. The connected set (CON_{set}) is the collection of all connected sensor nodes in the network. A sensor node S_i is an element of CON_{set} iff $\forall G_j \exists \{G_j : Dist(S_i, G_j) \le R_s\}$.
(b) *Unconnected nodes and unconnected sets*: An unconnected node is that node which does not have any gateway node within its sensing radius. The unconnected set (UNCON_{set}) is the collection of unconnected sensor nodes in the

network. A sensor node S_i is an element of $\text{UNCON}_{\text{set}}$ iff $\forall G_j \nexists \{G_j : Dist(S_i, G_j) \leq R_s\}$.

(c) *Agent nodes and agent sets*: An agent node for an unconnected node S_i is the connected nodes within its sensing radius. The agent set AGNT_{set} of an unconnected sensor S_i is the set of all agent nodes of S_i. A sensor node S_j is an element of the agent set of S_i, i.e., ($\text{AGNT}_{\text{set}}(S_i)$) iff $S_j \in \text{CON}_{\text{set}}$ & $Dist(S_i, S_j) \leq R_s$.

10.4 Proposed DFMCA Algorithm

Working of the DFMCA algorithm may be divided into two phases: the cluster setup phase followed by the steady phase. In the cluster setup phase, all sensor nodes select their CHs among different possible choices of CHs/gateways and then form clusters. The cluster setup phase is followed by the steady phase in which the CMs send their packets to their respective CHs during their specified time slot. A time slot is assigned to each CM by their CH using time-division multiplexing access (TDMA). The CH aggregates the received packets and sends them to the base station. The steady phase runs for a fixed number of rounds. After that, the cluster setup phase is executed again. The cluster setup and steady phase are repeated till the nodes in the WSN are alive. During the steady phase, if any gateway node/CH dies, then the fault-tolerant algorithm is executed to recover from the fault.

10.4.1 Cluster Setup Phase

In cluster setup phase, the following types of control messages are used:

(a) *CH advertisement message* (*CH_ADV_MSG*): Each gateway node broadcasts this message within sensor communication range (R_s) to advertise its status. This message holds sensor node's ID, its residual energy, and its distance from the base station. This information is used by the sensor nodes to calculate the cost value of the gateway node.

(b) *Cluster join message* (*CH_JOIN_MSG*): A sensor node sends this message to the gateway node, which is selected as CH based on the cost value by that sensor. This message contains the sensor node's ID.

(c) *HELP message* (*HELP*): A senor node S_i, $S_i \in \text{UNCON}_{\text{set}}$, broadcasts this message to find an agent node for relaying packets to a gateway. This message contains the sensor node's ID.

(d) *Acknowledgment message* (*ACK*): A senor node S_i, $S_i \in \mathrm{CON}_{\mathrm{set}}$, sends this message if it receives any *HELP* message. This message contains the sensor node's ID, its residual energy, and the distance from its CH.

In cluster setup phase, each gateway node broadcasts a CH_ADV_MSG within sensor's communication radius (R_s). Upon receiving this message, a sensor node S_i becomes a member of $\mathrm{CON}_{\mathrm{set}}$. If the sensor node S_i is within the communication range of multiple gateways, it will receive multiple CH_ADV_MSG from those gateway nodes. The node S_i computes the cost value for each candidate CH to select the best CH among these gateways. The calculation of cost value depends on multiple parameters. Our algorithm helps a node in selecting its CH among the multiple gateways based on the multiple parameters using a multi-objective optimization technique.

On the other hand, if a sensor node S_j does not receive any CH_ADV_MSG message for a certain period of time, it declares itself as a member of $\mathrm{UNCON}_{\mathrm{set}}$. The members of the $\mathrm{UNCON}_{\mathrm{set}}$ are not covered by any gateway nodes due to random deployment of the nodes. So these nodes broadcast a *HELP* message within their communication radius to establish a path to some gateway node. If a sensor node S_i receives the *HELP* message from some node S_j such that $S_i \in \mathrm{CON}_{\mathrm{set}}$, then the node S_i replies an *ACK* message to the node S_j. Upon replying such *ACK* message, the node S_i becomes an element of $\mathrm{AGNT}_{\mathrm{set}}$ of the node S_j. Now, for the sensor node S_j, $S_j \in \mathrm{UNCON}_{\mathrm{set}}$, if $\mathrm{AGNT}_{\mathrm{set}}(S_j) \neq \emptyset$, then the node S_j selects one of the members from the $\mathrm{AGNT}_{\mathrm{set}}$ (S_j) to relay its packet to a gateway node. The selection of agent node is based on a cost value. Thus, an unconnected sensor node is assigned to a gateway node via multi-hop communication.

A. Cost value calculation

The important steps in this process are the element matrix, normalize matrix, and cost matrix calculations, and these matrices are discussed as follows:

(i) *Element matrix*:

It is a $p \times q$ matrix, where 'p' is the number of candidate nodes and 'q' is number of parameters, which are important for decision. Each element $e_{i,j}$ represents the jth parameter for the ith candidate. It is given as follows:

$$EM = \begin{bmatrix} e_{1,1} & e_{1,2} & \cdot & \cdot & e_{i,q} \\ e_{2,1} & e_{2,2} & \cdot & \cdot & e_{2,q} \\ \cdot & \cdot & \cdot & & \\ \cdot & \cdot & \cdot & \cdot & \\ e_{p,1} & e_{p,2} & \cdot & \cdot & e_{p,q} \end{bmatrix}$$

(ii) *Normalized matrix*:

The element matrix is transformed into the normalized matrix as follows. Each element of the normalized matrix $n_{i,j}$, lies between 0 and 1, where 0 means the worst and 1 means the best value choice. The elements of the normalized matrix are calculated as follows:

$$n_{i,j} = \left(\frac{e_{i,j} - w_j}{b_j - w_j}\right)$$

where w_j = worst value of $e_{k,j}$ and b_j = best value of $e_{k,j}$, $k = 1, 2, \ldots, p$.

Thus, the normalized matrix NM is given as follows:

$$NM = \begin{bmatrix} n_{1,1} & n_{1,2} & . & . & n_{i,q} \\ n_{2,1} & n_{2,2} & . & . & n_{2,q} \\ . & . & . & & \\ . & . & . & . & \\ n_{p,1} & n_{p,2} & . & . & n_{p,q} \end{bmatrix}$$

(iii) *Cost Matrix/Vector*:

The cost matrix is indeed a vector that is calculated by multiplying the normalized matrix with a weight matrix (vector), and it is given as follows:

$$\begin{bmatrix} c_1 \\ c_2 \\ . \\ . \\ c_p \end{bmatrix} = \begin{bmatrix} n_{1,1} & n_{1,2} & . & . & n_{i,q} \\ n_{2,1} & n_{2,2} & . & . & n_{2,q} \\ . & . & . & & \\ . & . & . & . & \\ n_{p,1} & n_{p,2} & . & . & n_{p,q} \end{bmatrix} \begin{bmatrix} w_1 \\ w_2 \\ . \\ . \\ w_p \end{bmatrix}$$

The maximum value of the cost vector indicates the best choice. The sensor node S_i, $\forall S_i \in \text{CON}_{\text{set}}$, can calculate its cost matrix/vector to select the best gateway node as its CH based on the local information only. After the CH has been selected, the sensor S_i sends the CH_JOIN_MSG message to the selected CH to inform about its membership in the cluster. So, all sensor nodes S_i, $\forall S_i \in \text{CON}_{\text{set}}$, join their respective CHs to form clusters.

Now, the sensors $S_j, S_j \in \text{UNCON}_{\text{set}}$, join CHs via multi-hop communication. The sensor S_j selects a CH agent from the set $\text{AGNT}_{\text{set}}(S_j)$ to relay its packet to the gateway. To select the best CH agent S_j, it forms the element matrix by using residual energy of agent nodes, distance of S_j from the agent nodes, and distance of the agent nodes from their respective CHs, as its critical parameters. From the element matrix, the node S_j calculates the cost matrix and then selects the best CH agent to establish a link with gateway node. In this way, all sensor nodes become members of a cluster in the network. The DFMCA method is given in algorithmic form as follows:

Algorithm: DFMCA Clustering: *Input:*

- *Set of gateway nodes* $G = \{G_1, G_2 \ldots G_m\}$
- *Set of sensor nodes* $S = \{S_1, S_2 \ldots S_n\}$

Output:

- *An assignment* $S \rightarrow G$

1. *For each gateway node* $G_i \in G$
   Broadcast a CH_ADV_MSG message to announce its existence in network.
   End //for
2. *For each sensor node* S_i

   If S_i *receives CH_ADV_MSG from gateway* G_k *then*
   Cand_CH (S_i)= *Cand_CH* $(S_i) \cup G_k$ // S_i *becomes a member of* CON_{set} *set*
   Else
   $UNCON_{set} = UNCON_{set} \cup S_i$ // S_i *becomes member of* $UNCON_{set}$.
   End
   End //for
3. *For each sensor* $S_i, (S_i \in CON_{set})$
   3.1. *Construct element matrix (EM) for CH selection using gateway's residual energy, distance of gateway from sink and distance of sensor* S_i *from gateway.*
   3.2. *Calculate normalized matrix, NM from element matrix EM.*
   3.3. *Calculate cost matric/vector C from normalized matrix NM.*
   3.4. *Select gateway node corresponding to maximum value in cost matrix C as CH for node* S_i.
   End //for
4. *For each sensor* $S_j, S_j \in UNCON_{set}$
   4.1. *Broadcast HELP message.*
   4.2. *While* S_j*is receiving reply from* $S_k, S_k \in CON_{set}$
   $AGNT\_CH\ (S_j) = AGNT\_CH(S_j) \cup S_k$ // S_k *become a member of* $AGNT\_CH\ (S_j)$ *set.*
   End //while.
   4.3. S_j *Constructs element matrix for agent CH selection from* $AGNT\_CH\ (S_j)$ *set using residual energy, distance of agent CH from its CH and distance of agent CH from* S_j.
   4.4. *Calculate normalized matrix from element matrix EM.*
   4.5. *Calculate cost matrix/vector C from normalize matrix NM.*
   4.6. *Select candidate node as agent CH for node* S_j*corresponding to maximum value in C. //* S_j*isassignedtosomegatewayviaagentCH*
   4.7. $CON_{set} = CON_{set} \cup S_j$& $UNCON_{set} = UNCON_{set} - S_j$. // S_j*becomes a member of* CON_{set},
   End //for

In our algorithm, for each sensor node $S_{i,}$ the rows of the element matrix for CH selection represent the gateways within the communication radius of the sensor S_i. The jth element of each row represents a parameter critical for CH selection. The critical parameters for CH selection are gateway node's energy, distance of gateway from the base station, and distance of the sensor node from the gateway node. All the sensor nodes which have received the CH_ADV_MSG message can construct their element matrix independently.

10.4.2 Steady Phase

After the completion of cluster formation phase, the steady phase begins in which the CHs set up a TDMA schedule for their CMs to transmit their packets to the CH without collision. The CHs aggregate the received packets to reduce the redundancy and send the aggregated packets to the base station directly. This completes one round of the network operation. The steady phase runs for a fixed number of rounds before the cluster setup phase is executed again. During the steady phase, if a CH dies, all its CMs run the fault-tolerant algorithm to recover from the fault.

10.4.3 Fault tolerance

The faulty CH can be detected by its CMs if they do not receive any acknowledgment from their CH for their sent packets. Once the fault has been detected, its CMs $S_i, \forall S_i \in \mathrm{CON_{set}}$, check their $\mathrm{CON_{set}}(S_i)$ after removing the faulty CH. If $\mathrm{CON_{set}}(S_i) \neq \emptyset$, then S_i selects CH using its cost matrix, as discussed earlier. On the other hand, if $\mathrm{CON_{set}}(S_i) = \emptyset$, the S_i declares itself as a member of $\mathrm{UNCON_{set}}$. The unconnected sensors joins a CH via multi-hop routing using agent nodes as discussed in cluster formation phase.

Algorithm: Fault-tolerance:

1. *For each Sensor S_i in faulty cluster set*

   $CON_{set}(S_i) = CON_{set}(S_i) - G_k$ *// Update its CON_{set} by removing faulty gateway G_k*
   End //for

2. *For each sensor $S_i$in faulty cluster set*

   If $CON_{set}(S_i) \neq \emptyset$ then
   Select CH of S_i // the best gateway for CH based on cost matrix calculated in clustering.
   Else $\forall S_{i,}(CON_{set}(S_{i,}) = \emptyset)$
   $UNCON_{set} = UNCON_{set} \cup S_i$

3. *For each sensor* S_i *(*$S_i \in UNCON_{set}$*)*

   3.1 *Broadcast HELP message.*

   While S_i *is receiving reply from* $S_k, S_k \in CON_{set}$
   $AGNT\_CH\ (S_i) = AGNT\_CH(S_i) \cup S_k$ *//* S_k *become a member of* $AGNT\_CH\ (S_j)$ *set.*
   End //while.

   3.2 S_i*Constructs element matrix for agent CH selection from* $AGNT\_CH(\ S_j)$ *set using residual energy, distance of agent CH from its CH and distance of agent CH from* S_i.
   3.3 *Calculate normalized matrix from element matrix EM.*
   3.4 *Calculate cost matrix/vector C from normalize matrix NM.*
   3.5 *Select candidate node as agent CH for node* S_i*corresponding to maximum value in C.*
   // S_i*isassignedtosomegatewayviaagentCH*
   3.6 $CON_{set} = CON_{set} \cup S_i$ *&* $UNCON_{set} = UNCON_{set} - S_i$. *//* S_i *becomes a member of* CON_{set}*,*
   End //for

10.5 Simulation Results

For performance evaluation, we test our proposed algorithm and compare it with the load-balanced clustering algorithm (LBCA) [10] and the fault-tolerant clustering algorithm (FTCA) [11]. We use the following parameters for performance evaluation: network lifetime and energy conservation. In our experiments, we have computed the network lifetime in terms of rounds when the first gateway node becomes dead (FGND). The energy conservation is defined as the average remaining energy of the sensor nodes in network. The simulations have been carried out using Dev C++ and MATLAB. The parameters with their typical values used in our simulations are given in Table 10.1.

We have computed network lifetime in terms of the number of gateway nodes alive with respect to the number of rounds as shown in Fig. 10.2. Here, we have initially considered 35 gateway nodes. It is evident from Fig. 10.2 that our DFMCA outperforms both the LBCA [10] and FTCA [11].

We have also computed the network lifetime in terms of number of rounds when the first gateway node becomes dead (FGND), as shown in Fig. 10.3, for different number of gateway nodes, i.e., 30, 35, 40, 45, and 50. Our DFMCA method performs better than both the LBCA [10] and FTCA [11], for different number of gateway nodes.

We have also evaluated our DFMCA in terms of average remaining energy of the sensors in network as shown in Fig. 10.4. As evident from Fig. 10.4, our algorithm conserves more energy of sensor nodes as compared to the LBCA [10] and FTCA [11].

Table 10.1 Parameters and their values

Parameter name	Notation	Value
Target area	A	100×100 m^2
Sink location	S	(50, 0)
No. of sensors	N	350
No. of gateways	G	30–50
Initial sensor energy	E_{res}	0.02 J
Initial gateway energy	E_{res_g}	0.1 J
Tx range	r	10 m
Control packet size	C_p	100
Data packet size	D_p	500
Free space (ε_{fs})	ε_{fs}	10 pJ/m^2/bit
Multi-path fading (ε_{mp})	ε_{mp}	0.0013 pJ/bit/m^4
Tx or *Rx* electronics	α_{tx} or α_{rx}	50 nJ/m^2/bit

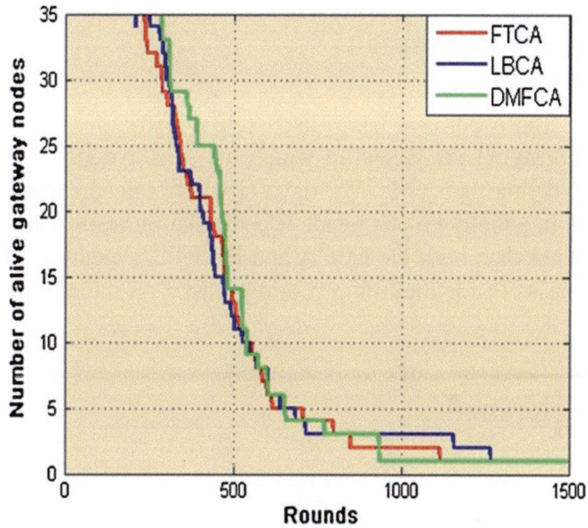

Fig. 10.2 Network lifetime of various algorithms ($G = 35$)

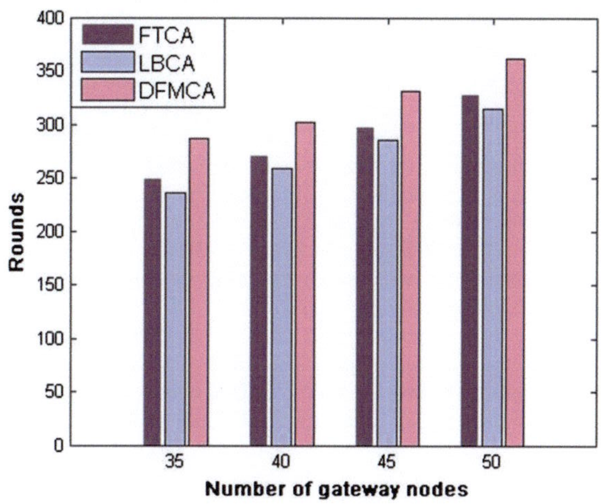

Fig. 10.3 Network lifetime (FGND)

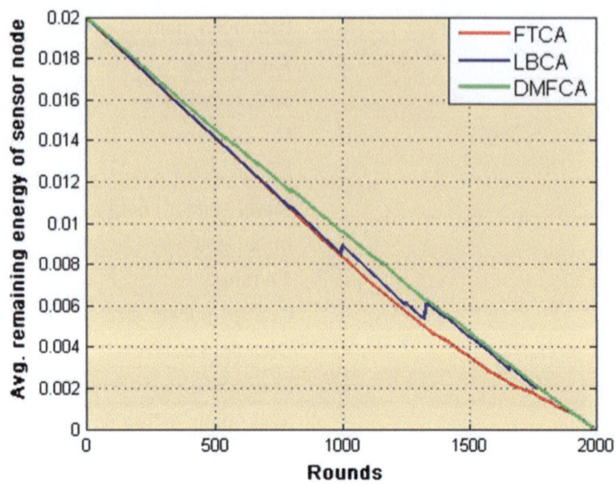

Fig. 10.4 Average residual energy of sensor nodes

10.6 Conclusion

In this paper, we have discussed a distributed fault-tolerant clustering algorithm (DFMCA) for WSNs in which the sensor nodes consider multiple parameters for selecting their CHs. We have also introduced a fault-tolerant algorithm to recover from a faulty cluster head. The simulation results show that our algorithm performs better than both the LBCA and the FTCA methods. In future work, we will extend this work for multi-hop routing, based on the modified DFMCA.

References

1. Akilidz IF, Su W, Sankarasubramaniam Y, Cayirci E (2002) Wireless sensor networks: survey. Comput Netw 38:393–422
2. Smaragdakis G, Matta I, Bestavros A (2004) SEP: a stable election protocol for clustered heterogeneous wireless sensor networks. In: Second international workshop on sensor and actor network protocols and applications SANPA
3. Duan C, Fan H (2007) A distributed energy balance clustering protocol for heterogeneous wireless sensor networks, wireless communications, networking and mobile computing. In: WiCom international conference, pp 2469–2473
4. Qing L, Zhu Q, Wang M (2006) Design of a distributed energy efficient clustering algorithm for heterogeneous wireless sensor networks. Comput Commun 29:2230–2237
5. Pratyay K, Jana PK (2012) Energy efficient load-balanced clustering algorithm for wireless sensor network. In: ICCCS, Procedia Technology (Elsevier), vol 6, pp 771–777
6. Heinzelman WB, Chandrakasan A, Balakrishnan H (2000) Energy-efficient communication protocols for wireless microsensor networks. In: Proceedings of Hawaii international conference on system sciences
7. Lindsey S, Raghavenda CS (2002) PEGASIS: power efficient gathering in sensor information systems. In: Proceeding of the IEEE aerospace conference. Big Sky, Montana
8. Younis O, Fahmy S (2004) HEED: a hybrid, energy-efficient, distributed clustering approach for ad hoc sensor networks. IEEE Trans Mobile Comput 3:366–379

9. Manjeshwar A, Grawal DP (2001) TEEN: a protocol for enhanced efficiency in wireless sensor networks. In: Proceedings of the 15th parallel and distributed processing symposium. IEEE Computer Society, 2011, San Francisco, pp 2009–2015
10. Gupta G, Younis M (2003) Load-balanced clustering of wireless sensor networks. In: IEEE International Conference ICC '03, vol 3, pp 1848–1852
11. Gupta G, Younis M (2003) Fault-tolerant clustering of wireless sensor networks. IEEE WCNC 3:1579–1584

Chapter 11
Comparing Energy Efficiency of DF Relay-Assisted Cooperative and Noncooperative Short-Range Wireless Systems

Biswajit Ghosh, Aniruddha Chandra and Ashis Kumar Mal

11.1 Introduction

Minimizing the energy consumption over short-range data transfer has become the prime concern for dense wireless sensor networks. Optimizing the transmission energy is one of the major procedures for energy-efficient communication techniques. It was investigated in [1] that the transmit power consumption is much significant compared to the power consumption of the RF circuits, when the nodes' distance are far away from each other.

Cooperative techniques are one of the important candidates to reduce the required transmit energy in WSNs. This is acquired by cooperation among the nodes. The channel fading has a significant outcome on the consistency of energy consumption and data transmission in cellular network. Cooperation among the relay nodes helps in mitigating the effects of channel fading. In destination node, the same message received through direct path is buffered and consequently combined with the relayed paths message. Different cooperative schemes have been developed for improved energy efficiency of different communication schemes. The authors in [1, 2] discussed on energy efficiency of MIMO and cooperative MIMO

B. Ghosh (✉)
IT Department, Future Institute of Engineering & Management,
Kolkata 700150, WB, India
e-mail: biswanerist@gmail.com

A. Chandra
Department of Radio Electronics, Brno University of Technology,
Brno, Czech Republic
e-mail: aniruddha.chandra@ieee.org

A.K. Mal
ECE Department, National Institute of Technology,
Durgapur 713209, WB, India
e-mail: akmal@ece.nitdgp.ac.in

V. Nath (ed.), *Proceedings of the International Conference on Nano-electronics, Circuits & Communication Systems*, Lecture Notes in Electrical Engineering 403,
DOI 10.1007/978-981-10-2999-8_11

techniques in wireless networks. The authors [2] compared the performance of the switch and stay combining (SSC) and maximum ratio combining (MRC) based on the transmit antenna selection (TAS) at the receiver, whereas we focus on comparing the performance of selection combing (SC) and maximum ratio combining (MRC). The related literature [3] discussed about the significance of cooperative transmission compared to without cooperative schemes in WSNs. They suggested that cooperative transmission technique is also more energy-efficient for small transmission range, but they did not investigate the effect of relay placement. Some other literature [4] investigate the energy efficiency of relay for different locations of relay, modulation level, and packet size. They consider the transmit power is fixed for cooperation and without cooperation scheme. The authors conclude that for short-distance direct transmission is more energy-efficient compared to cooperation. In [5], the authors suggested that transmit power of cooperative and noncooperative scheme can be minimized by establishing an acceptable limit for the packet loss. The authors [6] worked with decode and forward (DF) relay which is based on parallel coding and prepared a finite SNR analysis for the collaborative scheme based on maximum ratio combing (MRC), but did not investigated with other combing schemes. In [7, 8] energy-efficient relay placements were considered. The authors [7] investigate energy efficiency for multi-hop relayed communication among end nodes, instead of that our analysis is based on single relay (dual hop) case only.

In this paper, we extend our earlier model [9] by comparing the noncooperative scheme with cooperative schemes when operating over a Rayleigh fading channel and using M-ary quadrature amplitude modulation (MQAM), and try to figure out the best one in terms of energy efficiency. We also investigate effective relay placement regions and how different design parameters such as spectral efficiency (R), path loss (n), and distance (d) affect the placement region (Fig. 11.1).

The rest of this paper is organized as follows: Section 11.2 presents the energy consumption models. In Sect. 11.3, numerical results are discussed. Finally, Sect. 11.4 concludes the paper.

11.2 Energy Consumption Modeling

For a short-range wireless transmission, we can express the total power consumption as

$$P_C = P_{\mathrm{PA}} + P_{\mathrm{TX}} + P_{\mathrm{RX}} \tag{11.1}$$

where P_{PA} indicates power consumption by the power amplifier (PA) at transmitter (TX), while power for the rest of transmitter blocks is given by P_{TX}, and P_{RX} denotes the power consumption circuit components at receiver (RX). For detailed expressions of P_{TX} and P_{RX}, the readers are referred to [10, 11].

As per the notational conventions in [1], we may write P_{PA} as

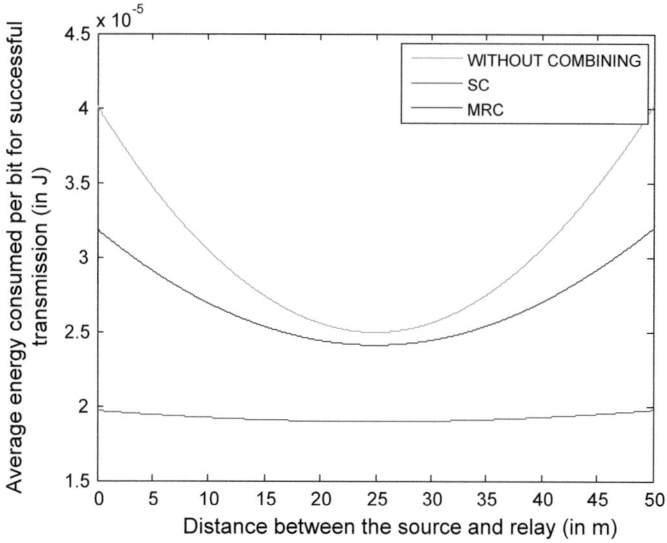

Fig. 11.1 Variation in energy consumption with relay position (d_{SR}) for cooperative and noncooperative schemes

$$\begin{aligned} P_{\mathrm{PA}} &= (\xi/\eta)P_{T,ij} \\ &= (\xi/\eta) \times \mathrm{term}_{ij} \times E_b \times 2R_b \end{aligned} \tag{11.2}$$

where $\mathrm{term}_{ij} = (4\pi/\lambda)^2 d_{ij}^n M_l N_f / G$. In the term specified, η is the drain efficiency of the PA and ξ is the peak-to-average power ratio (PAPR) which for MQAM may be written as $\xi = 3\left[(\sqrt{M}-1)/(\sqrt{M}+1)\right]$ and $P_{T,ij}$ is the transmitted power. The parameters d, n, λ, M_l, N_f, and G denote distance between source and destination, path loss exponent, carrier wavelength ($\lambda = c/f_c$; c = speed of light in free space, f_c = carrier frequency), link margin, noise figure, and combined TX-RX antenna gains, respectively.

The received signal at each transmission time slot can be expressed as

$$y_{ij} = \sqrt{E_b R_b} h_{ij} x + n_{ij} \tag{11.3}$$

where $i \in \{S, R\}$ and $j \in \{R, D\}$, h_{ij} represents the fading parameter of the channel, x represents signal to be transmitted, and n_{ij} is the noise vector. For a Rayleigh fading channel, the probability density function (PDF) of the received instantaneous signal-to-noise ratio (SNR) is given by Goldsmith [12]

$$f_\gamma(\gamma) = \frac{1}{\bar{\gamma}_{ij}} \exp\left(-\frac{\gamma}{\bar{\gamma}_{ij}}\right) \tag{11.4}$$

where $\bar{\gamma}_{ij}$ is the average SNR.

From (11.4), it is easy to show that the outage probability for single hop transmission in Rayleigh fading channel with threshold β is given by

$$O_o = 1 - \exp\left(-\frac{\beta}{\bar{\gamma}_{ij}}\right) \tag{11.5}$$

An outage arises at the receiver end when the received SNR falls below a threshold β (which permit error free decoding). This threshold is defined from Shannon's Capacity Theorem [12]

$$\beta = 2^{RL} - 1 \tag{11.6}$$

where L is number of hops, and R is the spectral efficiency.

The total energy consumption takes into account of the energy consumed by the PA, the consumption by the rest of the TX and RX circuitry, and the bit rate ($R_b = B \times R$; B is the bandwidth). Thus, the total energy consumption per bit on any link is given by

$$E_{bt,ij} = \frac{P_{PA,ij} + P_{TX} + P_{RX}}{R_b} \tag{11.7}$$

where $P_{\mathrm{PA},ij}$ is the power consumption by the power amplifier in the i–j link.

11.2.1 Noncooperative Scheme

When the destination receives the signal only via relay and ignores any direct signal available from source, outage occurs either if there is an outage in the *S-R* link or there is no outage in the *S-R* link but it occurs in the *R-D* link. If we express the outage probability along *S-R* link by O_{SR} and in the *R-D* link as O_{RD}, for similar fading scenarios in both the links, i.e., $O_{\mathrm{SR}} = O_{\mathrm{RD}} = O_{\mathrm{relay}}$, the end-to-end outage probability along the entire relayed path is given by

$$\begin{aligned} O_R &= O_{\mathrm{SR}} + (1 - O_{\mathrm{SR}})O_{\mathrm{RD}} \\ &= 2O_{\mathrm{relay}} + O_{\mathrm{relay}}^2 \\ &= 2\left[1 - \exp\left(-\frac{\beta}{\bar{\gamma}_R}\right)\right] + \left[1 - \exp\left(-\frac{\beta}{\bar{\gamma}_R}\right)\right]^2 \end{aligned} \tag{11.8}$$

whereas $O_{\mathrm{relay}} = 1 - \exp(-\beta/\bar{\gamma}_R)$ as derived from (11.5) and the average SNR per link, $\bar{\gamma}_R = -\beta/\ln(1 - O_R)$ is set to achieve a target outage probability O_R. The corresponding bit energy $E_{b,\mathrm{Relay}}$ at the receiver is simply $E_{b,\mathrm{Relay}} = \bar{\gamma}_R \times N_0$, where N_0 is the one-sided noise power spectral density.

It may be noted that O_{relay} also represents the percentage of time when energy consumption does not contribute to successful transmission. To transmit one bit through the relayed path, the energy consumption is

$$\begin{aligned} E_{bt,\mathrm{Relay}} &= O_{\mathrm{relay}}\left(\frac{P_{\mathrm{PA,SR,Relay}} + P_{\mathrm{TX}} + 2P_{\mathrm{RX}}}{R_b}\right) \\ &\quad + (1 - O_{\mathrm{relay}})\left(\frac{P_{\mathrm{PA,SR,Relay}} + P_{\mathrm{PA,RD,Relay}} + 2P_{TX} + 2P_{\mathrm{RX}}}{R_b}\right) \end{aligned} \tag{11.9}$$

where $P_{\mathrm{PA,ij,Relay}} = (\xi/\eta)P_{\mathrm{t,ij,Relay}}$ and $P_{t.ij,\mathrm{Relay}} = 2R_bE_{b,\mathrm{Relay}} * \mathrm{term}_{ij}$.

Finally, considering that the average number of retransmissions required is $1/(1 - O_R)$ to compensate for outage, the average energy consumption for every successful bit transmission is given by

$$E_{bt,R\_\mathrm{suc}} = \frac{E_{bt,\mathrm{Relay}}}{1 - O_R}. \tag{11.10}$$

11.2.2 Cooperation Through SC

In selection combining, all the various paths are active at the same time and the one with the highest SNR is being considered as the combined output. So for the combined output, outage occurs when the path with the highest SNR is in outage or in other words when all the paths present are in outage at the same time (since if the path with the highest SNR has SNR value less than the outage threshold, then for all the other paths it will be automatically less than the threshold). However, for energy consumption calculation, the energy consumption in all the paths has to be taken into account since all the paths are active at any point of time.

The end-to-end outage probability for the SC scheme can be written as

$$\begin{aligned} O_{\mathrm{sc}} &= (O_s + (1 - O_s)O_s) \times O_s \\ &= \left[1 - \exp\left(-\frac{\beta}{\bar{\gamma}_{\mathrm{SC}}}\right)\right]^2 + 1 + \exp\left(-\frac{\beta}{\bar{\gamma}_{\mathrm{SC}}}\right) \end{aligned} \tag{11.11}$$

In order to find the minimal transmit power required by the SC cooperative scheme, for a target total outage probability O_{sc}, we find the average SNR which is denoted by $\bar{\gamma}_{\mathrm{sc}}$ to find the bit energy, $E_{\mathrm{b,sc}} = \bar{\gamma}_{\mathrm{sc}} \times N_0$, at the receiver.

To transmit one bit through the relayed path energy consumption is given by

$$E_{bt,\mathrm{SC}} = O_s\left(\frac{P_{\mathrm{PA,SR,sc}} + P_{\mathrm{TX}} + 2P_{\mathrm{RX}}}{R_b}\right) + (1 - O_s)\left(\frac{P_{\mathrm{PA,SR,sc}} + P_{\mathrm{PA,RD,sc}} + 2P_{\mathrm{TX}} + 3P_{\mathrm{RX}}}{R_b}\right) \tag{11.12}$$

where $O_s = 1 - \exp(-\beta/\bar{\gamma}_{\mathrm{sc}})$, $P_{\mathrm{PA,ij,sc}} = (\xi/\eta)P_{t,ij,sc}$ and $P_{t.ij,sc} = 2R_bE_{b,sc} * \mathrm{term}_{ij}$. Considering the average number of retransmissions required to be $1/(1 - O_{\mathrm{sc}})$, the average energy consumption for successful transmission is

$$E_{bt,\mathrm{SC\_suc}} = \frac{E_{bt,\mathrm{SC}}}{1 - O_{\mathrm{sc}}}. \tag{11.13}$$

11.2.3 Cooperation Through MRC

MRC may also be used to combine the signals obtained over relayed path and direct path. The MRC scheme is a complex scheme and is based on desirable combining of all the accessible paths at the receiver. The SNR of the individual paths may be under the threshold value of SNR, but the combined SNR of different paths might be above it.

In our case, there are only two branches: one is the relayed path and the other one is the direct path. For MRC, the combined SNR present at the output is equivalent to the summation of the entire individual branch SNRs.

We consider the following two cases: first, when there is outage in the *S-R* link, and second, when there is no outage in the *S-R* link. When there is outage in the *S-R* link then due to the adaptive nature of the relay, the relay will not transmit anything to the destination. This results in only one path to be active, and hence the combined SNR will be same as that of the SNR of the direct path, i.e., $\gamma_{\mathrm{MRC}} = \gamma_D$ where γ_D represents the SNR along the direct path. Combined output depends on the SNR of the direct path so it will be in outage when the direct path is in outage. The joint probability of outage in the S-R link and the outage of the combined output is given by

$$\begin{aligned} p_A &= O_{\mathrm{SR}} \times O_{\mathrm{RD}} = O_o^2 \\ &= 1 - 2\exp\left(-\frac{\beta}{\bar{\gamma}_{\mathrm{MRC}}}\right) + \exp\left(-\frac{2\beta}{\bar{\gamma}_{\mathrm{MRC}}}\right) \end{aligned} \tag{11.14}$$

When there is no outage in the *S-R* link, then both the relayed path and the direct path will be active, and hence the combined SNR will be the summation of the SNRs of both the paths, i.e., $\gamma_{\mathrm{MRC}} = \gamma_D + \gamma_{\mathrm{RD}}$ where γ_{RD} is the SNR along the *R-D* link. As the relay is adaptive, we do not consider end-to-end SNR.

Now the outage of the combined output is calculated as

$$\begin{aligned}\Pr\{\gamma_{\mathrm{MRC}} \le \beta_{\mu}\} &= \Pr\{\gamma_{\mathrm{RD}} \le \beta_{\mu} - \gamma_D | \gamma_D\} \\ &= \int_0^{\beta_{\mu}} \Pr\{\gamma_{\mathrm{RD}} \le \beta_{\mu} - \gamma_D\} f(\gamma_D) \mathrm{d}\gamma_D \\ &= 1 - \exp\left(-\frac{\beta_{\mu}}{\bar{\gamma}}\right)\left(1 + \frac{\beta_{\mu}}{\bar{\gamma}}\right)\end{aligned} \tag{11.15}$$

where $\beta_{\mu} = 2^{2R} - 1$. The limits of the integration are taken from 0 to β_{μ} since if γ_D becomes greater than β_{μ}, then the summation of γ_D and γ_{RD} can never be less than β_{μ}.

The total end-to-end outage probability for maximal ratio combining is be given by

$$\begin{aligned}O_{\mathrm{mrc}} &= p_A + p_B \\ &= 1 - 2\exp\left(-\frac{\beta}{\bar{\gamma}_{\mathrm{MRC}}}\right) + \exp\left(-\frac{2\beta}{\bar{\gamma}_{\mathrm{MRC}}}\right) \\ &+ \left(1 - \exp\left(-\frac{\beta_{\mu}}{\bar{\gamma}_{\mathrm{MRC}}}\right)\left[1 + \frac{\beta_{\mu}}{\bar{\gamma}_{\mathrm{MRC}}}\right]\right) \times \exp\left(-\frac{\beta}{\bar{\gamma}_{\mathrm{MRC}}}\right)\end{aligned} \tag{11.16}$$

where $\bar{\gamma}_{\mathrm{MRC}}$ is the average SNR for maximum ratio combining and

$$p_B = \left(1 - \exp\left(-\frac{\beta_{\mu}}{\bar{\gamma}_{\mathrm{MRC}}}\right)\left[1 + \frac{\beta_{\mu}}{\bar{\gamma}_{\mathrm{MRC}}}\right]\right) \times \exp\left(-\frac{\beta}{\gamma_{\mathrm{MRC}}}\right) \tag{11.17}$$

Similar to SC, for a target total outage probability O_{mrc}, we can find the average SNR which is denoted by $\bar{\gamma}_{\mathrm{mrc}}$ to find the bit energy $E_{b\mathrm{mrc}}$ required at the receiver. Also, the energy consumption for single bit transmission will be

$$\begin{aligned}E_{bt,\mathrm{mrc}} &= O_{\mathrm{mrc}}\left(\frac{P_{\mathrm{PA,SR,MRC}} + P_{\mathrm{TX}} + 2P_{\mathrm{RX}}}{R_b}\right) \\ &+ (1 - O_{\mathrm{mrc}})\left(\frac{P_{\mathrm{PA,SR,MRC}} + P_{\mathrm{PA,RD,MRC}} + 2P_{\mathrm{TX}} + 3P_{\mathrm{RX}}}{R_b}\right)\end{aligned} \tag{11.18}$$

where $O_{\mathrm{mrc}} = 1 - \exp(-\beta/\bar{\gamma}_{\mathrm{MRC}})$, $P_{\mathrm{PA},ij,\mathrm{MRC}} = (\xi/\eta)P_{t,ij,\mathrm{MRC}}$ and $P_{t.ij,\mathrm{MRC}} = 2R_b E_{b,\mathrm{MRC}} * \mathrm{term}_{ij}$. Finally, the average energy consumption per bit for successful transmission with MRC is

$$E_{bt,\mathrm{MRC}\_suc} = \frac{E_{bt,\mathrm{MRC}}}{1 - O_{\mathrm{mrc}}}. \tag{11.19}$$

11.3 Results and Discussions

In this section, we compare the energy consumption for one successful bit transmission following noncooperative (without combining), SC, and MRC schemes. The figures which given here are simulated by MATLAB simulator. The relay node is assumed to be situated on a straight line joining the source node and the destination node. From Figs. 11.2, 11.3, 11.4, 11.5, 11.6, 11.7, and 11.8, the relay potion is in midpoint between source and destination. We assume that the target total outage probabilities are same (0.05) for all cooperative and noncooperative schemes. The system parameters [3, 10] are, $P_{TX} = 97.6$ mW, $P_{RX} = 112.4$ mW, $n = 2.3$, $d = 50$ m, $M = 4$, and $R = 2$ b/s/Hz, unless otherwise specified.

Figure 11.1 shows the variation in energy consumption with all these three schemes plotted against the distance of the relay from the source. The energy efficiency with SC is better than MRC when the path loss coefficient is low ($n = 2.3$). Also, from the implementation point of view, an MRC combiner increases complexity of the receiver at the destination.

Further, we observed that the optimum positions of relay for all these three schemes are close to the midpoint of the *S-D* joining line.

11.3.1 *Effect of S-D distance*

In this subsection, we compare the energy consumption when the relay is placed at exactly midpoint of the *S-D* path. As shown in Fig. 11.2, the noncooperative

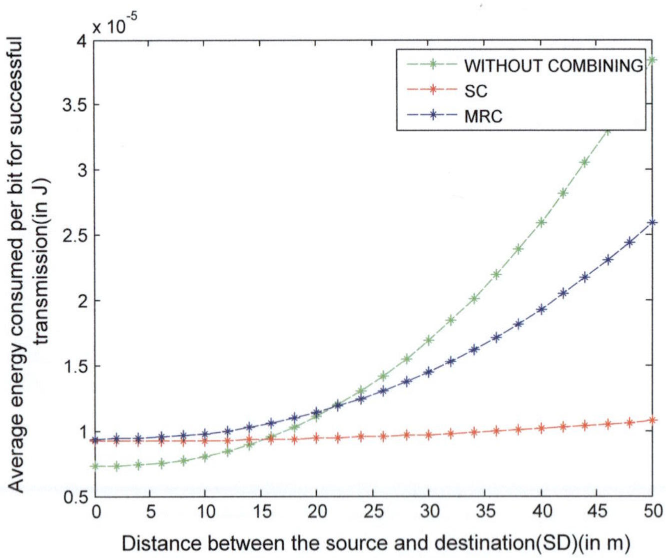

Fig. 11.2 Variation in energy consumption with distance between source and destination

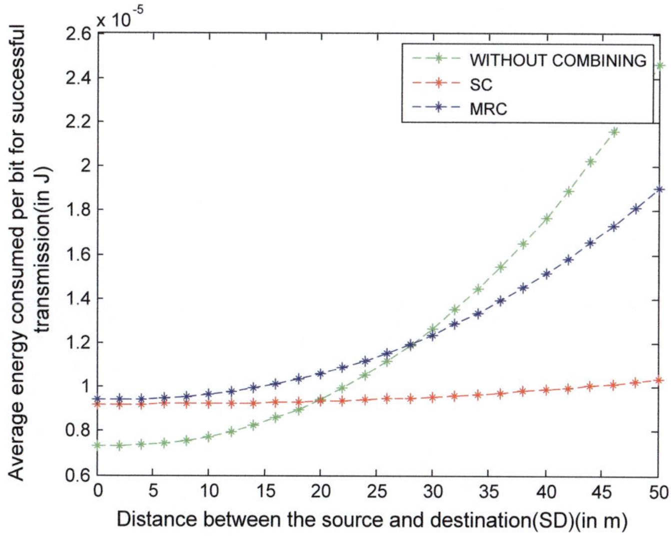

Fig. 11.3 Variation in energy consumption with distance between source and destination. The target outage probability is 0.009

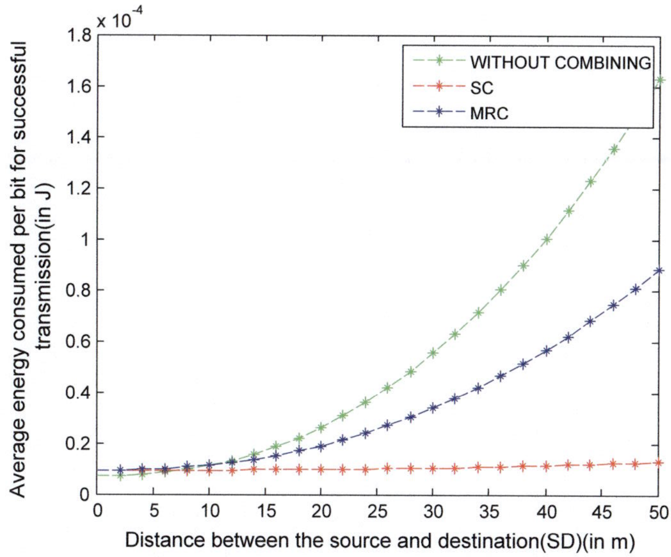

Fig. 11.4 Variation in energy consumption with distance between source and destination. The target outage probability is 0.001

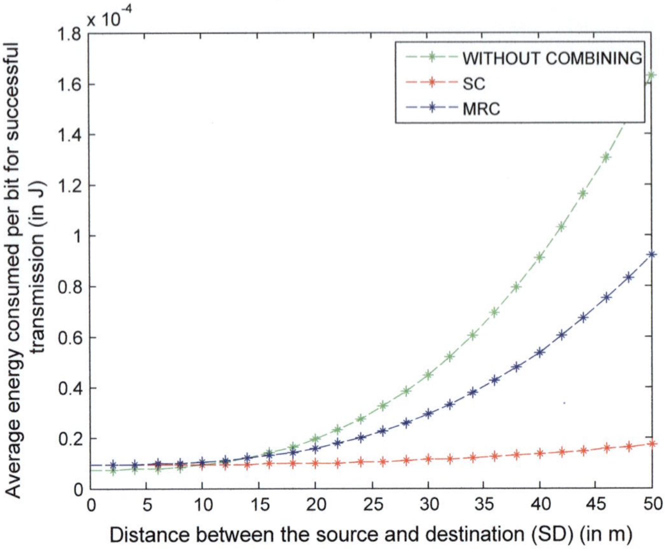

Fig. 11.5 Variation in energy consumption with distance between source and destination. The path loss coefficient is 2.8

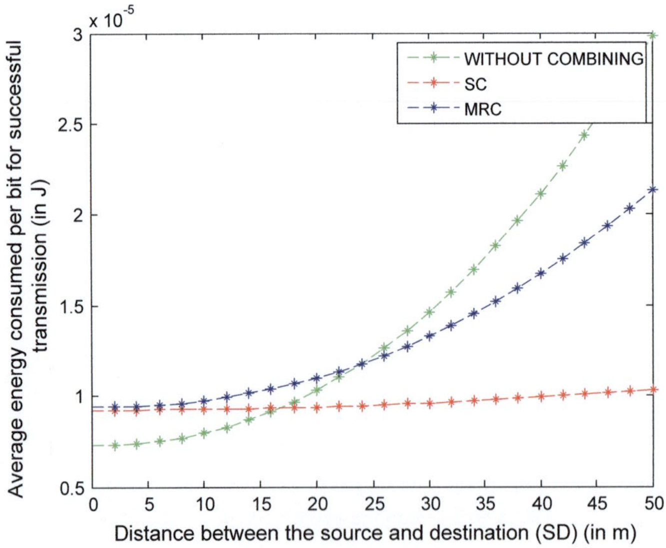

Fig. 11.6 Variation in energy consumption with distance between source and destination. The path loss coefficient is 2.2

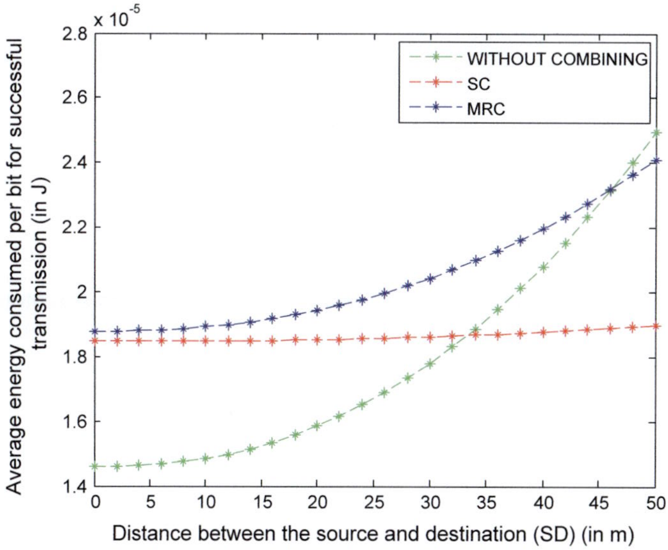

Fig. 11.7 Variation in energy consumption with distance between source and destination. The spectral efficiency is 1

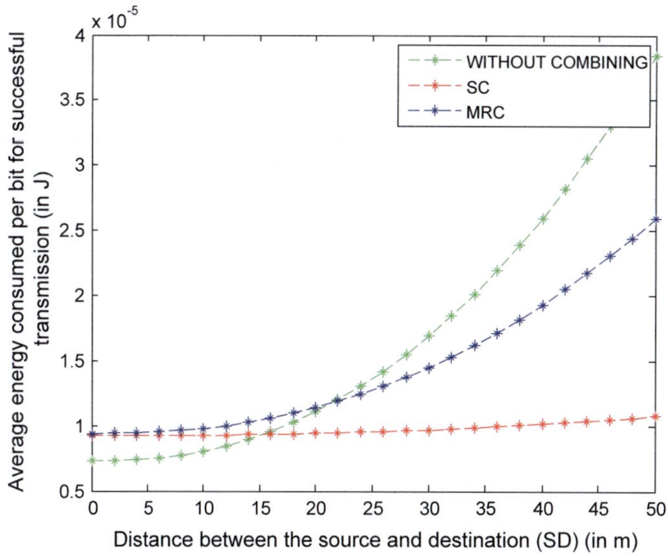

Fig. 11.8 Variation in energy consumption with distance between source and destination. The spectral efficiency is 2

scheme is more energy-efficient compared to SC scheme and MRC scheme, when *S-D* distance is less than 15 and 21 m, respectively. Thus, we conclude that for short distances, noncooperative scheme is a better choice than the cooperative schemes.

11.3.2 Effect of Target Outage (O_o)

Comparing Figs. 11.3 and 11.4, one may find that for increased target outage values, the distance range where noncooperative scheme outperforms the combining schemes increases.

11.3.3 Effect of Path loss coefficient (n)

Figures 11.5 and 11.6 depict that for high path loss exponents (greater than 2.8), combining schemes perform better than the without-combining scheme. We also observe from Fig. 11.6 that the without-combining strategy exhibits better energy efficiency when the distance of *S-D* path is less than 15 m.

11.3.4 Effect of Spectral Efficiency (R)

In Figs. 11.7 and 11.8, we observed that for higher spectral efficiencies in combining schemes turn out be more energy-efficient than the without-combining scheme. Particularly, for lower values of the spectral efficiency, there exists a significant range of *S-D* distance. Referring to Fig. 11.7, the noncooperative scheme is more energy-efficient than SC scheme and MRC scheme, at the time while SD distance is 34 and 46 m, respectively.

11.4 Conclusion

In this paper, we have considered a three-node short-range wireless transmission model, out of which one of the nodes is a relay. The source station transmits to the destination through the relay. The relay is assumed to operate in adaptive DF mode, and cooperative communication at destination may be realized to reduce the energy wasted in transmission. We considered two diversity schemes, SC and MRC, and provided the respective calculations for outage probabilities and energy consumptions. We conclude from results that the energy consumption with cooperation is lesser, but the energy efficiency is observed only after a certain transmission range. We also conclude that at lower spectral efficiencies, the energy required for transmission without combining is significantly lower than complex combining methods in short-range communications.

Acknowledgements This work was supported by the SoMoPro II programme, Project No. 3SGA5720, co-financed by the People Program me (Marie Curie action) of the Seventh Framework Programme of EU according to the REA Grant Agreement No. 291782 and by the South-Moravian Region.

References

1. Cui S, Goldsmith A, Bahai A (2004) Energy-efficiency of MIMO and cooperative MIMO techniques in sensor networks. IEEE J Sel Areas Commun 22(6):1089–1098
2. Brante GGO, Stupia I, Souza RD, Vandendorpe L (2013) Outage probability and energy efficiency of cooperative MIMO with antenna selection. IEEE Trans Wireless Commun 12:5896–5907
3. de Oliveira Brante GG, Kakitani MT, Souza RD (2011) On the energy efficiency of some cooperative and non-cooperative transmission schemes in WSNs. In: Proceedings of IEEE CISS, Baltimore, MD, pp 1–6, March 2011
4. Wang S, Nie J (2010) Energy efficiency optimization of cooperative communication in wireless sensor network. In: EURASIP Journal on Wireless Communications and Networking, pp 1–8
5. Sadek AK, Yu W, Liu KJR (2009) On the energy efficiency of cooperative communications in wireless sensor networks. ACM Trans Sensor Netw 6(1):1–21
6. Khormuji M, Larsson E (2009) Cooperative transmission based on decode-and-forward relaying with partial repetition coding. IEEE Trans Wireless Commun 8(4):1716–1725
7. Zhang R, Gorce J-M (2008) Optimal transmission range for minimum energy consumption in wireless sensor networks. In: Proceedings of the IEEE WCNC, Las Vegas, USA, pp 757–762
8. Wang S, Nie J (2010) Energy efficiency optimization of cooperative communication in wireless sensor networks. EURASIP J Wireless Commun Netw 162326:1–8
9. Chandra A, Ghosh B, Biswas N, Brante G, Souza RD (2013) Energy efficient relay placement for dual hop wireless transmission. Int J Electron Lett 1(4), 198–209 (Taylor & Francis)
10. Ghosh B, Ghosh A, Biswas N, Chandra A (2012) Placing the 'third' node: an energy efficiency perspective. In: Proceedings of IEEE CODEC, Kolkata, pp 1–4, Dec 2012
11. Biswas N, Ghosh B, Chandra A (2013) Energy efficiency relay node placement in aeta-mu fading channel. In: Proceedings of the IEEE ICT. IEEE, Kumaracoil, pp 1–5
12. Goldsmith A (2005) Wireless communications. Cambridge University Press, Cambridge

Chapter 12
Modified PSO-Based Equalizers for Channel Equalization

D.C. Diana and S.P. Joy Vasantha Rani

12.1 Introduction

Adaptive equalization [1] plays an important role in the high-speed digital transmission to remove and recover the problem of inter-symbol interference (ISI). The adaptive algorithms [2] such as steepest descent, least mean square (LMS), recursive least square (RLS), affine projection algorithm (APA), and their variants [3] reported in literature have the chance of getting trapped in local minima [4–6] while optimizing the equalizer weights. The performance of these algorithms further degraded in nonlinear channel conditions [6]. To overcome these problems, different derivative-free optimization algorithms are proposed, whereas PSO is one among them. For solving optimization algorithms, PSO is proven as an efficient method and was applied successfully in the area of adaptive equalization [6]. PSO stays as one of the best algorithms for channel equalization in the recent years [5–7]. And also it provides minimum mean square error (MSE) compared to genetic algorithms used in the channel equalization [6].

From the first introduction of PSO [8], several variants [9–16] are proposed. The inertia weight parameter 'w' is the first modification found in literature which plays a major role in convergence and improves the simulation time. Initially, Shi and Eberhart introduced inertia weight [10]. In their work, a range of constant w values are used and found that PSO shows a weak exploration for large w values, i.e., $w > 1.2$, and it tends to trap in local optima with $w < 0.8$. When w is in the range [0.8, 1.2], PSO shows the global optimum in least average

D.C. Diana (✉) · S.P. Joy Vasantha Rani
Department of Electronics Engineering, Madras Institute of Technology, Anna University, Chennai, India
e-mail: diana_ece@yahoo.co.in

S.P. Joy Vasantha Rani
e-mail: joy_mit@annauniv.edu

V. Nath (ed.), *Proceedings of the International Conference on Nano-electronics, Circuits & Communication Systems*, Lecture Notes in Electrical Engineering 403,
DOI 10.1007/978-981-10-2999-8_12

number of iterations. A random value is selected to track the optima in [11], as given in Eq. (1.1).

$$w = 0.5 + \frac{\text{rand}()}{2} \tag{12.1}$$

where rand() is a random value between [0, 1].

Time-varying inertia weight is a common approach used in PSO, which determines the inertia weight based on the current and total iteration. Time-decreasing inertia weight methods are used in literature to improve the convergence rate. A linearly decreasing inertia weight has been used for adaptive equalization in [6], based on the update law:

$$w_n = \left[(w_i - w_f)(m - n)/(m - 1)\right] + w_f \tag{12.2}$$

where w_i is the initial weight, w_f the is maximum weight, 'm' is the maximum iteration value, and 'n' is the current iteration index. A nonlinear decreasing inertia weight is proposed by Chatterjee and Siarry [12] based on the equation:

$$w_n = \left[(w_i - w_f)(m - n)^{np}/(m - 1)^{np}\right] + w_f \tag{12.3}$$

where np is the nonlinear modulation index. With different values of np, inertia weight gives different variations from w_i to w_f.

Feng et al. [13, 14] proposed another modification as given in Eq. (1.4).

$$w_n = \left[(w_i - w_f)(m - n)/m\right] + w_f * z \tag{12.4}$$

where $z = 4z(1 - z)$. Initially, z is the random value between (0, 1).

Lei et al. [15] choose a fuzzy complement function as inertia weight defined by Eq. (1.5).

$$w = \frac{1 - \beta}{1 - s\beta} \tag{12.5}$$

where β is n/m and s is greater than −1.

The other nonlinear approach [16] is given as:

$$w = \left(\frac{2}{n}\right)^{0.3} \tag{12.6}$$

Zheng et al. [17] proposed an increasing inertia weight. Similarly, Jiao et al. [18] proposed another nonlinear increasing inertia weight as in Eq. (1.7),

$$w = w_{\text{initial}} \times u^n \tag{12.7}$$

Here w_{initial} is usually between [0, 1] and u is [1.0001, 1.005]. As per [18], u is set as 1.0002.

This paper analyzes the above existing modifications and proposes novel time-varying inertia weight methods for adaptive equalization with minimum mean square error. The following section describes the methodology used in adaptive equalization. Section 12.3 explains the structure and training of the equalizer filter using PSO. Section 12.4 discusses the results.

12.2 Methodology

Figure 12.1 depicts a basic block diagram used in adaptive equalization [1].

The equalizer input is the convolution sum of the random input Bernoulli sequence $\{x(n)\} = \pm 1$ (with zero mean and variance 1) and channel model. The channel output is added with the random additive white Gaussian noise (*AWGN*). The noise sequence has zero mean and variance 0.001.

Initially, the channel model is assumed as a linear channel [2] with three paths as given in Eq. (12.8).

$$\begin{aligned} h_n &= \tfrac{1}{2}\left[1 + \cos\left(\tfrac{2\pi}{W}(n-2)\right)\right], \quad && n = 1, 2, 3 \\ h_n &= 0 && \text{otherwise} \end{aligned} \tag{12.8}$$

The factor W controls the amount of distortion. The effect of amplitude distortion is analyzed and given in Table 12.3. The effect of nonlinearities generated in the transmitter is modeled as three different nonlinear equations in (12.9), (12.10), and (12.11). The nonlinearity is introduced by relating output $y(n)$ and input $x(n)$ as in [19],

$$\text{Channel 1}: y(n) = \tanh(x(n)) \tag{12.9}$$

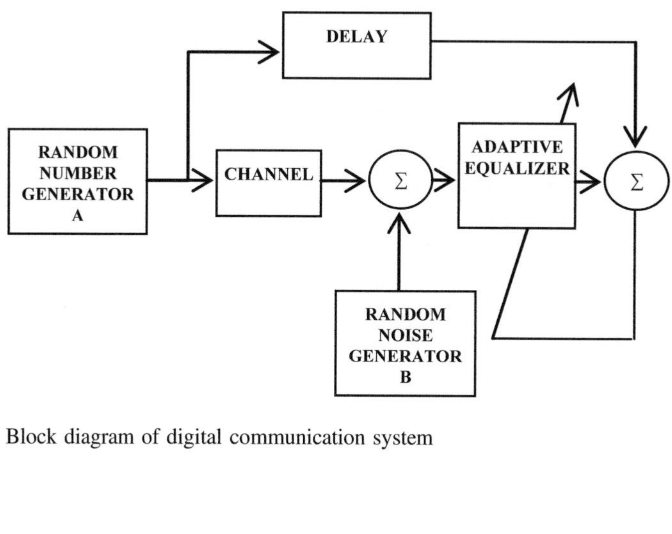

Fig. 12.1 Block diagram of digital communication system

$$\text{Channel 2:}\, y(n) = x(n) + 0.2x^2(n) - 0.1x^3(n) \tag{12.10}$$

$$\text{Channel 3:}\, y(n) = x(n) + 0.2x^2(n) - 0.1x^3(n) + 0.5\cos(\pi x(n)) \tag{12.11}$$

The nonlinear effect is introduced in the input $x(n)$ and it is denoted as $y(n)$. The channel output is represented as

$$r(n) = d_{\text{distorted}}(n) + v_n$$

Here $d_{\text{distorted}}(n)$ is the distorted version of the desired signal. The distortion is introduced by applying the input signal through any one of the nonlinear equations (Eqs. (12.9), (12.10), and (12.11)) and then convolved with the linear channel in Eq. (12.8).

$$d_{\text{distorted}}(n) = y(n) * h_n$$

The signal can be distorted by bandwidth limitation, multipath effect, and the nonlinearities introduced in the transmitter. v_n is the noise component modeled as white Gaussian noise with variance σ_n^2.

The error $e(n)$ is calculated as

$$e(n) = d(n) - r(n)$$

The adaptive algorithm updates the equalizer weights iteratively to minimize $e^2(n)$. Since $e^2(n)$ is positive and gives the instantaneous power, selected as cost (fitness) function.

12.3 System model

The system model [1] used for equalization is the simple linear transversal equalizer as in Fig. 12.2 and decision feedback equalizer shown in Fig. 12.3.

12.3.1 Linear Transversal Equalizer (LTE)

In this type of structure, the present and old received samples $r(t-kT)$ are appropriately weighted by the coefficients c_q and added to generate the output. The weights are trained to optimum value using adaptive algorithms. The output Z_k becomes

$$Z_k = \sum_{q=0}^{N-1} c_q r(t - qT) \tag{12.12}$$

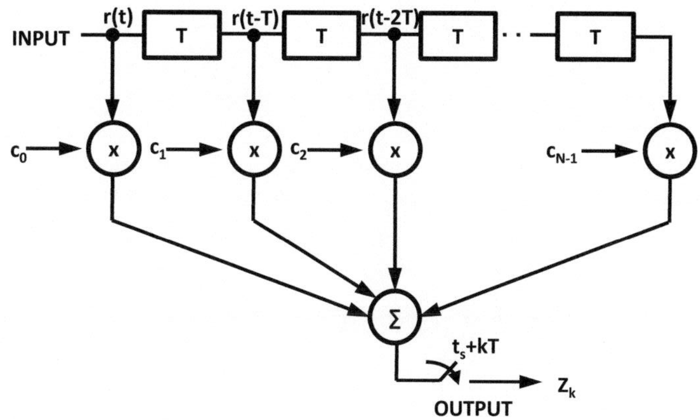

Fig. 12.2 Linear transversal equalizer structure

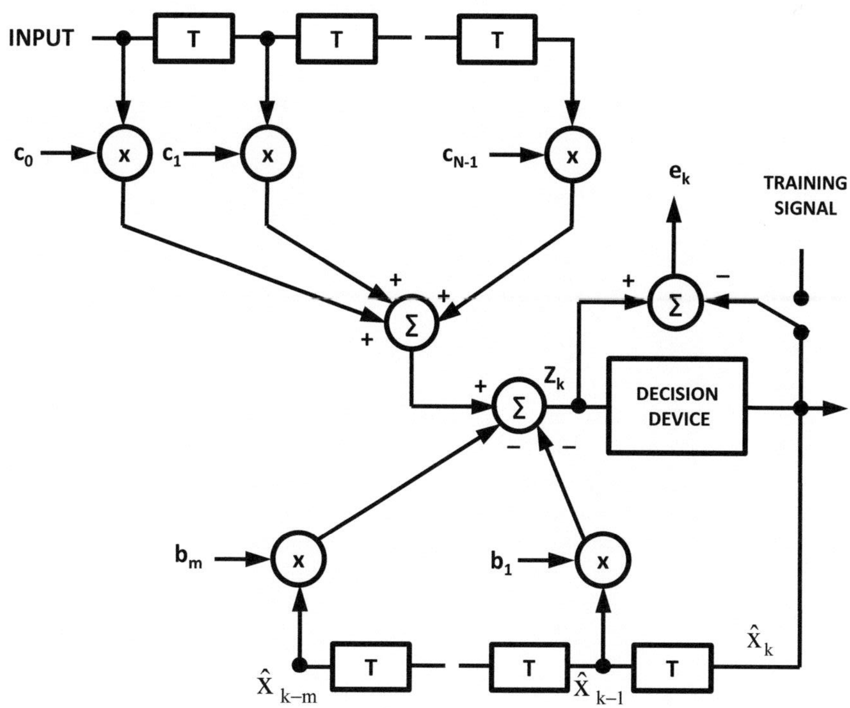

Fig. 12.3 Decision feedback equalizer structure

12.3.2 Decision Feedback Equalizer (DFE)

This type of nonlinear equalizer uses a forward and feedback filter. The forward and feedback filter outputs are summed to find the output of the equalizer. The forward filter output is sent back via the feedback filter. The ISI is canceled by deducting past symbol values from the equalizer output. The output of DFE is calculated as in Eq. (12.13)

$$Z_k = \sum_{q=0}^{N-1} c_q r(t - qT) + \sum_{i=1}^{m} b_i X_{k-m} \tag{12.13}$$

12.3.3 Basic PSO

PSO [8] starts by initializing particles in the random search space and then considers the social and cognitive behavior of the particles. The candidate solutions called 'particle' move around the n-dimensional search space with a velocity, which is adjusted based on its own experience and its neighbors' experiences. The own experience of a particle is denoted as P_{best}; all the particles' experience is denoted as G_{best}. A new G_{best} is generated in each update process, making other particles to fly toward G_{best}. The new velocity and position are calculated as:

$$V_i(t+1) = w * V_i(t) + ac_1 * \text{rand}_1 * (P_{\text{best}} - X_i) + ac_2 * \text{rand}_2 * (G_{\text{best}} - X_i) \tag{12.14}$$

$$c_i(t+1) = c_i(t) + V_i(t+1) \tag{12.15}$$

where ac_1 and ac_2 are positive constants, called as cognitive and social acceleration coefficients, respectively. The two random functions rand_1 and rand_2 are in the interval [0, 1]. $V_i(t)$ and $c_i(t)$ are velocity and position of particle i, respectively, in tth iteration. Location of the best solution (best tap weights) for particle i is P_{best} and G_{best} represents the best solution among all particles. In Eq. (12.14), w is the inertia weight, which controls the local search and global search.

12.3.4 Training by PSO

The PSO-based equalizer [6] is optimized the tap weights based on the following steps:

For LTE:

- T number of tap weights is assigned for equalizer.
- 'ws' samples of data are passed from channel output (distorted signal) to equalizer which generates ws numbers of estimated samples.
- Error is estimated by comparing delayed version of each input sample with equalizer output
- The mean square error function of each particle P is

$$\mathrm{MSE}(P) = \frac{\sum_{i=1}^{K} e_i^2}{ws}$$

- Fitness value MSE (P) is minimized using PSO-based optimization.
- If the MSE of a particle is less than its previous value, then term it as current local best value and its corresponding weight values as P_{best}
- The minimum of MSE of all particles in every iteration is taken as global best value.
- If the global best value is better than the previous one, select the corresponding tap weights to G_{best}.
- Calculated the change in position (Tap weights) of each particle using Eq. (12.14)
- Moved each particle (Tap weights) c_k in Eq. (12.12) to new position by Eq. (12.14).
- Repeated the above steps for the number of iterations specified or stopped when the algorithm converges to an optimum value with least MSE value.

For DFE:

- The coefficients are initialized randomly for forward and feedback filter.
- In the first iteration, only forward filter is active and after calculating the error, the output of the forward filter is fed back through feedback filter.
- The mean square error is calculated by subtracting the output of forward and feedback filters.
- The forward and feedback filter coefficients c_k and b_i in Eq. (12.13) are updated based on Eqs. (12.14) and (12.15).

12.3.5 Proposed Strategies

In PSO, higher values of inertia weight enhance global search while smaller values improve local search. Generally, the inertia weight decreases linearly from 1 to 0 over the entire iterations. The inertia value is high at first, which allows all particles to move freely in the search space during the initial steps, while it decreases over time in the later steps. The later search requires a low inertia value to gradually shift the search from global search to local search. The speed of convergence and

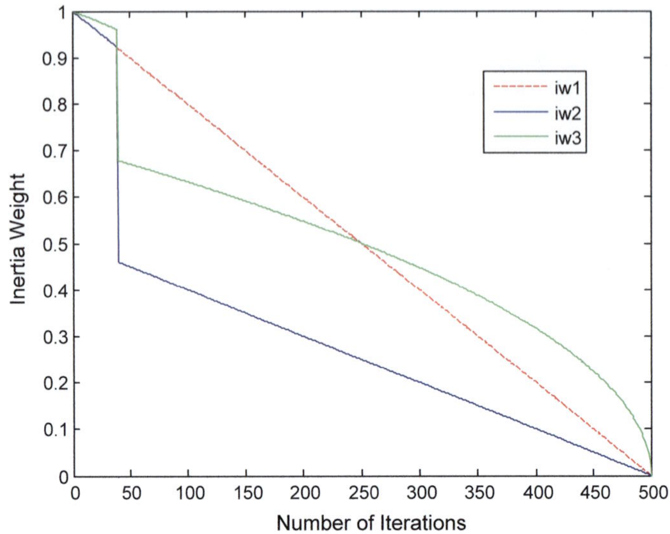

Fig. 12.4 Proposed inertia weight strategies

minimum MSE is improved when the inertia weight is suddenly changed from high to low after a particular iteration. Hence, the inertia weight is updated in three different ways as given in Eqs. (12.16), (12.17), and (12.18).

$$iw_1 = \eta \tag{12.16}$$

$$iw_2 = \begin{cases} \eta & \text{if } n \leq \text{N} \\ \eta/2 & \text{if } n > \end{cases} \tag{12.17}$$

$$iw_3 = \begin{cases} \sqrt{\eta} & \text{if } n \leq \text{N} \\ \sqrt{\eta/2} & \text{if } n > \text{N} \end{cases} \tag{12.18}$$

Figure 12.4 shows that, the inertia weight iw_1 decreases linearly from 1 to 0, whereas in the case of iw_2 and iw_3 the inertia weight decreases linearly until N iteration after that there is a sudden reduction in inertia weight. The common factor used in all time-varying inertia weight algorithms is $\eta = (m-n)/m$, where m denotes maximum iteration and n denotes current iteration. This factor η changes linearly from 1 to 0. If Eq. (12.14) is modified with a decreasing control function, it gives an effective time-varying inertia weight strategy as shown in Eqs. (12.17) and (12.18). The term N in Eqs. (12.17) and (12.18) is the intermediate iteration value used to reduce the inertia weight suddenly after Nth iteration. This reduction produces optimum performance compared to existing inertia weight-modified methods in terms of convergence speed and MSE.

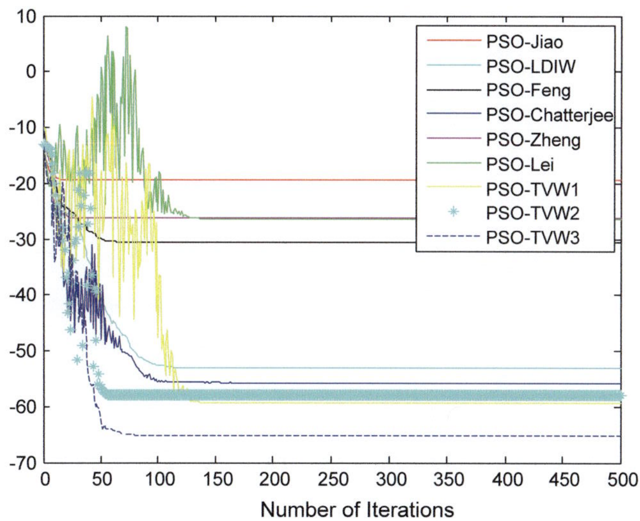

Fig. 12.5 Performance of proposed and other time-varying strategies in linear channel for LTE

12.4 Simulation Results

The simulations are performed in MATLAB 7.7 version. The parameters are set as amplitude distortion $W = 2.9$ which gives an eigenvalue spread 6, particle size $P = 40$, input window size $ws = 20$, the acceleration coefficients $ac_1 = 1$ and $ac_2 = 1$, tap weights $T = 7$, and the initial and final inertia weights $w_i = 0.9$ and $w_f = 0.3$, respectively. The term np used in Eq. (12.3) is taken as 0.7 and w_{initial} in Eq. (12.7) is selected as 0.5. The parameters considered above are minimum values to minimize the computational complexity in the algorithm. If the parameter values such as P, ws, ac_1 and ac_2 are increased, it can improve the performance and the same is analyzed in Sect. 4.1.

The PSO variants considered are PSO-LDIW [9], PSO-Chatterjee [17], PSO-Feng [18, 19], PSO-Lei, PSO-Zheng and PSO-Jiao. The proposed techniques, PSO with time-varying inertia weights, PSO-TVW, PSO-TVW1, PSO-TVW2, and PSO-TVW3 based on Eqs. (12.14), (12.15), and (12.16), are compared with the above variants in Fig. 12.5. Among all it is seen that the PSO variant PSO-LDIW and PSO-Chatterjee gives the best MSE value with convergence in 80th iteration. PSO variant PSO-LDIW is compared with proposed techniques in Table 12.1. Figure 12.6 depicts the convergence and MSE analysis of the proposed algorithms for LTE structure. The proposed PSO-TVW3-based algorithm converges fast with minimum MSE value.

Table 12.1 gives the comparison and effect of different channels using least mean square (LMS) algorithm, PSO-LDIW [9], PSO-TVW2, and PSO-TVW3 for LTE and DFE structures. From Table 12.1 and Figs. 12.5 to 12.6, it is shown that the proposed modifications outperform the other existing modifications based on

Table 12.1 Comparison of convergence rate and MSE with different channel models

Channel used	LMS		PSO-LDIW		PSO-TVW2		PSO-TVW3	
	MSE in dB	Convergence rate	MSE in dB	Convergence rate	MSE in dB	Convergence rate	MSE in dB	Convergence rate
Linear channel LTE	−53	200	−50	80	−58	50	−67	50
Linear channel DFE	−50	250	−48	90	−57	90	−67	100
Nonlinear channel 1-LTE	−15	200	−40	45	−37	50	−46	50
Nonlinear channel 1-DFE	−14	250	−40	60	−35	80	−46	80
Nonlinear channel 2-LTE	−40	200	−57	70	−55	50	−78	50
Nonlinear channel 2-DFE	−37	250	−55	80	−55	90	−62	100
Nonlinear channel 3-LTE	−15	600	−15	30	−18	50	−28	50
Nonlinear channel 3-DFE	−14	700	−15	50	−18	70	−23	50

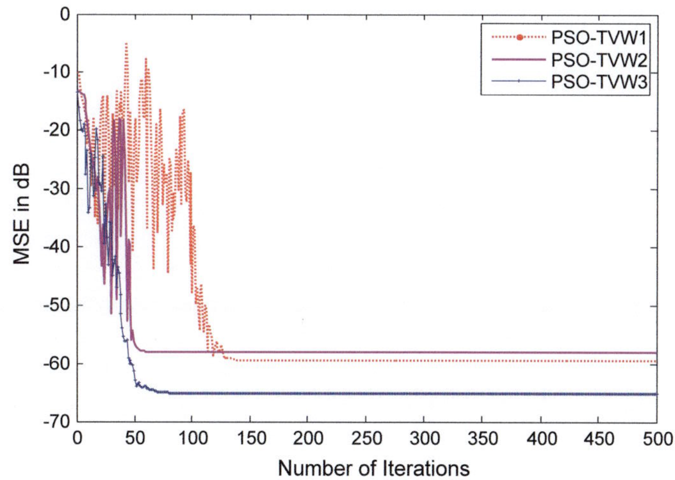

Fig. 12.6 Proposed PSO enhancements in linear channel for LTE

convergence and MSE. The PSO-TVW3 algorithm shows the best performance in all channel conditions. The LTE and DFE structures give approximately same MSE value, but differ in convergence rate which is shown in Table 12.1.

12.4.1 Sensitivity Analysis

The simulation-based sensitivity analysis is carried out to select optimum parameters in the algorithm. The parameter's values and choices have high impact on the efficiency of the method, and few others have less or no effect. The analysis is done with respect to six key parameters, namely the intermediate iteration value N, data window size ws, acceleration constants ac_1 and ac_2, the population size P, number of tap weights T, and distortion factor W. The effect of the basic PSO parameters swarm size or number of particles, window size, number of tap weights and acceleration coefficients are analyzed in [6]. The same is analyzed for PSO-TVW3 and is given in Table 12.2.

Table 12.2 Effect of PSO parameters on PSO-TVW3

Population P		Tap weights T		Window size WS		Acceleration coefficients c_1 and c_2		
P	MSE in dB	T	MSE in dB	WS	MSE in dB	c_1	c_2	MSE in dB
10	−52	5	−53	20	−65	1	1	−65
20	−64	7	−65	200	−68	1	2	−58
40	−70	9	−70	500	−69	2	1	−56
60	−72	13	−74	100	−70	2	2	−63

On average, an increase in the number of particles will always provide faster convergence. In contrast, the computational complexity can increase linearly with increase in population size. In Table 12.2, population size of 40 gives better convergence. So a problem-dependent minimum population size is good for better performance.

Setting the acceleration coefficients to a minimum value slows down the convergence speed. The local search and global search are best when the summation of acceleration coefficients becomes $ac_1 + ac_2 < 4$ in adaptive equalization. The acceleration coefficients greater than 1 also seem to give the best performance. For equal value of acceleration constants, the algorithm reaches its minimum in least number of iterations. The MSE calculated on each iteration is the average of the MSE over the window; a large window size increases the complexity per iteration and time consumption. In Table 12.2, window size does not make any greater changes in the MSE value. If the window size is small, the complexity can be reduced.

The tap weights are problem dependent. As given in Table 12.2, the increase in tap weights above a certain limit does not make much difference in MSE value, but it may increase the complexity. Figure 12.7 shows the analysis for different intermediate iteration N for PSO-TVW2. Table 12.3 compares the convergence rate and MSE for PSO-TVW2 and PSO-TVW3 with reference to N. An increase in the value of N increases the number of iterations required for convergence. Decreasing N value degrades the MSE performance. N value between 30 and 40 exhibits minimum MSE with faster convergence.

Table 12.4 explains the effect of amplitude distortion parameter W in linear channel. The MSE is computed with different amplitude distortion that leads to

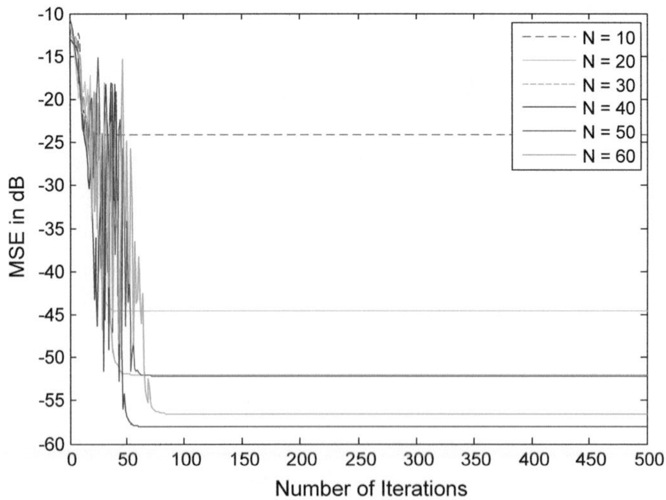

Fig. 12.7 Effect of different intermediate iteration value N on PSO-TVW2

Table 12.3 Comparison of convergence rate with different intermediate iteration value N

Intermediate iteration N	Convergence rate (iterations)	MSE in dB for PSO-TVW2	MSE in dB for PSO-TVW3
10	30	−24	−28
20	40	−45	−63
30	45	−52	−65
40	50	−58	−67
50	60	−52	−65
60	70	−56	−64

Table 12.4 Effect of amplitude distortion $\boldsymbol{W}$ on PSO-TVW2 for LTE and DFE

SNR in dB	LTE			DFE		
	For $W = 3.7$ MSE in dB	For $W = 3.1$ MSE in dB	For $W = 2.9$ MSE in dB	For $W = 3.7$ MSE in dB	For $W = 3.1$ MSE in dB	For $W = 2.9$ MSE in dB
5	−17	−18	−20	−17	−18	−19
10	−24	−26	−27	−29	−27	−25
15	−25	−27	−33	−32	−30	−32
20	−27	−35	−40	−35	−37	−37
25	−33	−44	−53	−38	−45	−53
30	−37	−61	−67	−46	−62	−67

different eigenvalue spread. Increase in amplitude distortion degrades the MSE performance. The performance degradation is not severe in proposed PSO-based algorithms compared to existing algorithms.

12.5 Conclusion

In this work, two new strategies are proposed to update in PSO algorithm for adaptive equalization. The results are discussed in linear and nonlinear channels for LTE and DFE structures. The proposed time-varying PSO algorithms, PSO-TVW2 and PSO-TVW3, show better performance than the existing algorithms in linear and nonlinear channels. Also it shows performance better for LTE and DFE structures. The PSO-TVW3 algorithm outperforms other modifications based on convergence and MSE. The sensitivity analysis is done to find the optimum parameter values. The performance is degraded for severe amplitude distortions in linear channel. The DFE structure delays the convergence but it performs well in severe amplitude distortion conditions.

References

1. Qureshi S (1985) Adaptive equalization. Proc IEEE 73:1349–1387
2. Haykin S (1996) Adaptive filter theory, 3rd edn. Prentice-Hall, Inc
3. Shin HC, Saved AH (2004) Mean-square performance of a family of affine projection algorithms. IEEE Trans Signal Process 52:90–101
4. Karaboga N, Cetinkaya B (2011) A novel and efficient algorithm for adaptive filtering: artificial bee colony algorithm. Turk J Electr Eng Comput Sci 19(1):175–190
5. Krusienski DJ (2004) Enhanced structured stochastic global optimization algorithms for IIR and nonlinear adaptive filtering. Ph.D. thesis, Department of Electrical Engineering, The Pennsylvania State University, University Park, PA
6. Al-Awami AT, Zerguine A, Cheded L, Zidouri A, Saif W (2011) A new modified particle swarm optimization algorithm for adaptive equalization. J Digital Signal Process 21(2): 195–207
7. Das G, Pattnaik PK, Padhy SK (2014) Artificial neural network trained by particle swarm optimization for non-linear channel equalization. J Exp Syst Appl 41(7):3491–3496
8. Kennedy J, Eberhart R (1995) Particle swarm optimization. In: Proceedings of the IEEE international conference on neural networks, vol 4, pp 1942–1948
9. Nickabadi A, Ebadzadeh MM, Safabakhsh R (2011) A novel particle swarm optimization algorithm with adaptive inertia weight. J Appl Soft Comput 11(4):3658–3670
10. Shi Y, Eberhart R (1998) A modified particle swarm optimizer. In: Proceedings of the international conference on evolutionary computation, IEEE world congress on computational intelligence, pp 69–73
11. Eberhart RC, Shi Y (2000) Comparing inertia weights and constriction factors in particle swarm optimization. In: Proceedings of the 2000 congress on evolutionary computation, vol 1, pp 84–88
12. Chatterjee A, Siarry P (2006) Nonlinear inertia weight variation for dynamic adaption in particle swarm optimization. Comput Oper Res 33(3):859–871
13. Feng Y, Teng G, Wang A, Yao YM (2007) Chaotic inertia weight in particle swarm optimization. In: Second international conference on innovative computing, information and control (ICICIC 07), pp 475–1475
14. Feng Y, Yao YM, Wang A (2007) Comparing with chaotic inertia weights in particle swarm optimization. In: International conference on machine learning and cybernetics, pp 329–333
15. Lei K, Qiu Y, He Y (2006) A new adaptive well-chosen inertia weight strategy to automatically harmonize global and local search ability in particle swarm optimization. In: ISSCAA
16. Fan S, Chiu Y (2007) A decreasing inertia weight particle swarm optimizer. Eng Optim 39 (2):203–228
17. Zheng Y, Ma L, Zhang L, Qian J (2003) Empirical study of particle swarm optimizer with an increasing inertia weight. In: IEEE congress on evolutionary computation
18. Jiao B, Lian Z, Gu X (2008) A dynamic inertia weight particle swarm optimization algorithm. Chaos Solitons Fractals 37(3):698–705
19. Patra JC, Pal RN, Baliarsingh R, Panda G (1999) Nonlinear channel equalization for QAM signal constellation using artificial neural networks. IEEE Trans Syst Man Cybern 29(2):262–271

Chapter 13
Investigation of Electrostatic Actuation Scheme for Low Voltage MEMS Switch

Sudhanshu Kumar, Neela Chattoraj, Manish K. Sinha and Niharika Danu

13.1 Introduction

Micro actuators are essential components in micro electro mechanical system (MEMS) for producing physical movement at the micro and nanometer level. The micro actuators are based on several actuation principles such as electrostatic actuation [1], electromagnetic actuation [2], Piezoelectric actuation [3] and thermal actuation [4]. At micro level, surface area to volume ratio increases. Hence, surface effect is dominating over volume effect; electrostatic force used for actuation depends on surface area; therefore, complete investigation should be carried out for explaining electrostatic actuation phenomenon at micro and nano level.

MEMS switch provided an alternative choice over semiconductor switch. Simultaneously, it is operated at high actuation voltage, which makes it incompatible for microelectronic circuits. Therefore, most important designing task for MEMS switch is to reduce actuation voltage (V_{dc}). The actuation voltage can be reduced by choice of appropriate design and other parameters used for switching. Their geometry as well as materials is used to determine the actuation voltage, stiffness and lifetime of switch. In some cases, porous beam are preferred to reduce stiffness of beam which finally results in high beam deflection and reduction of residual stress [5]. Including this porous beam also help in removal of sacrificial layer easily. Before actual fabrication, devices are simulated, modeled and analyzed using simulation software to give a complete flow process of the design. For MEMS devices, there are several FEA (Finite Elemental Analysis)-based software such as

S. Kumar · M.K. Sinha
Department of Physics, Birla Institute of Technology, Mesra, Ranchi 835215, India
e-mail: sudsuman2013@gmail.com

N. Chattoraj (✉) · N. Danu
Department of Electronics and Communication Engineering, Birla Institute of Technology, Mesra, Ranchi 835215, India
e-mail: nchattoraj@bitmesra.ac.in

V. Nath (ed.), *Proceedings of the International Conference on Nano-electronics, Circuits & Communication Systems*, Lecture Notes in Electrical Engineering 403,
DOI 10.1007/978-981-10-2999-8_13

COMSOL multiphysics, CoventorWare and ANSYS. In this paper, a low actuation voltage MEMS switch is proposed and simulated. The overall switch structure is designed for microelectronics circuits.

13.2 MEMS Switch Design Consideration

The layout of proposed design of electrostatic-based MEMS switch is shown in Fig. 13.1. The upper plate of a cantilever beam having length 220 μm and width of 100 μm suspended over the anchor, lower plate consist of pull down electrode which behaves as parallel plate capacitor. The tip of cantilever beam is designed in such a way that with application of actuation voltage it short-circuits ‘signal in’ and ‘signal out’ ports; hence provide normal switching action in electronic circuits (Table 13.1).

The MEMS switch has been designed and investigated with the help few governing equations [6, 7]. Capacitance between upper plate and lower plate is given by (13.1),

$$C = \frac{\varepsilon A}{d} \tag{13.1}$$

where, ε denotes the dielectric permittivity of the medium between upper and lower plates, d represents the separation between plates, and A represents the area of individual plate.

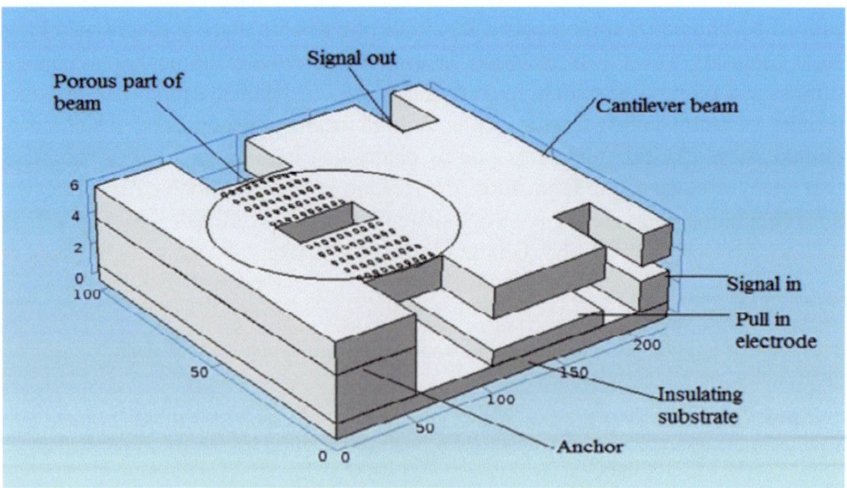

Fig. 13.1 Schematic diagram of proposed MEMS switch

Table 13.1 Dimensions of different parts of MEMS switch

Sl. No.	Parts of MEMS switch	Length (μm)	Width (μm)	Height (μm)
1.	Insulating substrate (lower plate)	220	100	1
2.	Anchor	50	100	3
3.	Pull in electrode (actuation pad)	75	100	1
4.	Cantilever beam	220	100	1
5.	Hole	Diameter = 2 μm		2

Electrostatic potential energy developed with the application of applied voltage (V) is given by (13.2),

Electrostatic potential energy,

$$U = \frac{CV^2}{2} \tag{13.2}$$

Electrostatic force experience by cantilever beam is the negative gradient of potential energy ($F = -\nabla U$), and it was found to be

$$F = \frac{\varepsilon A V^2}{2d^2} \tag{13.3}$$

But cantilever beam remains in equilibrium state till spring force ($F_d = k\Delta d$) is equal to the electrostatic force experience by the cantilever beam. Mathematically, it is represented as

$$|F| = |F_d| \tag{13.4}$$

Solving, Eq. (13.4) followed by differentiating obtained potential difference with cantilever tip deflection, pull in voltage is obtained to be,

$$V_{PI} = \sqrt{\frac{8kd^3}{27\varepsilon_0 A}} \tag{13.5}$$

13.3 Parametric Analysis

The performance variation of MEMS switch has been studied on the basis of electrical parameter as well as mechanical parameter. A very basic design of polysilicon-based MEMS switch, having length 220 μm and width 100 μm as shown in Fig. 13.2, is simulated and compared for different electromechanical parameters.

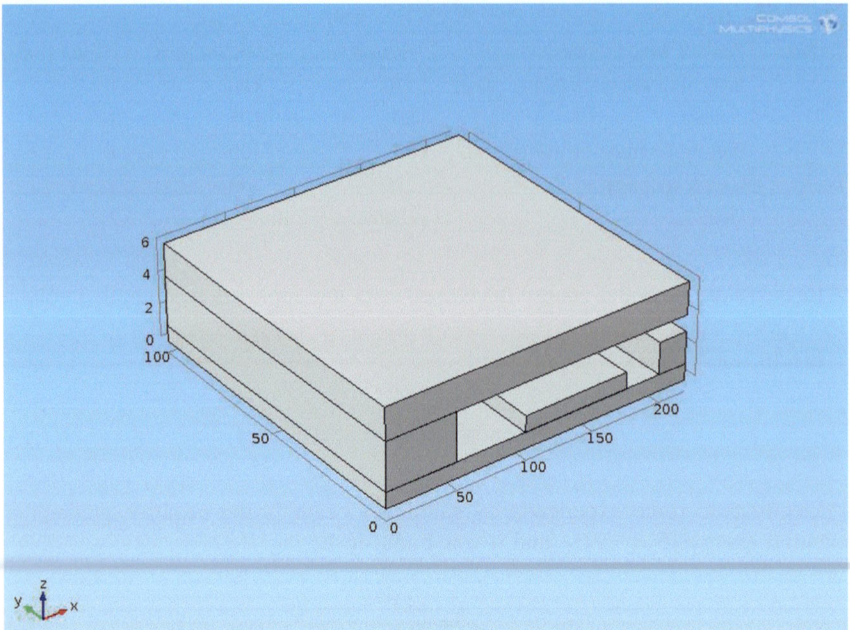

Fig. 13.2 A very basic design of cantilever-based MEMS switch

13.3.1 Effect of Position of Pull in Electrode

The variation in tip deflection of cantilever switch has been studied with change is position of pull in electrode as shown is Fig. 13.3. It was found that with the increase in separation of pull in electrode from anchor, tip deflection increases, and it is graphically represented by Fig. 13.4.

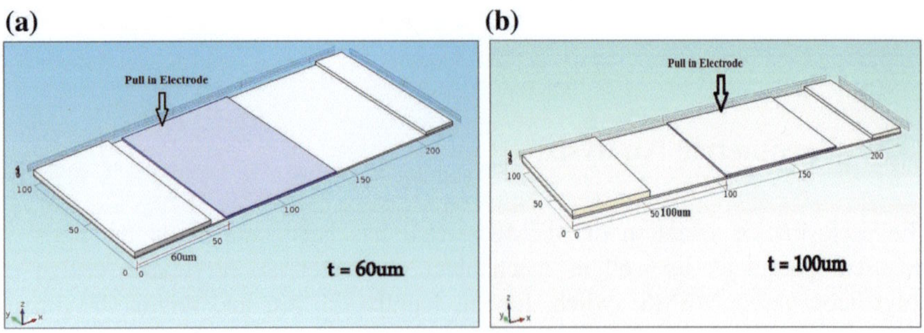

Fig. 13.3 **a**, **b** Different positions (*t*) of pull in electrode

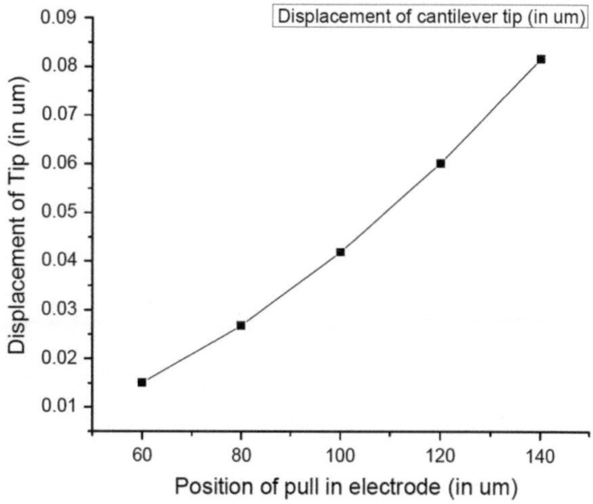

Fig. 13.4 Graph representing variation in tip displacement with pull in electrode's position

13.3.2 *Effect of Holes*

In MEMS switch as shown in Fig. 13.1, small holes are defined in suspended cantilever beam to reduce spring constant (stiffness/k) in order to improve beam deflection. Holes also reduce the residual stress and spread it over a wide area (Fig. 13.5b) rather than concentrated over a line (Fig. 13.5a).

Stress spread over large area improves lifetime of MEMS switch. Porous cantilever beam also helps in removing sacrificial layer by providing more area of contact between etchant and sacrificial layer. Because of the fringing effect [5], the capacitance between two plates of switch decreases and hence increases pull in voltage, which is not desirable; to avoid it, holes are defined only a particular portion of upper plate as shown in Fig. 13.1.

13.3.3 *Effect of Actuation Voltage*

13.3.3.1 Tip Deflection of Cantilever Beam

Electrostatic and mechanical forces are responsible for deflection of cantilever tip; whenever electrostatic force (F) exceeds spring force ($F_{\Delta d}$), there is a net deflection of cantilever tip from its equilibrium position and for a particular value of applied DC voltage (normally two-third of separation between upper and lower plate) MEMS switch turns ON. This particular voltage is known as pull in voltage as given by Eq. (13.5). Figure (13.6) shows very clear graphical study of applied DC voltage from 0 to 8 V. By applying DC voltage at different position through out the

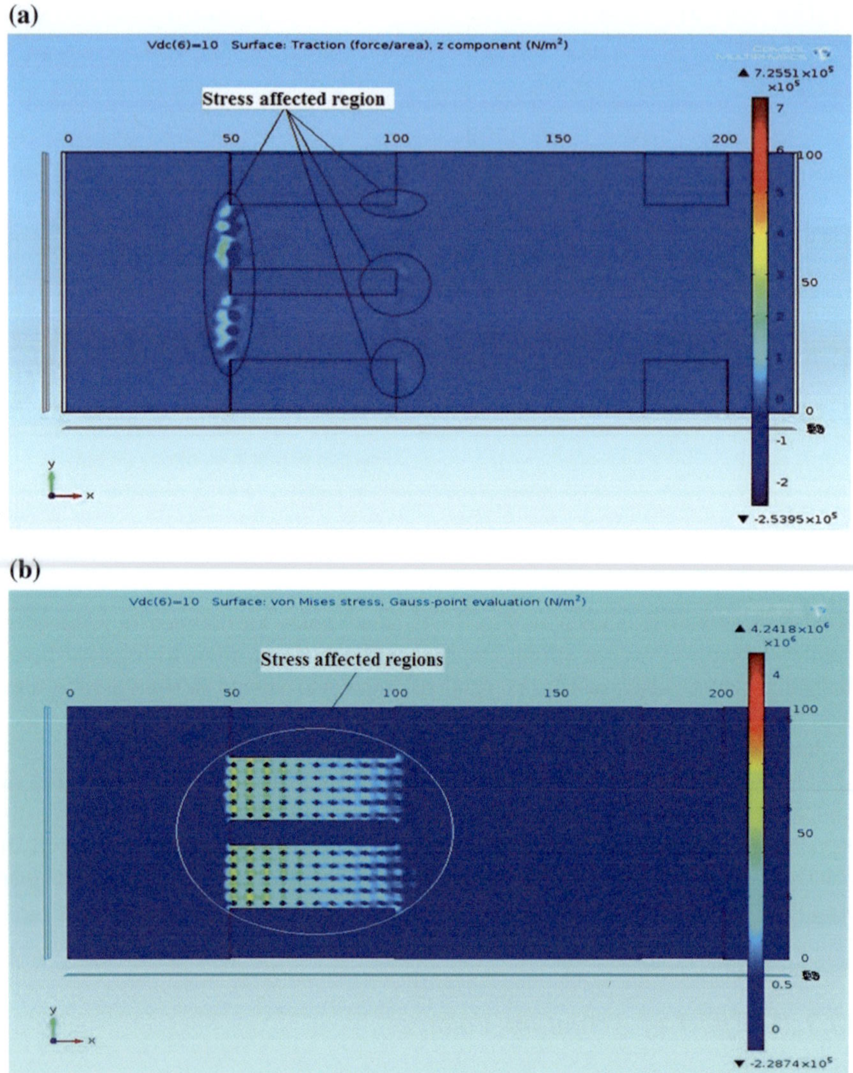

Fig. 13.5 **a** *Top* view of cantilever switch showing stress, **b** *top* view of a porous cantilever switch showing stress

length of the beam, deflection varies significantly, and it was maximum at tip and near anchor which is shown in Fig. 13.6.

Figure 13.7 clearly represents short circuiting (ON state) of switch when actuation voltage of 8 V is applied, it remains in ON state until actuation voltage is not removed. With the removal of actuation voltage, cantilever restores its equilibrium position (OFF state), this is because of mechanical restoring force (spring force/$F_{\Delta d}$) acted on it.

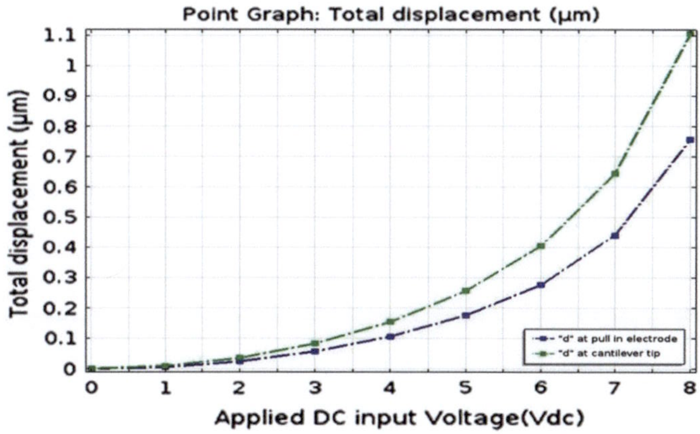

Fig. 13.6 A graph representing variation in displacement with applied potential

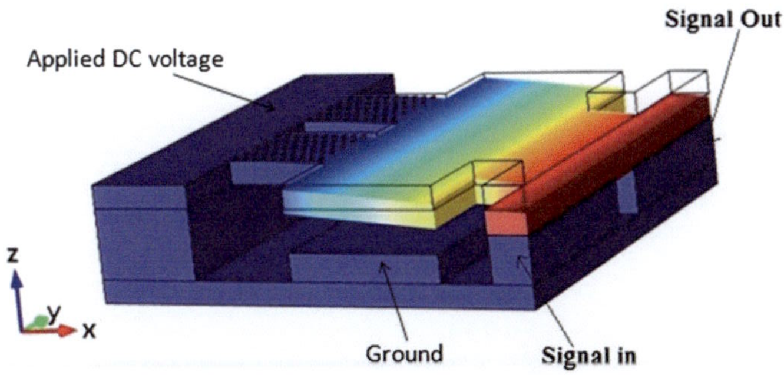

Fig. 13.7 Schematic diagram of MEMS switch in ON state when actuation voltage 8 V is applied

13.3.3.2 Capacitance Between Two Plates

The capacitance value of electrostatic micro actuator is correlated with actuation voltage by relation given by ($V \propto C$). This relationship between applied voltage and capacitance is derived from Eqs. (13.1–13.3). A graph obtained in Fig. 13.8 from simulation result satisfied the capacitance and actuation voltage proportionality relationship. Maximum value of capacitance (0.046 pF) is obtained for applied DC voltage of 7 V.

With the increase in applied actuation voltage, the deflection of cantilever increases which lead to decrease in value d (separation between upper and lower plate of micro actuator). Clearly, from Eq. (13.1),

$$C \propto \frac{1}{d} \tag{13.6}$$

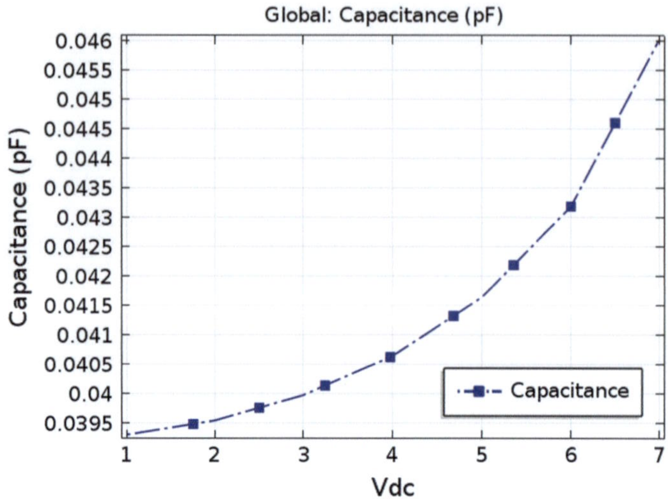

Fig. 13.8 A graphical representation of variation of capacitance with applied voltage

Fig. 13.9 A graphical representation of variation of cantilever tip deflection against capacitance value

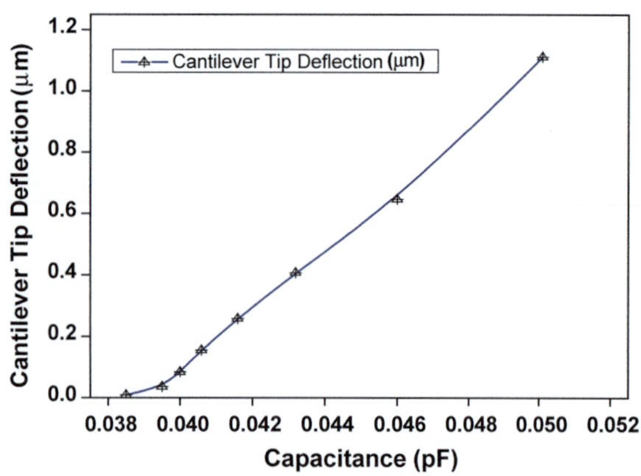

i.e., enhanced value of cantilever tip deflection is due to increase in capacitance value and vice versa, which is obtained by Fig. 13.9 as shown.

So, it was concluded from obtained result that with increase in actuation voltage, the separation between two plates of capacitor decreases [8], i.e., cantilever tip deflection increases which ultimately results in increase in capacitance value. The simulation results obtained in Figs. 13.8 and 13.9 satisfied the same pattern.

Table 13.2 Comparative analysis of the proposed switch

Sl. No.	Literature reviewed	Actuation voltage (V)	Capacitance	Material used	Dimension (μm)
1.	Spasos et. al [9]	54.4	–	–	–
2.	Vaibhav et al. [10]	14.3	1.8 pF	Niaro (Au–Ni)	325 × 180 × 20
3.	Bansal et al. [11]	20	–	Metal (not specified)	300 × 150 × 2.5
4.	Hu et al. [12]	13	–	P-silicon	300 × 300 × 1.5
5.	This work	8	0.046 pF	P-silicon	220 × 100 × 6

13.4 Comparative Analysis

The designed switch is capable of operating at an actuation voltage of 8 V with a dimension of (220 × 100 × 6) μm. RF-MEMS switch by Spasos et al. [9] describes RF-MEMS switch operated at 54.4 V. MEMS switch by Vaibhav et al. [10] designed using Niaro (Au–Ni) was capable of operating at actuation voltage of 14.3 V, which is comparatively low but not suitable for microelectronic circuits. Other than these, few other designs suggest actuation voltages of 20 V and 13 V for their operation. The comparison of the designed switch with the previous work has been shown below in Table 13.2.

Although this design does not operate in ultralow voltage range still with less design complexity, actuation voltage of 8 V is low enough to be integrated in microelectronic circuits.

13.5 Conclusion

The MEMS switch is designed and simulated for low voltage applications. Ideas used for reduction of actuation voltage are decrease in separation between upper plate and lower plate of MEMS switch, well-defined porous cantilever beam, proper position of pull in electrode, and major consideration of industrially available material (polysilicon) for low cost fabrication. The designed and simulated MEMS switch with dimensions (220 μm × 100 μm × 6 μm) can be fabricated using micromachining technique for industrial applications.

Acknowledgments The authors would like to thank Hemant Kumar and COMSOL India for their software support, B. Pathak, Aks, Esha, and N. Poddar for discussions and BIT Mesra for resources.

References

1. Lahiri SK, Saha H, Kundu A (2009) RF MEMS SWITCH: an overview at a glance. In: International conference on computer and devices for communication
2. Yan D, Khajepour A, Mansour R (2003) Modeling of two-hot-arm horizontal thermal actuator. J Micromech Microeng 13:312–322
3. Seki K, Ruderman M, Iwasaki M (2014) Modeling and Compensation for hysteresis properties in piezoelectric actuators. In: AMC2014-Yokohama 14–16 Mar 2014, Yokohama, Japan
4. Behlert R, Kunzig T, Schrag G, Wachutka G (2014) Theoretical and experimental investigations on failure mechanisms occuring during long-term cycling of electrostatic actuators. In: IEEE, 15th International conference
5. Fang DM, Li XH, Yuan Q, Zhang HX (2010) Effect of etch holes on the capacitance and pull-in voltage in MEMS tunable capacitors. Int J Electron 97(12):1439–1448
6. Srikar VT, Spearing SM (2003) Material selection for microfabricationed electrostatic actuators. Sens Actuators A 102(2003):279–285
7. Son M, Yin J, Xunjun H, Yue W (2008) Design and analysis of a novel low actuation voltage capacitive. RF MEMS switches. IEEE, Jan 6–9 2008, Sanya, China
8. Sem Jebar Kumar J, Tetteh EA, Paul Braineard E (2014) A study of why electrostatic actuation is prefered and a simulation of an electrostatically actuated cantilever beam. Int J Eng Sci Emerg Technol 6(5):441–446
9. Spasos M, Tsiakmakis K, Charalampidis N, Nilavalan R (2011) RF-MEMS switch actuation pulse optimization using Taguchi's method. Microsyst Technol 17(10):1351–1359
10. Kolte V, Mahajan PM (2014) Design and analysis of non-uniform shaped RF MEMS switch. J Comput Electron 13(14):547–554
11. Bansal D, Kumar A, Sharma A, Kumar P, Rangra KJ (2013) Design of novel compact anti-stiction and low insertion loss RF MEMS switch, vol 1(20). Springer, Berlin, pp 337–340
12. Hu F, Li Z, Qian Y, Yao J, Xiong X, Niu J, Peng Z (2012) A multi-electrode and pre-deformed bilayer spring structure electrostatic attractive MEMS actuator with large stroke at low actuation voltage. J Micromech Microeng 22(9):960–1317

Chapter 14
On Online Counting of Cigarette in Packets—An Image Processing Approach

Abahan Sarkar, Sourav Chakrabarty and B.K. Roy

14.1 Introduction

The fast-moving consumer goods (FMCG) industry is highly competitive. To increase the growth, dynamic business tactics with flexible strategies are to be incorporated which can be changed rapidly when new challenges are threatened. Packaging and brand value play a key role in marketing to determine the future of a product. It is the brand which carries faith and satisfaction among the consumers. To live up to the expectations of their consumers, companies need to put their best efforts to keep everything in place. They use automated machinery in every step of production leading to higher productivity and uniformity of products. However, in spite of careful production, some defects still creep in which may be in the form of incorrect number of cigarettes per packet, incorrect barcode being printed on packet, illegible or no labeling in the packets, etc.

Any of the aforesaid faults encountered after the product is delivered from the production floor can put company's reputation at stake. In the digital era of fast computing, we can always come up with a solution provided the algorithm is best chosen. Image processing on LabVIEW platform is such a powerful and reliable tool which has the capability to accurately perform inspection of products moving at very high speed and generating control decisions quickly cratering to industrial requirements. It is very important for the producing enterprise to guarantee that packaging bears the proper features since it plays a vital role for any product.

Counting objects is a simple task for small set of data. It becomes a challenging task for a human and can lead to human errors when the number increases. Thus to improve the precision of quantitative measurements, automating the process is very much necessary. Contact method which may disturb and delay the process is used for counting in most of the applications. In this chapter, an automated and

Abahan Sarkar (✉) · S. Chakrabarty · B.K. Roy
National Institute of Technology Silchar, Silchar 788010, Assam, India
e-mail: abahan55@gmail.com

V. Nath (ed.), *Proceedings of the International Conference on Nano-electronics, Circuits & Communication Systems*, Lecture Notes in Electrical Engineering 403,
DOI 10.1007/978-981-10-2999-8_14

non-contact method for counting the number of cigarettes in a cigarette box moving on a conveyor is proposed. In many industries, the count is checked in bulk by weighing the batch instead of checking each box. This may lead to decrease in accuracy of count in individual boxes. In this regard, the proposed work helps the industry by designing an automated counting system to check the count of the product.

The proposed technique uses some high-resolution cameras, which acquire pictures of cigarette packets at proper instants and process the images using appropriate algorithms for defects if any. If a defect is encountered at any stage, that particular packet will be dropped from the production line. Thus, product quality control is fulfilled.

Various techniques related to automated product inspection are proposed by many researchers in recent past. The domain of automated inspection for industrial products is found to be populated mostly by vision-based applications. Speaking of cigarette packet inspection in particular, Fang-xia et al. [1] proposed vision-based inspection system for flaw detection in external packaging of cigarette packets. Cui et al. [2] proposed a vision-based inspection system for cigarettes lying in open tin containers which can detect incorrect number of cigarettes, improper placing, and paper handle defects. Ping et al. [3] designed a capacitance-based technique for detection of improper number of cigarette packets in containers utilizing back-propagation neural networks. Park et al. [4, 5] used morphological image processing operation for the identification of cigarette and paper spoon in tin packages. Wenping et al. [6] proposed a method for fault detection utilizing vision-based microcontroller. Sorting balance problem is introduced by Sun et al. [7]. Xuguang et al. [8] proposed a self-owned intelligent sort instrument for cigarette draw resistance. Use of capacitance-based inspection techniques can be found in the works of Soleimani et al. [9] where they have used the concept of electrical capacitance tomography (ECT) for crack detection in common dielectric materials and White [10] for monitoring gas cavities in a packed cradle of solids. Zhang et al. [11] proposed a corn moisture detection technique based on the relationship between the capacitance and the dielectric constant of corn. Xia et al. [12] proposed capacitance-based pulverized coal concentration detection based on the concept of dielectric variation. Malamus et al. [13] surveyed the applications and state-of-the-art tools and techniques related to vision-based industrial inspection systems. Kumar in his survey paper presented the inspection techniques for fabric defect detection [14]. Applications related to quality inspection of thermal fuses [15], pharmaceutical tablets [16], bearing seals [17], bulb filament [18], PCB [19], bottle-printed labels [20], food products [21], and liquid level in bottles [22] are also reported. Antonini et al. [23] discussed some of the main tools and techniques employed in the area of automatic counting of objects in digital images. Semertzidis et al. [24] proposed an automated method of counting feeder fish by processing the video frames individually and independently utilizing image processing techniques. Abahan et al. [25] present defect identification in labels on cigarette packets using LabVIEW. Joze et al. [26] designed a color- and size-based automated visual inspection system for missing or broken tablet detection. A machine vision apple sorting method is reported [27]. Automatic counting of cigarettes in cigarette packets using LabVIEW is discussed [28]. Barcode detection and its implementation on chip are reported [29–33].

It is clear from the survey of reported work that although automated inspection is in research trend these days, relatively less work has been done in the area of automated product inspection for FMCG cigarette industry. In this chapter, an image processing approach-based techniques on LabVIEW platform is designed for the online counting of cigarettes present in a packet.

The chapter is organized as follows: Introduction to the research domain is presented in Sect. 14.1; Sect. 14.3 discusses the associated problem. Section 14.4 deals with proposed solution. Results and Discussions are given in Sect. 14.5. Finally, Sect. 14.6 concludes the chapter.

14.2 Experimental Setup

The setup process consists of a conveyor belt with Smart Camera NI-1744 [34] and proximity sensor on which the cigarette packets are moving with a constant speed as shown in Fig. 14.1.

Fig. 14.1 Experimental setup of the proposed technique

The conveyor belt is operated at a speed of 0.6 m/s. The speed of the motor is controlled by the user from the front panel window of the proposed LabVIEW [35–39] program.

Automation in the production line is a very essential activity of a process industry. The process of automation is also used in quality check of the produced product, but quantitative measurement is also an essential and important activity.

14.3 Problem Description

Automation in production line is a very essential feature of the FMCG industry. Automated techniques are extensively applied to check the quality of the finished product. In this chapter, packets containing cigarettes moving through conveyor belt need to be inspected for correct number of cigarettes. If a packet is found to contain less number of cigarettes, an appropriate control decisions leading to fault alarm are to be generated, so that the faulty packet can be removed from the production line.

14.4 Solution Methodology

The proposed solution addresses the problem discussed in the previous section by employing image processing techniques on LabVIEW platform. Vision Builder for Automated Inspection software is configured for counting purpose. More than 200 real time pictures of different cigarette packets are considered to validate the proposed scheme. The general method for object counting follows the following framework (Fig. 14.2).

14.4.1 Image Acquisition

This step intends to capture images through the camera. The quality of image depends on camera parameters, lighting conditions, size of objects, and distance from which image is taken. In this work, real-world images of cigarette packets are acquired with the help of a smart camera NI-1744 with exposure time 1.29 ms and gain 144. In online testing, an image is acquired and simultaneously processes in the smart camera itself before acquiring the next image. In this chapter, the camera is triggered by a laser beam switch whenever a cigarette packet comes directly under the field of view of the camera. For acquisition, **NI Vision Acquisition Express VI** is used.

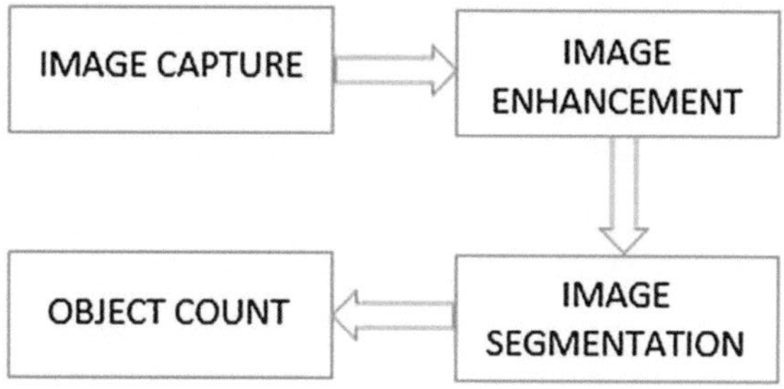

Fig. 14.2 Generalized framework for object counting using image processing

14.4.2 *Region of Interest (ROI)*

ROI is that portion of an image which contains the desired information and the rest portion of image is discarded to reduce the processing burden. Hence, ROI must contain the object of interest in the image. Here, we use rectangular-type ROI. For extracting ROI from the image, **ROI Descriptor**, **IMAQ Select Rectangle,** and **IMAQ Overlay rectangle** are used.

14.4.3 *Image Enhancement*

Resultant image after image enhancement is more appropriate compared to original for certain application. One or more features during this process are modified depending on the application. These include basic gray-level transformations, histogram modification, average, and median filtering. In this chapter, the following steps are performed for image enhancement.

14.4.3.1 Filtering

Filtering is done with a suitable mask to make the desired features more prominent. Smoothing, sharpening, and edge enhancement are the result of filtering. In this work, we prefer canny edge detection for its low error rate, good localization, and minimal response. Gaussian filter is used to filter out the available noise with kernel size 5. It is given below:

$$k = \frac{1}{100}\begin{bmatrix} 1 & 2 & 4 & 2 & 1 \\ 2 & 4 & 8 & 4 & 2 \\ 4 & 8 & 16 & 8 & 4 \\ 2 & 4 & 8 & 4 & 2 \\ 1 & 2 & 4 & 2 & 1 \end{bmatrix}$$

Then we find the intensity gradient of the noise with a pair of convolution masks x-gradient (F_X) and y-gradient (F_Y).

$$\Delta F = |(Z_3 + 2Z_6 + Z_9) - (Z_1 + 2Z_4 + Z_7)| + |-(Z_1 + 2Z_2 + Z_3) - (Z_7 + 2Z_8 + Z_9)|$$

and direction,

$$\theta = \arctan\left(\frac{F_X}{F_Y}\right)$$

For filtering, **IMAQ Canny Edge Detection VI** is used in LabVIEW.

14.4.3.2 Thresholding

The threshold level is selected accordingly which gives the best detection under the available lighting and camera aperture. We are looking for bright object with lower threshold value 241. The threshold equation is given below:

$$\begin{aligned} g_{\text{result}}(x, y) &= 1 \text{ if } f_{\text{original}}(x, y) \geq 241 \\ &= 0 \text{ if } f_{\text{original}}(x, y) < 241 \end{aligned}$$

For thresholding, **IMAQ Threshold VI** is used in LabVIEW.

14.4.4 Morphological Operation

The morphological operations such as erosion and dilation are employed to ensure no adjacent cigarette get counted as a single object during object counting. In this work, we use erosion with structuring element 7×7.

$$\text{Element} = \begin{bmatrix} 1 & 1 & 1 & 1 & 1 & 1 & 1 \\ 1 & 1 & 1 & 0 & 1 & 1 & 1 \\ 1 & 1 & 0 & 0 & 0 & 1 & 1 \\ 1 & 0 & 0 & 0 & 0 & 0 & 1 \\ 1 & 1 & 0 & 0 & 0 & 1 & 1 \\ 1 & 1 & 1 & 0 & 1 & 1 & 1 \\ 1 & 1 & 1 & 1 & 1 & 1 & 1 \end{bmatrix}$$

For morphological operations, **IMAQ Morphology VI** is used in LabVIEW.

14.4.5 Image Segmentation

Segmentation is used to divide an image into different regions. Each region has same type of pixels. Union of all regions covers the entire image.

14.4.6 Object Counting

The images after passing through the above steps are ready for counting. In this step, **IMAQ Count Object VI** is used on the corresponding images to count the numbers of cigarettes using LabVIEW and **NI Vision Assistant express VI** and **Detect objects are** used to detect the same using Vision Builder.

The block diagram of the proposed technique for counting the number of cigarettes in a packet is shown in Figs. 14.3 and 14.4.

On completion of counting, the **IMAQ Count Object VI and set inspection Status** matches the count with a predefined value which represents the ideal number of cigarettes of a packet. A fault alarm is generated if the estimated number is not matched with the predefined number. The flowchart of the proposed system is shown in Fig. 14.5.

14.5 Results and Discussion

Online testing is done to count the number of cigarette using more than 200 online images of the production line. Vision Builder consists of one custom overlay indicator which displays the inspection status with number of cigarettes. One numerical indicator is also used to display the number of cigarettes [34–39]. The results of online testing are shown in Figs. 14.6, 14.7, 14.8, 14.9, and 14.10.

The testing data set of 200 images consist of 160 good images and 40 defective images. The success of our algorithm can also be realized [2, 4] as follows in Table 14.1.

The performance study of the proposed algorithm compared with existing techniques is shown in Fig. 14.11, and it shows that our proposed algorithm provides better result with compare to the existing methods. High value of precision, sensitivity, specificity, accuracy, and F1 measure designates that the proposed method is proficient of detecting most of the defective cases as defective and most of the good cases as good.

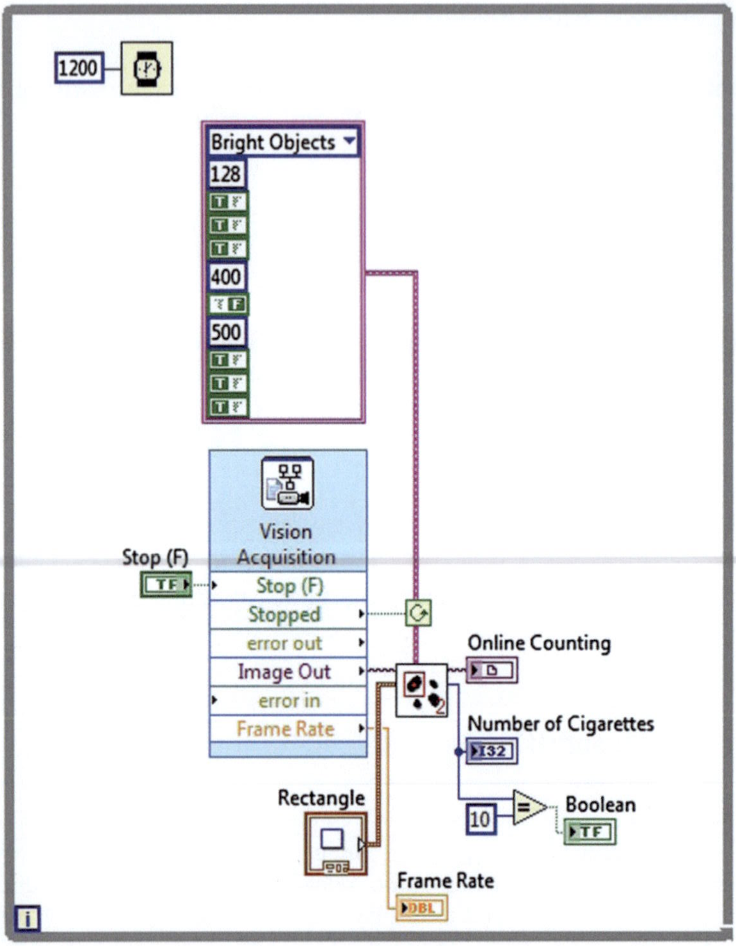

Fig. 14.3 Block diagram of the proposed scheme using LabVIEW

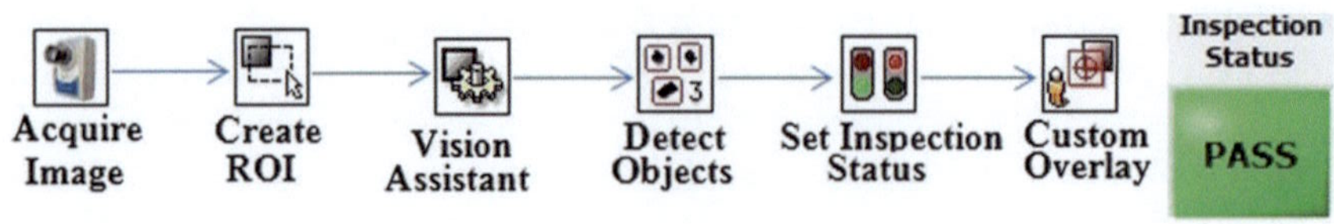

Fig. 14.4 Block diagram of the proposed scheme using Vision Builder

Fig. 14.5 The flowchart of the proposed scheme

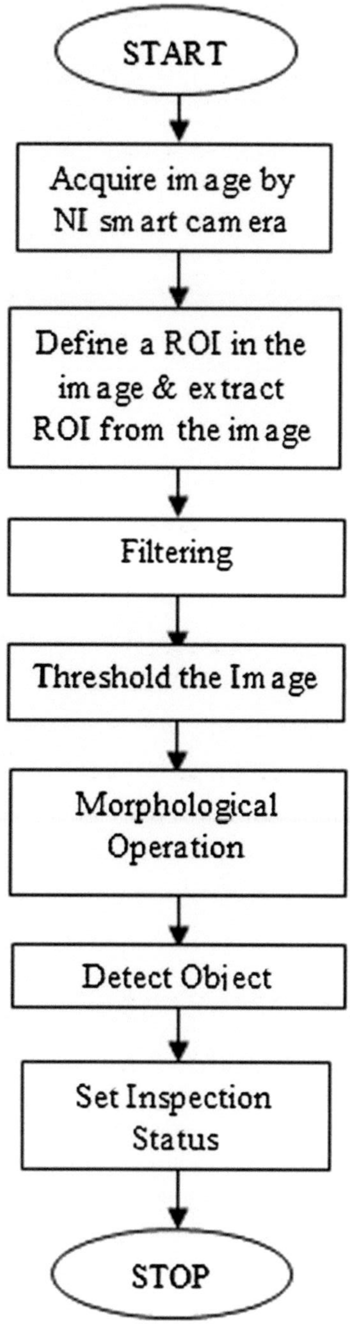

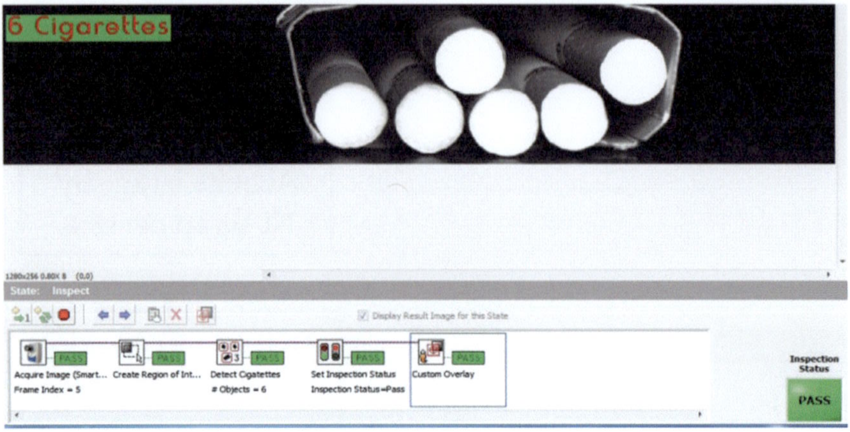

Fig. 14.6 Result using Vision Builder for 6 cigarettes

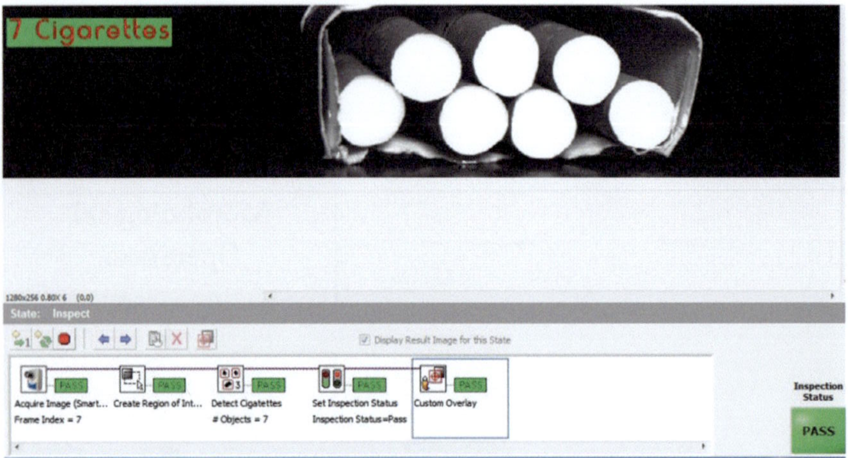

Fig. 14.7 Result using Vision Builder for 7 cigarettes

Results show that the proposed method is proficient of detecting most of the defects in cigarette packets. The inspection system provides high sensitivity, and specificity is shown in Table 14.1 and hence can be employed in cigarette-packaging industry efficaciously.

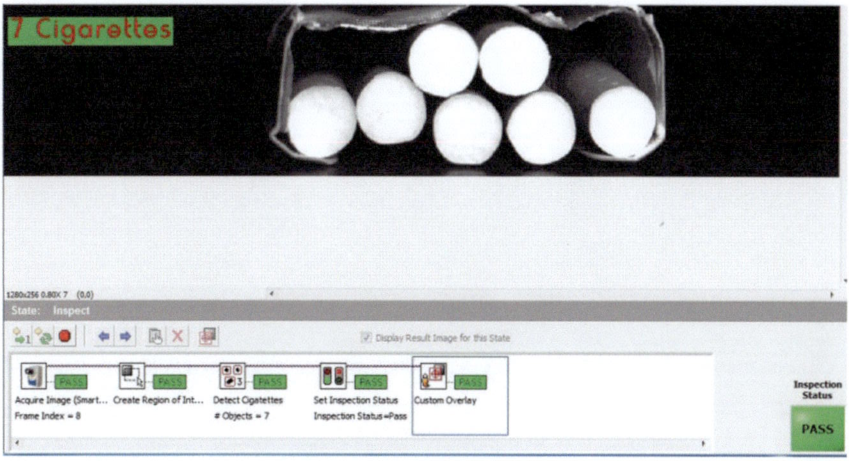

Fig. 14.8 Result with 7 cigarettes using Vision Builder

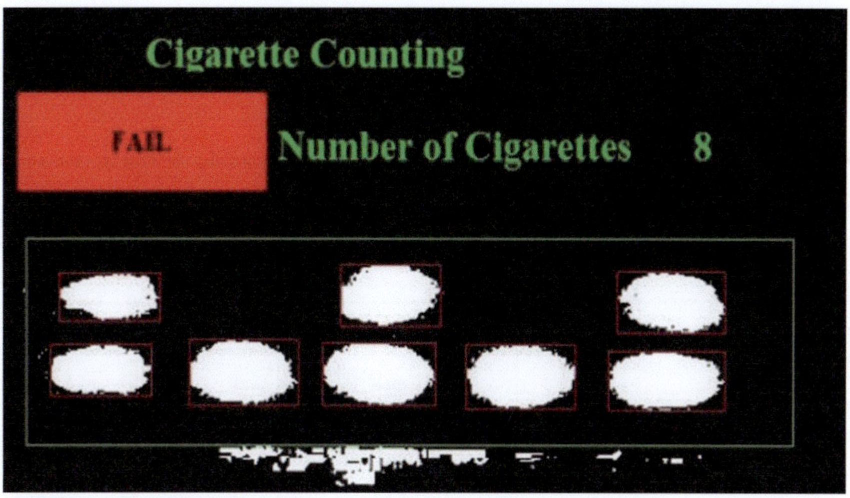

Fig. 14.9 Result using LabVIEW (8 cigarettes)

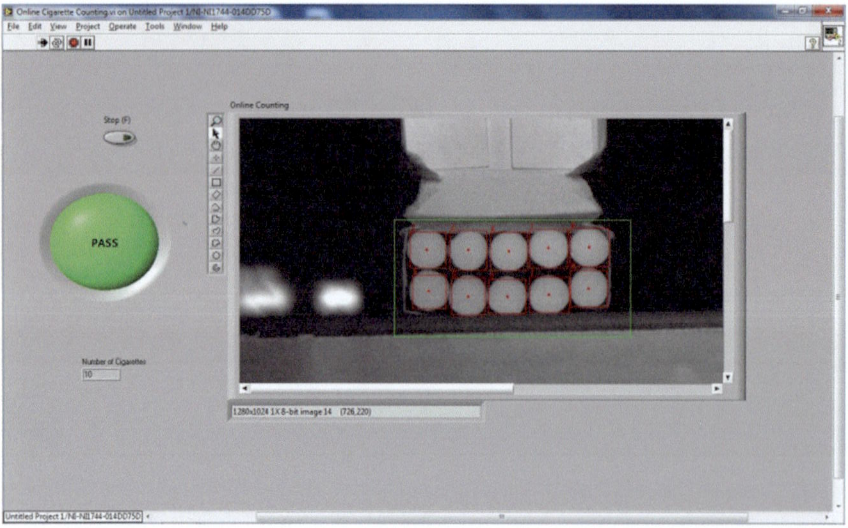

Fig. 14.10 Result using LabVIEW (10 cigarettes)

Table 14.1 Results of the proposed algorithm

	Detected as defective cases	Detected as good cases	
Defective cases (40)	40 (true positive)	0 (false negative)	Sensitivity = (40)/(40 + 0) = 1
Good cases (160)	4 (false positive)	156 (true negative)	Specificity = (156)/(4 + 156) = 0.975

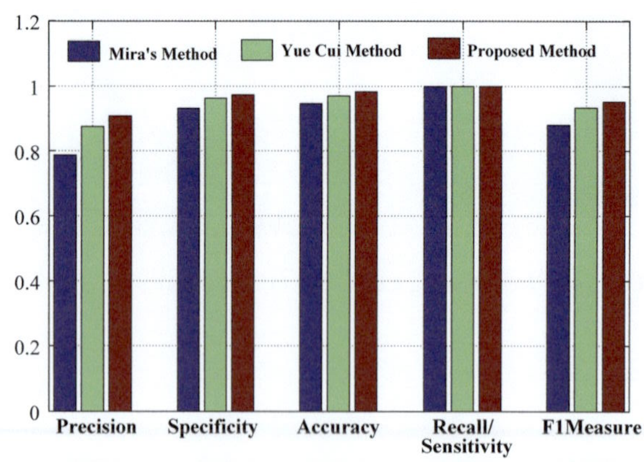

Fig. 14.11 Performance analysis of the proposed algorithm

14.6 Conclusion

In this chapter, an image processing-based online cigarette-counting technique is designed and implemented using LabVIEW and Vision Builder. The system competently counts the number of cigarettes. It is clear from the experimental results that the proposed technique has achieved the set objective satisfactorily and motivates for real-world implementation. Modification and extension of the proposed techniques can be easily implemented on similar FMCG products since the complete processing is carried out on software. Also it motivates us to work with real-time applications.

References

1. Fang-Xia H et al (2007) Study on the key image processing technology in the inspection of packing quality for small-pack cigarettes. In: Second workshop on digital media and its application in museum & heritages, Chongqing, pp. 67–71
2. Cui Y et al (2009) Automated pattern recognition and defect inspection system. In: Fifth international conference on image and graphics, ICIG'09, pp. 768–773)
3. Ping Z et al (2010) The design of a capacitance sensor for the detection of cigarette packets lack. In: 2nd international conference on signal processing systems, Dalian, pp. V2-707–V2-710
4. Park M et al (2000) Pattern recognition from segmented images in automated inspection systems. In: IEEE International symposium on ubiquitous multimedia computing, pp. 87–92
5. Park M et al (2009) Automated defect inspection systems by pattern recognition. Int J Signal Process Pattern Recogn 2:001
6. Wenping H et al (2008) Design of a video system to detect and sort the faults of cigarette package. In: IEEE international symposium on it in medicine and education, pp. 994–997
7. Sun CW et al (2011) The optimization configuration of cigarettes sorting replenishment workers based on genetic algorithm. In: International conference on information management, innovation management and industrial engineering (ICIII), Vol. 2, pp. 59–62
8. Xuguang W et al (2011) Study on sort instrument of cigarette draw resistance based on zero-drive chain. Third Int Conf Measuring Technol Mechatronics Autom (ICMTMA) 3:518–521
9. Soleimani M et al (2011) Crack detection in dielectric objects using electrical capacitance tomography imaging. Insight-Non-Destructive Test Condition Monit 53(1):21–24
10. White RB (2002) Using electrical capacitance tomography to monitor gas voids in a packed bed of solids. Meas Sci Technol 13(12):1842
11. Zhang H et al (2013) Corn moisture measurement using a capacitive sensor. J Comput 8 (6):1627–1631
12. Xia C et al (2009) A pulverized coal concentration measurement system based on capacitance sensor in international technology and innovation conference. Xian, China
13. Malamas EN et al (2003) A survey on industrial vision systems, applications and tools. Image Vis Comput 21(2):171–188
14. Kumar J (2008) Computer-Vision-Based Fabric Defect Detection: A Survey. IEEE Trans Industr Electron 55(1):348–363
15. Sun T-H et al (2014) Automated thermal fuse inspection using machine vision and artificial neural networks. J Intell Manuf. doi:10.1007/s10845-014-0902-y

16. Možina M et al (2013) Automated visual inspection of imprint quality of pharmaceutical tablets. Mach Vis Appl 24(1):63–73
17. Shen H et al (2012) Bearing defect inspection based on machine vision. Measurement 45 (4):719–733
18. Ng CK et al (2011) A real time quality monitoring system for the lighting industry: a practical and rapid approach using computer vision and image processing (CVIP) tools. Int J Eng Bus Manag. doi:10.5772/45670
19. Mar NSS et al (2011) Design and development of automatic visual inspection system for PCB manufacturing. Robot and Comput-Integr Manufact 27(5):949–962
20. Grosso E et al (2011) Automated quality control of printed flasks and bottles. Mach Vis Appl 22(2):269–281
21. Brosnan T, Sun D-W (2004) Improving quality inspection of food products by computer vision-a review. J Food Eng 61(61):3–16
22. Pithadiya KJ, Modi CK, Chauhan JD (2011) Selecting the most favourable edge detection technique for liquid level inspection in bottles. Int J Comput Info Syst Ind Manage Appl (IJCISIM) ISSN, 2150–7988
23. Antonini G, Thiran JP (2006) Counting pedestrians in video sequences using trajectory clustering. IEEE Transact Circuits Syst Video Technol 16(8):1008–1020
24. Semertzidis T et al (2010) Video sensor network for real-time traffic monitoring and surveillance. IET Intel Transport Syst 4(2):103–112
25. Sarkar A, Dutta T, Roy BK (2014) Counting of cigarettes in cigarette packets using LabVIEW. In International conference on communications and signal processing (ICCSP) pp. 1610–1614
26. Derganc J, Likar B, Tomaževič D, Pernuš F (2003) Real-time automated visual inspection of color tablets in pharmaceutical blisters. Real-Time Imaging 9(2):113–124
27. Bennedsen BS, Peterson DL (2005) Performance of a system for apple surface defect identification in near-infrared images. Biosyst Eng 90(4):419–431
28. Sarkar A, Dutta T, Roy BK (2014) Counting of cigarettes in cigarette packets using LabVIEW. In: International conference on communications and signal processing (ICCSP), pp. 1610–1614
29. Narayan GR, James MV (2012) Barcode recognition from video by combining image processing and xilinx. Procedia Eng 38:2140–2146
30. Houni K, Sawaya W, Delignon Y (2008) Spatial resolution of 1D image-based barcode reading: an information theoretical approach. In: 3rd international symposium on communications, control and signal processing, ISCCSP 2008, pp. 87–92
31. Yang H, Kot AC, Jiang X (2012) Binarization of low-quality barcode images captured by mobile phones using local window of adaptive location and size. IEEE Trans Image Process 21(1):418–425
32. Wakaumi H, Nagasawa C (2006) A ternary barcode detection system with a pattern-adaptable dual threshold. Sens Actuators A: Physical 130:176–183
33. Wakaumi H (2010) A high-density ternary barcode detection system with a fixed-period delay method. Eurosensor XXIV Conf Procedia Eng 5:252–255
34. Smart Camera for Embedded Machine Vision, National Instruments, 2009
35. James Kevin (2000) PC interfacing and data acquisition: techniques for measurement. Newnes Publishers Instrumentation and Control, Boston
36. Gupta S, John J (2005) Virtual Instrumentation using LabVIEW. Tata McGraw Hill Publishing Co Ltd, USA
37. LabVIEW (2009) User manual, National instruments
38. Jeroma J (2000) Virtual Instrumentation by LabVIEW
39. Wells LK, Travis J (1997) LabVIEW for everyone. Prentice Hall, New Jersey

Chapter 15
Fuzzy Min-Max Neural Network-Based Intrusion Detection System

Azad Chandrashekhar and Jha Vijay Kumar

15.1 Introduction

Development in the field of communication technology, computer system and the World Wide Web provides more and more benefits to the corporate and individual users but also the development is misused by the some ill-intended social animal as a medium for the crime, i.e., is the cybercrime. Today, cybercrime is the global problem for the www development community, and it is increased day by day. According to the security experts, the yearly losses due to the cybercrime is in billons, but this is estimated, not the actual amount, and actual amount may be more because nobody knows how many cybercrime cases are unreported [1]. The concept behind the popularity of the www is the easy accessibility and the low cost of thc information processing and the less time consuming in the fast moving era. Computer virus, malware, Trojan, unauthorized usage, theft of data and services, DoS attack, etc. are the contributors to the cybercrime; due to these cyberworld losses, the integrity, availability, and reliability of the data and services hosted on the web. Therefore, the important www security methods used in the today's scenarios are access control, firewall, IDS, antivirus, authentication applications, encryption mechanisms so that the cope of cybercrime are reduced [2] or to provide cent percent security but today none of the systems provide cent percent security. According to [3], 72% of the organizations use intrusion detection system as security tool, so that they may escape from the cope of the cybercrimes. The main

A. Chandrashekhar · J. Vijay Kumar (✉)
Department of Computer Science and Engineering,
Birla Institute of Technology Mesra, Ranchi 835215, India
e-mail: vkjha@bitmesra.ac.in

A. Chandrashekhar
e-mail: csazad@bitmesra.ac.in

V. Nath (ed.), *Proceedings of the International Conference on Nano-electronics, Circuits & Communication Systems*, Lecture Notes in Electrical Engineering 403,
DOI 10.1007/978-981-10-2999-8_15

reasons behind the popularity of the intrusion detection system (IDS) are as follows: (i) it may provide the security on host as well on the whole network; (ii) network monitoring; and (iii) maintaining the consistency, availability, reliability, and the integrity of the data and services hosted on the host or network. Intrusion detection systems are the hardware or the software or both that provide security to the data and the services on the host or on the network and also monitor the network according to the defined policies [4]. If any violation of the security policies is caught, then the system administrator is alerted. The unauthorized user can access the computer resources through the web or from the inside the organizations. They access the system by violating the established security rules. IDS are implemented on the two fundamental concepts: the detection of the anomalous behaviors against the known normal behavior which is called anomaly detection-based IDS, and misuse detection by the known signatures of the system vulnerabilities that are called misuse detection-based IDS [5]. IDS basically works on the following steps to alert the system administrator on violation of security policies: (i) network traffic monitoring, (ii) abnormal traffic detection, and (iii) reporting to system administrator by raising alarm. Intrusion detection systems are divided into two categories, according to the installation parameters: (i) host-based IDS (HIDS), and (ii) network-based IDS(NIDS) [6, 7]. HIDS are installed on the single host to monitor the host on which it resides. The NIDS are installed on a particular system like server to monitor the entire network [8, 9]. The common functions of the NIDS and the HIDS are the log monitoring and analysis, data and services security, raising alarm to the administrator, and policy implementation [10, 11].

The development of the intrusion detection system includes the data collection, data preprocessing (feature extraction, feature selection, outlier detection, etc.) learning, intrusion detection, and report and response generation. All the components of the IDS have their own significance, and the role in the intrusion detection cannot be ignored. In today's scenario, the IDS must be an adaptive system so that the changing traffic pattern and the different types of the network vulnerabilities to be caught in at host and the network level. The complete suite of the intrusion detection system is a need for the automated system; it is due to the unbalanced data distribution, high volume of the data, large numbers of the attributes, availability of the noise, and the outliers. CERT [12] reports that the number of network attack increases day by day, and also the amount of data on the internet increases exponentially [13]. Attack on the network services not only gives financial loss, but also violates the security rules of the individuals or the organizations. Today, there are access control, malware protection, email security, firewalls, and IPS- and IDS-related concepts, and devices that exist failed to provide 100% security to the data and services hosted on the World Wide Web [4]. From the literature available in the [4–7], it is clear that the field of the intrusion detection is gaining importance as the size of the network log data increases exponentially day by day, and the manual analysis is approximately impossible or might it will take so much time and money. So there is a need of network intrusion detection system which can automatically extract useful information form large volume of the data which is previously unknown; these features are available in the data mining or machine learning [14, 15]. The data mining can do prediction, classification [16], clustering [17], and association rule mining [18] between the data items. So the field of intrusion detection system based on the data mining is the best encapsulation of the technological framework to overcome the cope of the intrusion in the World Wide Web.

15.2 Related Work

In [19], a hybrid IDS is proposed which is based on the RBF/Elman neural network, and it has the capability to detect the both anomaly and the misuse type of the intrusions. The RBF is used for the classification, and the Elman is used to restore the memory of the past event. The IDS is trained and tested using the KDD Cup dataset, and the results show that it has superior detection rate and low false alarm rate. In [20], a fuzzy clustering and ANN-based intrusion detection is proposed. In this, fuzzy clustering method is used to generate the training subsets, the different ANN are learned on those subsets, and finally fuzzy aggregation module to aggregate the results from the learned ANNs. The system is learned and evaluated using the KDD Cup dataset. In [21], two ICLN- and SICLN-based intrusion detection systems are proposed. ICLN is the unsupervised classification method, and the SICLN is the supervised classification method. The experimental result of the paper shows that the ICLN and the SICLN provide the superior performance, and the SICLN gives better result than the traditional clustering methods. In [22], artificial neural network is used for intrusion detection. They use feed forward neural network and the back propagation network for the detection of unusual traffic in computer network. The system is evaluated using the DARPA dataset, and the result of the paper shows that the system provides the promising result. In this [23], authors presented a multilevel hierarchal Kohenan net-based IDS. The system is trained and evaluated using the randomly selected data instances from the KDD Cup dataset. In [24], authors presented neural network-based intrusion detection system for computer network monitoring. In this, they presented the network data analysis, development of the feature selection method, and generation of the learning dataset, and the system is trained using the error back propagation and Levenberg–Marquardt neural network. In [25], authors use the asymmetric cost of the false positive and negative error for the performance enhancement of the neural network-based IDS. Their proposed method accomplishes security and system's performance.

In [26], NeC4.5 intrusion detection system is proposed, and the system uses the ensembles of neural network for the training set generation, and the C4.5 classification is used for the final classification task. The result shows that the NeC4.5

provides better results in comparison with the C4.5 and the ensembles of the neural network. In [27], a light-weight intrusion detection system is proposed for anomaly detection in the computer network. In this, authors worked on the data preprocessing, feature selection, and the classification; for feature selection, they used genetic algorithm, and for the classification, they used the Neurotree classifier. The training set is first preprocessed using the ensembles of the neural network classifiers, and then the labeled new dataset is used for the final classification using the Neurotree classifier. The result shows that the system provides the promising performance as compared to the REP Tree, Naive Bayes, decision stump, random forest, and the random tree.

In today's scenario, the field of the intrusion detection attracted the researchers from the last decade and not yet the systems developed who provide the cent percent security. Neural network-based intrusion detection system in the literatures like back propagation, and MLP, which takes multiple scanning of the training set for the classification knowledge-based formation, lack of adaptability, high learning time, lack of nonlinear separability. In this paper, a novel intrusion detection system is proposed which has the capability of online adaption, nonlinear separability, lesser learning time, and the single scanning of the learning dataset.

15.3 Materials and Method: Fuzzy Min-Max Neural Network

Fuzzy min-max neural network (FMMNN) classification [28] is based on the concept of the hyperbox fuzzy sets. The FMM neural network is a collection of the N input vectors as an input to the neural network, B numbers of hyperbox as a hidden layer, and the M class labeled as an output to the neural network. Each input is connected to all the hyperboxes, and each hyperbox is connected to a particular class label. A hyperbox represents a region in the n-dimensional pattern space that contains pattern vectors those who have the full class membership value. The hyperbox is defined by its min–max point and the hyperbox membership function. The membership function gives a belongingness value to each hyperbox fuzzy set corresponding to each input vector to which a pattern is associated with the hyperbox. In this neural network, the learning is performed by a series of expansion and contraction process, and the testing of the patterns is carried out by the finding the membership value of each input pattern corresponding to each hyperbox fuzzy sets. The membership value lies between 0 and 1,1 means the full class membership, and the 0 means no membership.

The main motivation behind the selection of Fuzzy min-max neural network for the intrusion detection are the online adaption, nonlinear separability, overlapping classes handling capability, single scanning of the training set, less training time requirement, easy verification and validation, hard and soft decision, less parameter tuning requirements, etc.

15.3.1 Fuzzy Set Hyperbox

A fuzzy set hyperbox represents a region in the n-dimensional pattern space. The hyperbox is defined by its min point, max point and the membership function. The hyperbox contained only those patterns which have the highest membership value against the all hyperboxes. The hyperbox fuzzy set is defined in Eq. 15.1:

$$B_j = \{A, v_j, w_j, F(A, v_j, w_j)\} \quad \forall A \in I^n. \tag{15.1}$$

Here, B_j is the jth hyperbox, A is the n-dimensional input vector or pattern, v_j is the min point of the hyperbox B_j, w_j is the max point of the hyperbox B_j, and $F(A, v_j, w_j)$ is the membership function of the hyperbox B_j. Suppose $A(a_1, a_2, a_3 \ldots a_n)$ is an input pattern, then the min and the max points of hyperbox B_j are defined as in Eqs. 15.2 and 15.3:

$$v_j = (v_{j1}, v_{j2}, v_{j3}, \ldots v_{jn}). \tag{15.2}$$

$$w_j = (w_{j1}, w_{j2}, w_{j3}, \ldots w_j n). \tag{15.3}$$

Now from the above-defined hyperbox B_j, we can conclude that the aggregate of the hyperbox fuzzy set of the kth pattern class C_k is defined as the union of the K hyperboxes that belong to the pattern class k and is defined in Eq. 15.4:

$$C_k = \cup_{j \in K} B_j. \tag{15.4}$$

15.3.2 Membership Function

The membership function of the fuzzy hyperbox is used to measure the degree of belongingness of input pattern to a hyperbox. The value of the membership function of a hyperbox B_j for the input pattern A_j is lies between 0 and 1. When the membership value is closer to 1, it means that the pattern is more contained in the hyperbox; 1 represents the full membership. The membership function is the sum of the two's complement of the average of the min point violation and the average of the max point violation. The membership function is defined in Eq. 15.5:

$$b_j(X_h) = \frac{1}{2n} \sum_{i=1}^{n} \left[\max(0, 1 - \max(0, \gamma \min(1, x_{hi} - w_{ji}))) \right.$$
$$\left. + \max(0, 1 - \max(0, \gamma \min(1, v_{ji} - x_{hi}))) \right]. \tag{15.5}$$

Here, $b_j(A_h)$ membership value of the input pattern A_h corresponding to the hyperbox b_j. γ is the sensitivity parameter that regulates the membership value, i.e., when the distance between hyperbox b_j and the input pattern X_h increases, the membership value decrease and vice versa.

$X_h = (x_{h1}, x_{h2} \ldots x_{hn})$, is the hth input pattern, $v_j = (v_{j1}, v_{j2}, v_{j3} \ldots v_{jn})$ is the min point of the hyperbox b_j and $w_j = (w_{j1}, w_{j2}, w_{j3}, \ldots w_j n)$ is the max point of the hyperbox b_j.

15.3.3 *FMM-Based Learning Methodology for IDS*

The hyperbox learning process is the combination of the three subprocess that are expansion, overlap test, and the contraction process. The fundamental prerequisite of the learning process is the dataset. The data is the ordered pair of the n attributes and the one decision attribute or the class attribute. In the intrusion detection, learning process first selects the input pattern along with the class label and finds the appropriate hyperbox for the same class, and if the chosen hyperbox accommodates the selected input pattern, then min and the max points of the hyperbox are adjusted, otherwise create the new hyperbox and assign the corresponding min and the max point. Due to expansion of the existing hyperboxes or the new hyperbox creation, process may create the problem of overlapping with the existing hyperboxes. Overlapping is eliminated with the help of contraction process. The details of the expansion, overlap test, and the contraction process are explained in the subsequent sections.

15.3.3.1 Hyperbox Expansion Phase

The goal of the hyperbox expansion process is to assign the input pattern to the already existing hyperboxes that have the highest degree of the membership value. If the membership function cannot accommodate it, then create new hyperbox. The size of the hyperbox depends on the θ, and its value lies between 0 and 1; both 0 and 1 are included. Let (x_h, c_h) be an ordered pair of the input pattern, $x_h = (x_{h1}, x_{h2}, \ldots x_{hn})$ be the set of the n input attributes, c_h be the class label of the input pattern. The input pattern x_h is included in the box b_j if,

$$n\theta \geq \sum_{i=1}^{n} \left(\max\left(w_{ji}, x_{hi}\right) - \min\left(v_{ji}, x_{hi}\right)\right). \tag{15.6}$$

If Eq. (15.6) is satisfied, then update the min and the max points of the fuzzy hyperbox b_j, which are given below in Eqs. (15.7) and (15.8); otherwise create, assign, and label the hyperbox, and also assign the min and max points.

$$v_{ji}^{\text{new}} = \min\left(v_{ji}^{\text{old}}, x_{hi}\right) \quad \forall i = 1,\ 2,\ 3 \ldots n. \tag{15.7}$$

$$w_{ji}^{\text{new}} = \max\left(w_{ji}^{\text{old}}, x_{hi}\right) \quad \forall i = 1, 2, 3 \ldots n. \tag{15.8}$$

15.3.3.2 Hyperbox Overlap Test Phase

The goal of the hyperbox overlap test phase is to finding out the overlap between the hyperboxes that have the different classes. Suppose that the box B_1 is expanded in the previous step and the box B_m is the box in the pattern space representing another class. Therefore, testing the possible overlap between the B_1 and the B_m is carried out by dimension-by-dimension comparison, and the index of the dimension of the smallest overlap along any dimension is saved for contraction process. If $\delta^{\text{old}} - \delta^{\text{new}} > 0$. Then, $\Delta = i$ and $\delta^{\text{old}} = \delta^{\text{new}}$. Here, Δ is the dimensional index where overlap exists; overlap test is carried out for all the dimensions. If the overlap does not exist, then the minimum overlap is set to indicate the contraction is not necessary, i.e., $\Delta = -1$. The overlap between the hyperboxes is tested by the below mention cases, initially $\delta^{\text{old}} = 1$.

Case 1. $v_{ji} < v_{ki} < w_{ji} < w_{ki}$

$$\delta^{\text{new}} = \min\left(w_{ji} - v_{ki}, \delta^{\text{old}}\right)$$

Case 2. $v_{ki} < v_{ji} < w_{ki} < w_{ji}$

$$\delta^{\text{new}} = \min\left(w_{ki} - v_{ji}, \delta^{\text{old}}\right)$$

Case 3. $v_{ji} < v_{ki} < w_{ki} < w_{ji}$

$$\delta^{\text{new}} = \min\left(\min\left(w_{ki} - v_{ji}, w_{ji} - v_{ki)}, \delta^{\text{old}}\right)\right.$$

Case 4. $v_{ki} < v_{ji} < w_{ji} < w_{ki}$

$$\delta^{\text{new}} = \min\left(\min\left(w_{ji} - v_{ki}, w_{ki} - v_{ji)}, \delta^{\text{old}}\right)\right.$$

15.3.3.3 Hyperbox Contraction Phase

The objective of the contraction process is to adjust the dimension of the hyperboxes so that the overlap between the boxes is eliminated. The contraction process keeps in mind to minimally disturb the size of the hyperboxes. The contraction process depends on the Δ, If $\Delta > 0$, then Δth dimension is adjusted that will minimally disturb the shape of the hyperboxes. The contraction process is covered in the following below-mentioned cases.

Case 1. $v_{j\Delta} < v_{k\Delta} < w_{j\Delta} < w_{k\Delta}$

$$w_{j\Delta}^{new} = v_{k\Delta}^{new} = \frac{\left(w_{j\Delta}^{old} + v_{k\Delta}^{old}\right)}{2}$$

Case 2. $v_{k\Delta} < v_{j\Delta} < w_{k\Delta} < w_{j\Delta}$

$$w_{k\Delta}^{new} = v_{j\Delta}^{new} = \frac{\left(w_{k\Delta}^{old} + v_{j\Delta}^{old}\right)}{2}$$

Case 3. (a) $v_{j\Delta} < v_{k\Delta} < w_{k\Delta} < w_{j\Delta}$ and $(w_{k\Delta} - v_{j\Delta}) < (w_{j\Delta} - v_{k\Delta})$

$$v_{j\Delta}^{new} = w_{k\Delta}^{old}$$

Case 3. (b) $v_{j\Delta} < v_{k\Delta} < w_{k\Delta} < w_{j\Delta}$ and $(w_{k\Delta} - v_{j\Delta}) > (w_{j\Delta} - v_{k\Delta})$

$$w_{j\Delta}^{new} = v_{k\Delta}^{old}$$

Case 4. (a) $v_{k\Delta} < v_{j\Delta} < w_{j\Delta} < w_{k\Delta}$ and $(w_{k\Delta} - v_{j\Delta}) < (w_{j\Delta} - v_{k\Delta})$

$$w_{k\Delta}^{new} = v_{j\Delta}^{old}$$

Case 4. (b). $v_{k\Delta} < v_{j\Delta} < w_{j\Delta} < w_{k\Delta}$ and $(w_{k\Delta} - v_{j\Delta}) > (w_{j\Delta} - v_{k\Delta})$

$$v_{k\Delta}^{new} = w_{j\Delta}^{old}$$

15.4 Learning Algorithm for IDS

Input: Training set, Testing set, θ(theta), γ (sensitivity parameter), v_{ij}, w_{ij}
Output: FMM neural Networks

Process:

Step 1. Start
Step 2. Load the dataset
Step 3. Specify the initial parameters θ(theta), γ and (sensitivity parameter).
Step 4. If the input pattern is the first input pattern in the learning process, then assign the v_{ij} and w_{ij}.
Step 5. For $i = 2$ to n
Step 5.1. [Hyperbox expansion] For the *i*th input pattern, identify a same class hyperbox that can accommodate it according the membership function given in Eq. 15.5 and the expansion criteria given on the Sect. 15.3.3.1. If not, then create a new hyperbox and assign the min and max (v_{ij} and w_{ij}.) of the hyperbox and label the box.
Step 5.2. [Overlap Test] Check the overlap between the newly created and expanded hyperbox according to the cases given in the Sect. 15.3.3.2 with existing. If any overlap, goto step 5.3 else go to step 5.4.
Step 5.3. [Contraction process] Eliminate the overlap by minimally adjusting the dimensions of the hyperboxes according to the cases given in the Sect. 15.3.3.3.
Step 5.4. Increment *i*.
Step 5.5. Stop if the criterion is matched, then goto step 6 else goto step 5.
Step 6. Stop

15.5 Results and Discussion

This section describes the result of the proposed fuzzy min-max Neural Network-based intrusion detection system (FMM IDS) and also the comparison with the other classifier-based systems. The FMM IDS is tested with the KDD Cup dataset. The dataset contains wide variety of intrusion and normal data collected by the DARPA. The KDD Cup 99 dataset provide both the training and the test datasets. The training data include the two types of the network data: normal and attack patterns. The attack pattern is of the different categories like DoS, U2R, R2L, Probe and Normal. The test dataset of used which is the subset of the KDD Cup dataset that includes some attacks that are not in the training set. The dataset contains a total of 41 attributes and one class attribute or the decision attribute, which describes the category of the traffic pattern whether it is normal or attack. In this experiment, a total of 11850 instances are taken for experimentation, 7821 instances for training, and 4029 instances for testing the FMM IDS. The dataset is first preprocessed to remove the duplicate entries. Outlier and extreme values are filter-based on interquartile ranges concept of WEKA [29]. The FMM-based IDS is tested on sensitivity parameter $\gamma = 1$ and on the following θ values 0.03, 0.06, 0.12, 0.24, and 0.36; here, θ is the size of the hyperbox.

In FMM-based intrusion detection system training data set is used to train the system and testing dataset to test the system. For checking the effectiveness, the classification accuracy and the classification error are taken as performance parameters. Table 15.1 shows the results of FMM-based IDS on different θ values 0.03, 0.06, 0.12, 0.24, and 0.36. FMM-based IDS is implemented in the Windows 8, 4 GB RAM, Core i5 processor and MATLAB 2014 environment. The system gives best performance at $\theta = 0.36$. At $\theta = 0.36$, FMM-based IDS provides the 95.1770% accuracy and 4.8230% classification error. The system gives 94.4162% accuracy and 5.5838% classification error at $\theta = 0.12$, which is the second best result. Table 15.2 shows the comparison of the proposed system with the other existing systems. The FMM-based system is compared with the MLP classifiers, multilayer perceptron, RBF classifier, RBFN classifiers, SMO, voted perceptron, naïve Bayes, LibSVM, simple logistic regression, KDD Cup Winner and KDD Cup Runner UP. The result of the MLP classifiers, multilayer perceptron, RBF classifier, RBFN classifiers, SMO, voted perceptron, naïve Bayes, LibSVM, simple logistic regression are evaluated on WEKA with default values, and the results as shown in Table [15.2] shows that the proposed system gives the best results.

Table 15.1 Result of FMM IDS

θ	Accuracy	Error
0.03	91.1168	08.8832
0.06	91.3706	08.6294
0.12	94.4162	05.5838
0.24	92.8934	07.1066
0.36	95.1770	04.8230

Table 15.2 Comparison of the FMM IDS with other systems

Methods	Accuracy	Error
MLP classifier	89.5658	10.4342
Multilayer perceptron	93.5964	06.4036
RBF classifiers	90.3946	09.6054
RBFN classifiers	81.8069	18.1931
SMO	91.6356	08.3644
Voted perceptron	77.8605	22.1395
Naïve Bayes	67.3616	32.6384
LibSvm	91.4867	08.5133
Simple logistic regression	91.5612	08.4388
KDD cup winner [30]	91.8000	08.2000
KDD cup runner up [31]	91.5000	08.5000
Proposed FMMNN IDS	95.1770	04.8230

15.6 Conclusion

In this paper, a novel intrusion detection system has been proposed, which is based on the fuzzy min max neural network. The main component of the fuzzy min-max neural network is the fuzzy hyperbox to find the decision boundary between the various overlapping classes. The proposed system has been the online adaption facility, nonlinear separability. The most important property of the proposed system is the lesser time requirement for the learning against the tradition neural network like back propagation neural network, and Boltzmann neural network. The traditional network takes the multiple passes in learning process. The system is trained and evaluated using KDD Cup 99 dataset. The proposed system provides superior performance as compared to MLP classifiers, multilayer perceptron, RBF classifier, RBFN classifiers, SMO, voted perceptron, naïve Bayes, LibSVM, simple logistic regression, KDD Cup Winner and KDD Cup Runner UP.

References

1. Rao HR (ed) (2007) Managing information assurance in financial services. IGI Global
2. Vokorokos L, Ennert M, Radušovský J (2014) A survey of parallel intrusion detection on graphical processors. Cent Eur J Comp Sci 4(4):222–230
3. http://www.isaca.org/Journal/Past-Issues/2006/Volume-1/Pages/JOnline-Network-Intrusion-Detection-Know-What-You-Do-Not-Need.aspx#f1. Accessed on 18 Feb 2015
4. Azad C, Jha VK (2013) Data mining in intrusion detection: a comparative study of methods, types and data sets. Int J Info Technol Comput Sci (IJITCS) 5(8):75
5. Liao HJ, Lin CHR, Lin YC, Tung KY (2013) Intrusion detection system: a comprehensive review. J Network Comput Appl 36(1):16–24
6. Yang H, Li T, Hu X, Wang F, Zou Y (2014) A survey of artificial immune system based intrusion detection. Sci World J
7. Elshoush HT, Osman IM (2011) Alert correlation in collaborative intelligent intrusion detection systems—a survey. Appl Soft Comput 11(7):4349–4365
8. Chari SN, Cheng PC (2003) BlueBox: a policy-driven, host-based intrusion detection system. ACM Trans Info Syst Sec (TISSEC) 6(2):173–200
9. Gautam SK, Om H (2015) Multivariate linear regression model for host based intrusion detection. In: Computational intelligence in data mining, vol 3. Springer, India, pp 361–371
10. Vigna G, Kemmerer RA (1999) NetSTAT: a network-based intrusion detection system. J Comput Sec 7(1):37–71
11. Joshi SA, Pimprale VS (2013) Network intrusion detection system (NIDS) based on data mining. Int J En Sci Innovative Technol (IJESIT) 2
12. Ragsdale DJ, Carver CA, Humphries JW, Pooch UW (2000) Adaptation techniques for intrusion detection and intrusion response systems. In: IEEE international conference on systems, man, and cybernetics, vol. 4, pp 2344–2349
13. http://en.wikipedia.org/wiki/Internet_traffic
14. Liao S-H, Chu P-H, Hsiao P-Y (2012) Data mining techniques and applications–a decade review from 2000 to 2011. Expert Syst Appl 39(12):11303–11311
15. Julisch K (2002) Data mining for intrusion detection. Applications of data mining in computer security. Springer, US, pp 33–62

16. Azad C, Jha VK (2014) Data mining based hybrid intrusion detection system. Indian J Sci Technol 7(6):781–789
17. Chen T, Zhang X, Jin S, Kim O (2014) Efficient classification using parallel and scalable compressed model and its application on intrusion detection. Expert Syst Appl 41(13):5972–5983
18. Gu B, Guo H (2014) The intrusion detection system based on a novel association rule. In: International conference on information science, electronics and electrical engineering (ISEEE), vol. 2, pp 1313–1316
19. Tong Xiaojun, Wang Zhu, Haining Yu (2009) A research using hybrid RBF/Elman neural networks for intrusion detection system secure model. Comput Phys Commun 180(10):1795–1801
20. Wang G, Hao J, Ma J, Huang L (2010) A new approach to intrusion detection using artificial neural networks and fuzzy clustering. Expert Syst Appl 37(9):6225–6232
21. Lei JZ, Ghorbani AA (2012) Improved competitive learning neural networks for network intrusion and fraud detection. Neurocomputing 75(1):135–145
22. Shun J, Malki HA (2008) Network intrusion detection system using neural networks. In: Fourth international conference on natural computation ICNC'08, vol. 5, pp 242–246
23. Sarasamma ST, Zhu QA, Huff J (2005) Hierarchical Kohonenen net for anomaly detection in network security. Syst Man Cybern Part B: Cybern IEEE Trans 35(2):302–312
24. Linda O, Vollmer T, Manic M (2009) Neural network based intrusion detection system for critical infrastructures. In: International joint conference on neural networks, pp 1827–1834
25. Joo D, Hong T, Han I (2003) The neural network models for IDS based on the asymmetric costs of false negative errors and false positive errors. Expert Syst Appl 25(1):69–75
26. Zhou ZH, Jiang Y (2004) NeC4. 5: neural ensemble based C4. 5. IEEE Trans Knowl Data Eng 16(6):770–773
27. Sindhu S, Siva S, Geetha S, Kannan A (2012) Decision tree based light weight intrusion detection using a wrapper approach. Expert Syst Appl 39(1):129–141
28. Simpson PK (1992) Fuzzy min-max neural networks. I. classification. IEEE Trans Neural Networks 3(5):776–786
29. Hall M, Frank E, Holmes G, Pfahringer B, Reutemann P, Witten IH (2009) The WEKA data mining software: an update. ACM SIGKDD explorations newsletter 11(1):10–18
30. Pfahringer B (2000) Winning the KDD99 classification cup: bagged boosting. ACM SIGKDD Explorations Newsletter 1(2):65–66
31. Levin I (2000) KDD-99 classifier learning contest: LLSoft's results overview. SIGKDD Explorations 1(2):67–75

Chapter 16
Effect of Static Synchronous Series Compensator on Power Balancing in Wind Farms

Abhishek Gandhar, Balwinder Singh and Rintu Khanna

16.1 Introduction

Even today, there are many locations in the world which are not connected with grid because of cost and some other factors. Electrical energy can be provided to these remote areas by isolated power systems, but this is not a cost-effective solution; therefore, the cost of electricity can be made comparable to existing system by using renewable sources such as wind and hydro depending upon the sitting locations of available energy sources. However, the main hurdle in using renewable energy sources for stand-alone power system applications is that the availability of power has daily and seasonal patterns that may not match the load demand. A group of induction generators of small ratings are used to generate electric power for stand-alone operations [1]. Wind power is a sustainable and clean method to generate electrical power. More than half of the people in the world use wind power for electricity generation. It is clean, pollution-free, and cost-effective solution of electric power generation [2]. But the biggest demerit of wind energy is the continuous fluctuations in the wind speed, which affects the different parameters of electrical power. Nowadays, wind power generation is increasing day by day, so for controlling the power quality, some control mechanism is required for farms. In recent years, flexible AC transmission systems (FACTS) are used because of their ability to control almost every parameter of power systems. These FACTS devices

A. Gandhar (✉) · B. Singh · R. Khanna
PEC University of Technology, Chandigarh, India
e-mail: abhishek.gandhar@gmail.com

B. Singh
e-mail: balwindersingh@pec.ac.in

R. Khanna
e-mail: rintukhanna@pec.ac.in

V. Nath (ed.), *Proceedings of the International Conference on Nano-electronics, Circuits & Communication Systems*, Lecture Notes in Electrical Engineering 403,
DOI 10.1007/978-981-10-2999-8_16

have been investigated to achieve the best utilization of existing power systems [3–5].

In this paper, an important FACTS controller SSSC has been incorporated with test system to control the parameters of the system. In the present study, the wind farm is analyzed for the quality of power generation property with and without SSSC. A combination of two 1.5 MVA wind turbine generators is modeled as a compensation methodology for the fault contingency. It is also clearly presented that voltage profiles can also be enhanced with the application of SSSC in the system.

In the proposed study the voltage profiles of connected loads are presented after the insertion of SSSC in the test system. Software simulations have been carried out in a very effective Simulink toolbox of MATLAB to study the effect of SSSC of the voltage profile under different loading conditions. The present work has been organized into the following sections:

1. Basic information of SSSC,
2. Test system,
3. Simulation results of test system, and
4. Conclusion.

16.2 Basic Configuration of SSSC

The SSSC is a static series synchronous controller and is very similar to a series compensator. On the contrary, this controller has the same structure as that of STATCOM. But in SSSC operation, the transformer is connected in series with the supply line, while in the case of STATCOM, that coupling transformer is connected in parallel with the line. The SSSC inserted a series voltage into the series combination of line with a control over magnitude and direction. This sinusoidal voltage is in synchronism with supply voltage [6, 7]. Then, the resultant output voltage is having a phase difference of 90° with the line current, and it is also independent of line current. It can be varied according to the requirement of the system. Thus, the objective of the controller is to vary the system power by varying the imaginary part of voltage drop. A fraction of inserted voltage that is in phase of line current is responsible for the losses in the converter. The basic configuration of a SSSC is shown in Fig. 16.1 [7–9].

Here, phase-locked loop (PLL) is an integral part of control block of SSSC which matches with the +ve sequence current. The control of SSSC is shown in Fig. 16.2. The yield of the phase-locked loop (angle = ωt) is used to design the direct-axis and quadrature-axis fractions of the three-phase voltages and currents (nominated as V_d, V_q or I_d, I_q). The quantification block is evaluating the quadrature part of alternating +ve components of voltages V_1 and V_2 (V_{1q} and V_{2q}) also the direct-current voltage V_{dc}. AC and DC voltage controller which calculates the two fractions of this voltage V_{d_conv} and V_{q_conv}) demanded to get the needed DC

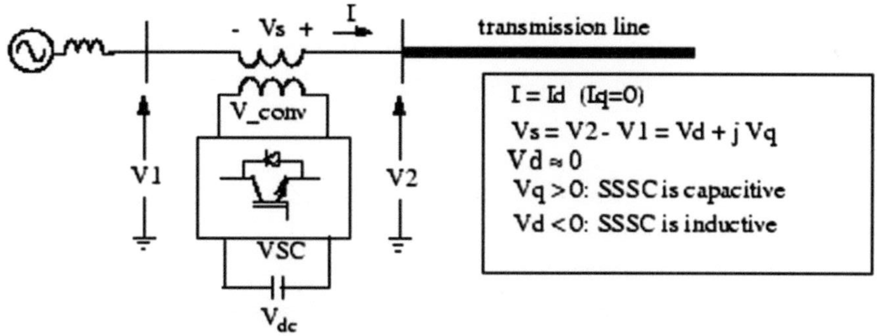

Fig. 16.1 The basic configuration of SSSC

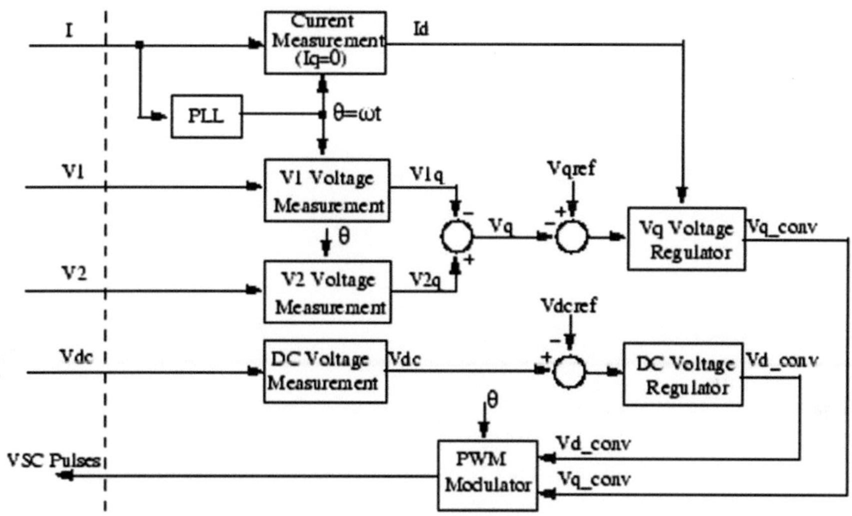

Fig. 16.2 Control system block diagram of SSSC [13]

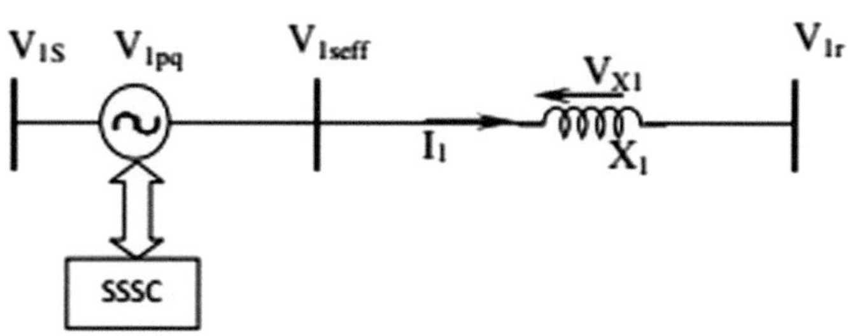

Fig. 16.3 Effect of injected voltage

voltage (V_{dcref}) and the inserted voltage (V_{qref}). The V_q voltage controller is helped by another controller which designs the V_{conv} voltage from the I_d calculations. The effect of injected voltage is illustrated in Fig. 16.3. The governing equations are given below:

$$P = \frac{V_s V_r}{X_L} \sin(\delta_s - \delta_r) = \frac{V^2}{X_L} \sin \delta \tag{16.1}$$

$$Q = \frac{V_s V_r}{X_L} [1 - \cos(\delta_s - \delta_r)] = \frac{V^2}{X_L} (1 - \cos \delta) \tag{16.2}$$

$$\delta = (\delta_s - \delta_r) \tag{16.3}$$

$$|V_s| = |V_r| = |V| \tag{16.4}$$

$$P_q = \frac{V^2}{X_{\text{eff}}} \sin \delta = \frac{V^2}{X_L \left[1 - \frac{X_q}{X_L}\right]} \tag{16.5}$$

$$Q_q = \frac{V^2}{X_{\text{eff}}} [1 - \cos \delta] = \frac{V^2}{X_L \left[1 - \frac{X_q}{X_L}\right]} [1 - \cos \delta] \tag{16.6}$$

16.3 Model of Test System

Here, the test system shown in Fig. 16.6 consists of wind farm with 2 wind turbines, each rated at 1.5 MW, and the wind turbines are connected in two groups containing one wind turbine generator each; induction generator of Group 1 is connected at bus ‘WT’, and induction generator of Group 2 connected at bus ‘D’ is shown in Fig. 16.2. Three different loads L-1, L-2, and L-3 are connected with A IEEE 4-bus system. The FACT controller SSSC has been implemented between these sections connected at bus WT and bus D. Thus, the SSSC controller is effectively controlling the different parameters of wind-powered distributed energy resource. The particular for the design is presented in the Appendix. In this paper, the investigations executed are based on the assumption that all the wind turbines in the wind farm are identical and have the same operating condition. This combination is representing a wind power system. In this paper, the investigations are executed on the assumption that the wind turbine is operating at a rated speed (9 m/s). The controlled parameters are presented in the following section (Figs. 16.4 and 16.5).

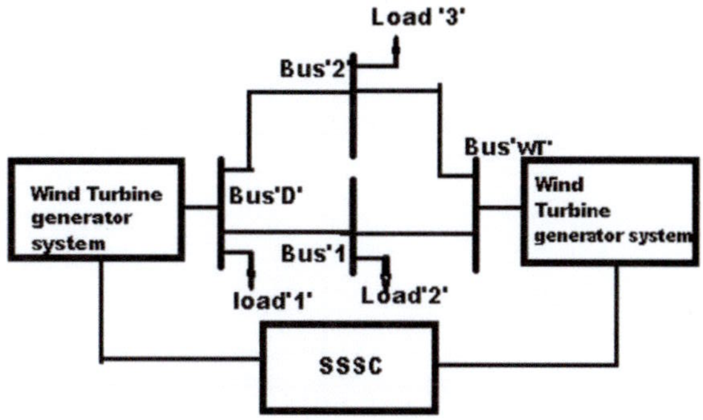

Fig. 16.4 Block diagram of test system [10–13]

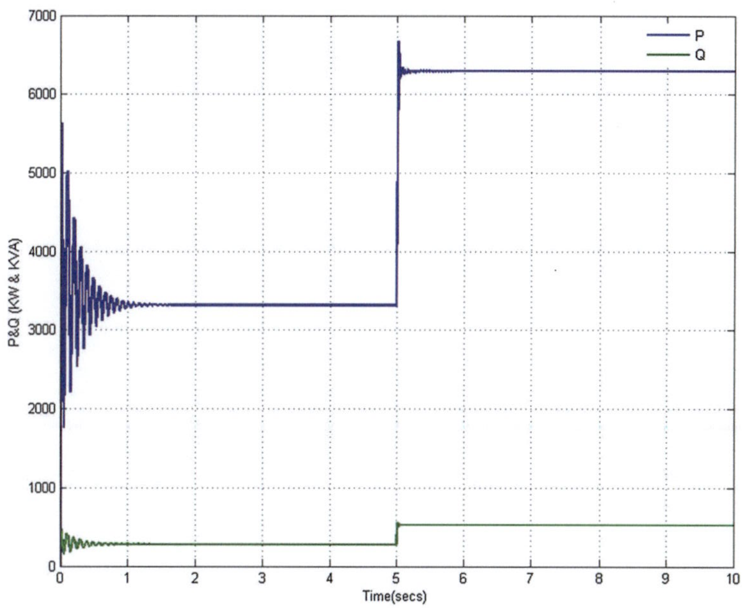

Fig. 16.5 P and Q profiles of load '1'

16.4 Simulation Results with SSSC

The active power and reactive curves of two groups of wind systems, where each group consists of six wind turbine induction generators connected with three loads that are presented here. A significant amount of reactive power and active power is required by wind generator of group '1', whereas wind induction generator of group '2' is having surplus active and reactive power. In reality, the fluctuation in the wind velocity brings the changes in the requirement of reactive power. Therefore, fixed capacitors are not capable of providing compensation to such dynamic

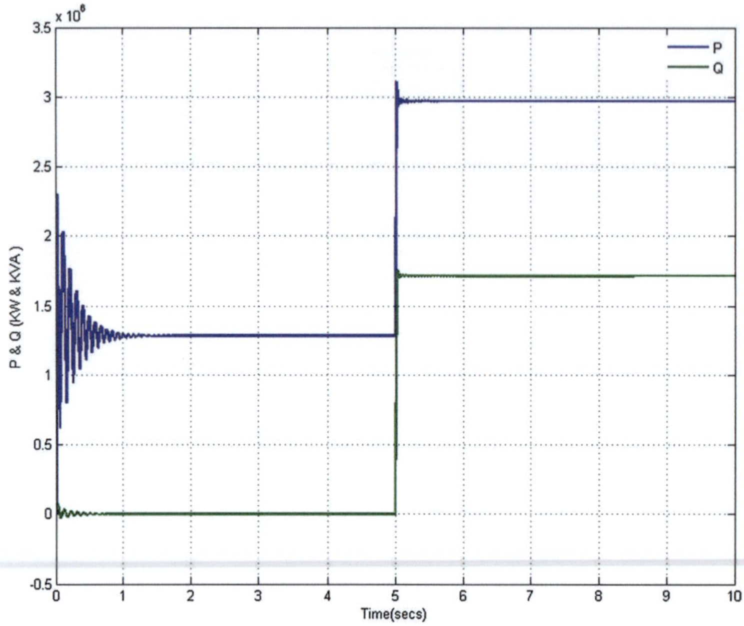

Fig. 16.6 P and Q profiles of load ‘2’

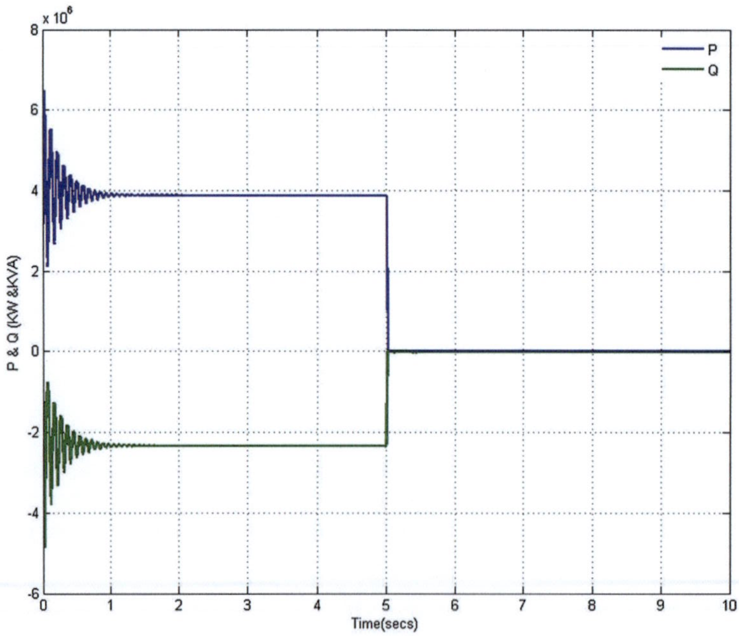

Fig. 16.7 P and Q profiles of load ‘3’

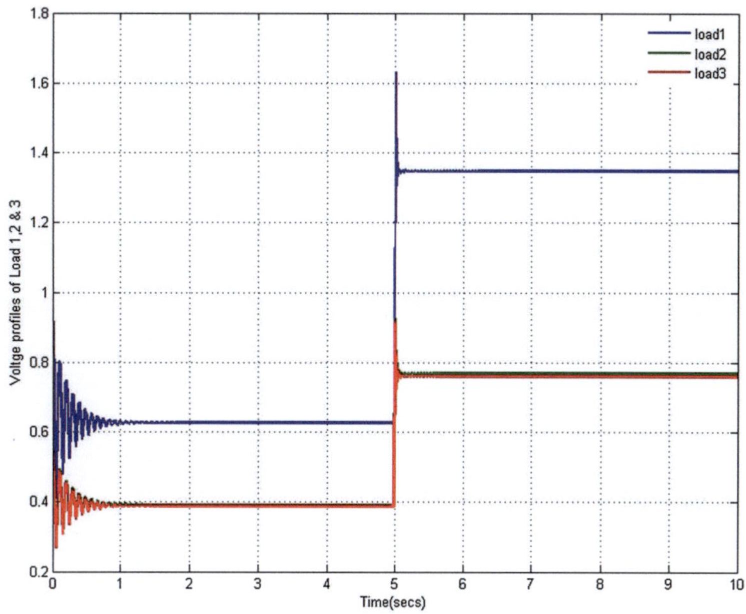

Fig. 16.8 Voltage profiles of load '1, 2 and 3'

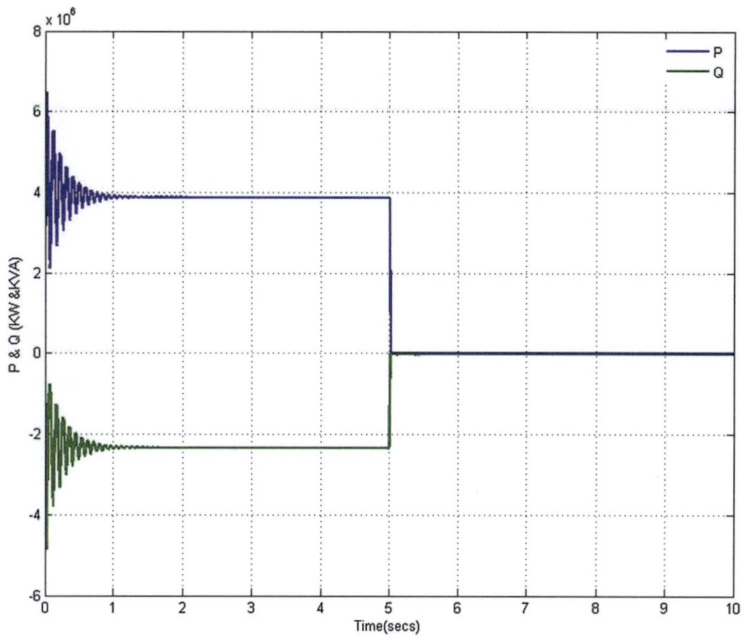

Fig. 16.9 P and Q profiles of wind generator '1'

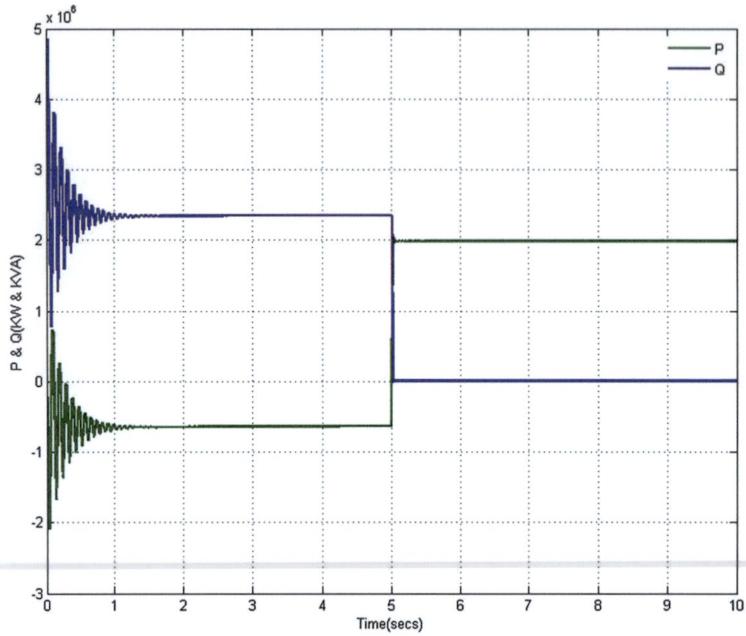

Fig. 16.10 P and Q profiles of wind generator ‘2’

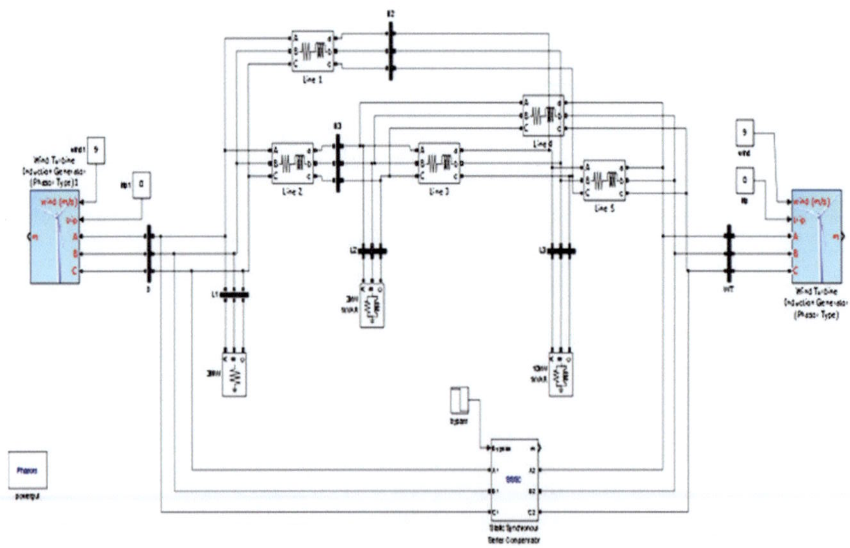

Fig. 16.11 Simulation of test system

Table 16.1 Voltage profile incremental change

Sr. no.	Loads	$V_{STATCOM}$ (p.u.)	V_{SSSC} (p.u.)
1	Load '1'	0.31	0.38
2	Load '2'	0.30	0.38
3	Load '3'	0.28	0.97

conditions. Then, this continual VAR demand in the system will be provided by a dynamic controller SSSC, and the range of compensation can be from no load to full load. Thus, a SSSC has been employed in between these two groups. It is shown in Figs. 16.7, 16.8, 16.9, 16.10 and 16.11 that SSSC is balancing active nd reactive VAR powers of the studied test system. The voltage profiles of three loads '1, 2 and 3' are also presented. A comparative analysis on the basis of voltage profile improvement at the time of induction of STATCOM and SSSC is also presented in Table 16.1. The SSSC is inserted at '5' s, and within no time, it significantly increases the voltage profiles of these loads shown in Fig. 16.10 (load '1' with green curve, load '2' with red curve, and load '3' with blue curve); thus, SSSC is playing a pivotal role of controlling powers of given test system and is proved as one of the most important and powerful FACT controllers of the family.

16.5 Conclusion

In this paper, the effective SSSC in improving the voltage profiles of different loads is analyzed. In the presented study SSSC helps in balancing the active and reactive VAR powers of the studied test system. It is clearly presented that SSSC helps in the mitigation of the active and reactive power imbalance for the considered system. So, here results clearly show the requirement of dynamic reactive power compensation in such conditions, and it can be achieved by using the FACTS controllers like SSSC.

Appendix A

See Tables 16.2, 16.3, 16.4, 16.5 and 16.6.

Table 16.2 Load data

Load	kW	kVA
1	10	1.0
2	03	1.0
3	03	–

Table 16.3 Wind turbine/generator data

1	Stator resistance	0.004843 pu
2	Rotor resistance	0.004377 pu
3	Stator leakage reactance	0.01248 pu,
4	Rotor leakage reactance	0.0179 pu
5	Magnetizing reactance	6.77 pu
6	Inertia constant	5.04 s
7	Friction factor	0.01 pu
8	Pole pairs	03
9	Base wind speed	9 m/s
10	Mechanical o/p power	9 MW

Table 16.4 Pitch angle regulator

Pitch angle regulator		
1	Kp	05
2	Ki	25
3	Max pitch angle	45°

Table 16.5 Series controller data

1	kVA	50 kVA
2	Voltage	480 V
3	Frequency	60 Hz
4	Max injected voltage	0.10 pu
5	Rse	0.16 pu
6	Xse	0.016 pu
7	Friction factor	0.01 pu
8	Pole pairs	03
9	Base wind speed	9 m/s
10	Mechanical o/p power	9 MW

Table 16.6 Injected voltage regulator

1	Kpse	0.03
2	Kise	1.5

References

1. Al-Majed SI, Fujigaki T (2010) Wind power generation: an overview. In: Proceedings of the international symposium modern electric power systems (MEPS), IEEE Xplore Press, Wroclaw, Poland, pp 1–6, 20–22 Sept 2010
2. Ahilan T, Mohammed KP, Arumugham S (2009) A critical review of global wind power generation. Am J Appl Sci 6:204–213
3. John Paserba J (2003) How FACTS controllers benefit AC transmission systems. IEEE Trans Power Delivery 1:949–957

4. Hingorani NG, Gyugyi L (2000) Understanding FACTS, concepts and technology of flexible AC transmission systems. IEEE Press, New York
5. Breuer W, Povh D, Retzmann D, Teltsch E (2004) Role of HVDC and FACTS in future power systems. In: Shanghai Power Conference, CEPSI- 2004, Shanghai
6. Panda Sidhartha (2010) Modelling, Simulation and optimal tuning of SSSC-based controller in a multi-machine power system. World J Model Simul UK 6(2):110–121
7. Gandhar Abhishek, Singh Balwinder, Khanna Rintu (2012) Impacts of FACTS technology-a state of art review. Int J Innovative Technol Exploring Eng (IJITEE) 1(4):28–31
8. Gyugyi L, Schauder CD, Sen KK (1997) Static synchronous series compensator: a solid-state approach to the series compensation of transmission lines. IEEE Trans Power Delivery 12 (1):406–417
9. Reddy KRS, Padhy RN, Patel RN (2006) Congestion management in deregulated power system using FACTS devices. IEEE Trans Power Syst 21(2)
10. Sode-Yome, Mithulananthan N, Kwang Lee Y (2007) A comprehensive comparison of FACTS devices for enhancing static voltage stability. IEEE Trans Power Delivery 06:1292–1298
11. Gupta S, Tripathi RK, Shukla RD (2010) Voltage stability improvement in power systems using facts controllers: state-of-the-art review. In: IEEE International conference on power control and embedded system (ICPCES), pp 1–8, Jan 2010
12. Thirumalaivasan R, Janaki M, Prabhu N (2013) Damping of SSR using subsynchronous current suppressor with SSSC. IEEE Trans Power Syst 28(1)
13. Divan D, Johal, H (2007) Distributed facts—A new concept for realizing grid power flow control. IEEE T Power Electr 22(6) 2253–2260

Chapter 17
Characterisation of TFT Sensors for Chemical Sensing Applications

A. Ravishankar Holla, Ananth Noorithaya and M. Uttarakumari

17.1 Introduction

The birth of flexible electronics took place due to its need in the space exploration activities. Solar cells were required to support the space vehicles. Hence, flexible solar cells were fabricated with silicon as the materials in them. Then, a necessity arose to make them thin. This was to improve their efficiency. The solar cells initially used in space crafts and satellites were not efficient. Later on, as more research was carried out in the field, flexible solar cells were introduced [1]. As more developments were made on this, flexible electronics became centred to the use of thin-film transistors (TFT). They are being used in a variety of applications which includes displays, sensors, and use-and-throw electronics [2]. TFTs are a unique group of field-effect transistors manufactured by depositing thin films of semiconducting materials, dielectric layers, and metallic contacts over a substrate which is the supporting base. The commonly used substrate materials are glass, plastic, etc. This is because in liquid crystal displays and flexible electronics, where in TFT's are primarily used, transparent substrates are usually preferred [3]. The TFTs differ from the conventional transistor with respect to the semiconductor material such as a silicon wafer typically is the substrate.

A.R. Holla (✉) · M. Uttarakumari
Department of Electronics & Communication,
R.V.C.E, Bangalore, India
e-mail: ravishankarholla@rvce.edu.in

M. Uttarakumari
e-mail: uttarakumari@rvce.edu.in

A. Noorithaya
VLSI Design & Embedded Systems,
Department of Electronics & Communication,
R.V.C.E, Bangalore, India
e-mail: ananthnoorithaya@hotmail.com

V. Nath (ed.), *Proceedings of the International Conference on Nano-electronics, Circuits & Communication Systems*, Lecture Notes in Electrical Engineering 403,
DOI 10.1007/978-981-10-2999-8_17

The configurations of TFT are based on the positions of source–drain contacts relative to the position of the gate:

1. Top-gate structure:

The top-gate structure is shown in Fig. 17.1. The gate material is deposited on the substrate. This is followed by the deposition of the dielectric layer on top of the gate deposited area. The organic or the active material is deposited next on the dielectric. The source and drain contacts are then etched by masking process. In top contact configuration, the charge injection is done vertical into the active layer, while the current flows through the channel in a lateral direction.

2. Bottom-gate structure:

The bottom-gate configuration is shown in Fig. 17.1. The gate is first deposited on the substrate. Next, the dielectric layer deposition is done, as in the previous configuration. The source and drain contacts are formed next. The active layer is deposited at last. So the source and drain contacts are partially concealed in this case. In this configuration, the charge injection is from the side plane into the active layer [4].

The cross-sectional views of above views are also shown in Fig. 17.1. In staggered configuration, the electrodes are placed over active layer. In coplanar, the electrodes of the diffusion layer are in the same plane as that of the active layer. For sensor applications, the use of the top contact structure is preferred because fabrication is a lot easier. Doping the organic layer under the contacts is used to facilitate the charge injection.

17.2 Literature Survey

Most of the flexible electronics today is implemented using TFTs. They were first manufactured by Brody et al. in 1968. In the last quarter century, there has been a substantial amount of research carried out in this field of study. There have been

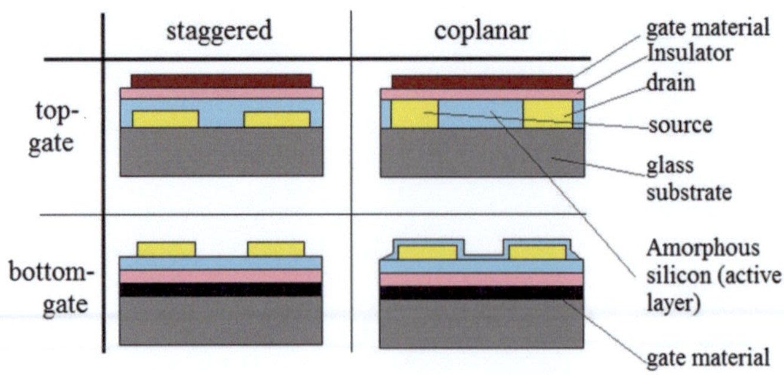

Fig. 17.1 TFT structure configurations [4]

new prospective materials found and tested. Also, there have been new techniques of fabrication experimented. In the early 1980s, the manufacture of liquid crystal display (LCD) was done using silicon in amorphous form. The advent of the flexible displays began in the 1990s. Flexible polyimide substrates were used to build the circuits containing TFT by Constant et al. in 1994. Since this work, many research organisation and companies have demonstrated the use of flexible substrates using various materials like a-Si:H (inorganic), zinc oxide, organic materials, and hybrid organic/inorganic materials [1].

TFTs have been fabricated using hydrogenated amorphous silicon (a-Si:H) as a main semiconductor material in the manufacture of displays. This is because it was found that the hydrogen atoms were helpful in rendering a lot of defects and dangling bonds as passive. This was mainly used for the manufacture of flat panel displays in the industry. The photo current produced in a semiconductor under light is of significance in the applications involving light such as photovoltaic devices. The light absorption coefficient in the domain of visible wavelength is relatively large for this material. This allows a massive recombination of electrons and holes in the device resulting in photo signals [5–8]. The a-SiGe:H, an alloy of this material, has a unique property of lowering the energy band gap, which enhances the spectral response in terms of photo currents in the range of the IR wavelengths [9, 10]. In a related work, the effects of photo illumination were analysed. This was done using TFTs based on a-SiGe:H when they were used as photo sensors. The generation rate of photo current was evaluated on the basis of the direction of photon incidence in the TFT structure. From the results, a highly sensitive a-SiGe:H TFT was implemented and was integrated to a touch-photo sensor in a display panel application [11].

When we consider the inorganic materials, the hydrogenated amorphous silicon is the forerunner in the industry. TFT which are manufactured by using this material are the chief ingredients in the active matrix in the LCD display panels today. These LCD panels are built using TFTs which have been fabricated at higher temperatures. However, most flexible substrates are processed at low temperatures. Therefore, this incompatibility has driven many companies and research groups to progress in the use of a-Si:H TFT backplanes in flexible displays. Although there is widespread use of a-Si:H in the industry, there are many devices in the market which have been substituted by different materials. For instance, organic LEDs (OLEDs) use organic materials instead of a-Si:H. The main drawback of using a-Si:H in TFT is that the TFT experiences a threshold voltage shift (Vt). Electrical stress is created due to continuous conduction of majority charge carriers in the active layer. As a result of this, the threshold voltage of the TFTs increases with time, and hence the current drive of it is reduced. In order to reduce the threshold, Vts, the processing temperature has to be increased. Hence, there is a need to find out suitable substrates which can withstand such high temperatures.

Polysilicon TFTs are fabricated at relatively higher temperatures using crystallisation of a-Si:H material. The crystalline structure gives rise to increased mobility. The threshold voltages of TFTs fabricated using polysilicon as active material are found to be stable.

TFT fabricated using amorphous In-Ga-Zn-O (α-IGZO) as active material has proven to be suitable to be fabricated at low temperatures as a substitute for silicon-based TFTs. Also, α-IGZO TFTs are transparent and show superior electrical performance. These can also be fabricated on plastic substrates which makes them a potential candidate for flexible electronic applications [11]. α-IGZO TFTs have already been used for flat panel displays [12, 13], The $I_{ds} - V_{gs}$ characteristics are the basis for the analysis of use of any active material in a TFT. When the gate bias is applied to the TFTs which use α-IGZO as active material, the characteristic shifts in the positive direction and the sub-threshold slope becomes steep. When light is irradiated onto the TFT, the characteristic shifts in the negative direction and the sub-threshold slope becomes lesser. This is due to the fact that the oxygen in the compound is responsible for generating flat trap states in the energy band gap [12].

The development of fabrication processes that use polymers for surface micromachining has increased the flexibility and capabilities for MEMS, especially in the biomedical and microfluidic fields. However, one major drawback to typical polymer MEMS processes is the lack of electrical signal conditioning or active sensing capabilities. Polymer MEMS devices currently rely on external circuits for signal conditioning during actuation or sensing, and typical processes can only fabricate passive structures without any transduction capabilities. Although embedded organic TFT circuits in polymer MEMS have been demonstrated [13], the fabrication technique requires micro stereo lithography which does not scale well for mass production when compared to the well-established amorphous silicon TFT technology [14] currently used for flat panel displays.

TFTs can also be used as sensors. In a recent work, a combined pH and impedance system was developed. There were related works which involved the development of such systems using microcontrollers. However, these systems required a lot of power. The use of TFTs served as an alternative [15]. This work involves the use iridium oxide (IrO_x)-thin films as an active material on a flexible substrate. Silver was used for metallic contacts [16].

In most cases where the organic materials are used for fabricating the TFTs, enhancement p-type organic devices were fabricated. The devices are operated in accumulation mode. Therefore, they have a positive "threshold voltage". An issue concerning the widespread use of organic transistors is the amount of leakage current that flows in the "off" state when a drain-to-source voltage exists. Rogers et al.'s [17] work showed that "the bottom-contact devices have low 'off' currents and better overall characteristics than the top contact devices". According to this observation, a fact in support of this is the use of pentacene in fabricating transistors using organic materials consistently results in stable devices with high on/off current ratio, these being bottom-contact types. After deposition, the substrate is allowed to cool slowly. This is called annealing. It is another great technique used to minimise the "off" currents in such devices. There has been a lot of work done to ascertain the factors affecting the conductivity in transistors fabricated using organic materials. In one such work, Vissenberg et al. [18] developed a field-effect mobility model to explain the conductivity in transistors fabricated using amorphous organic materials. They showed that the electrical conductivity in such devices depends on the molecular ordering in the morphology of the active layer.

Gas sensors were fabricated with metal-oxide-semiconducting (MOSs) materials. These devices find their applications in detecting gas leakages in industry and home. Gas sensors have been developed to detect ethanol, liquefied petroleum gas (LPG), carbon monoxide, etc. [19].

17.3 Method of Implementation

SILVACO is a private company well known as a provider of electronic design automation (EDA) software. It provides software for TCAD simulations. The EDA software licensed by this company is known to be a great tool for process and device simulation. It provides from semiconductor fabrication process simulation to design automation solutions for various fabrication technologies like CMOS and bipolar. The software when purchased fully as a suite becomes a virtual wafer fabrication tool. It includes various tools like ATLAS, TONYPLOT, TONYPLOT-3D, DECKBUILD, and DEVEDIT [20].

Athena is a tool provided by SILVACO which is used for two-dimensional simulation of a semiconductor device. It can also be used for numerical analysis. This tool has a modular structure of operation. This tool is special in a way that it provides the user to experience, through simulation, basic deposition and etching steps. It also provides a coding solution to design a basic structure of a semiconductor device. The doping concentration, the thickness of each region, and other critical attributes can be specified as desired. This software can be used to simulate silicon processing steps. This includes ion implantation, diffusion of dopants into the bulk material, and oxidation to observe the changes and manipulations in the device's physical parameters. This software supports the simulation of a wide range of techniques such as deposition, etch and reflow which are used in modern IC technologies. This tool includes the lithography simulation with advanced features like 2D aerial imaging, non-planar photoresist exposure, and post exposure bake and development. One can use this software for optimisation and analysis of different semiconducting structures.

While Athena is more concerned with the device construction and analysis, ATLAS is a physically based device simulator which is used to analyse the electrical characteristic of the device. This tool is so user-friendly and easy to use that the bias conditions and other physical and electrical constraints can be well specified by the user itself. This process is achieved by approximation. The operation of a device under consideration is mapped onto a two- or three-dimensional grid, with a number of intersections called grid points. These grid points form nodes and loops of an electrical circuit. The current and voltage equations are solved by continuous iterative procedure. A set of Maxwell's laws and differential equations is applied to the grid to approximately mimic the transport of charge carriers. Therefore, we can model the structure according to the DC, AC, or transient modes of operation without having to set any extra configurations.

Since the physical models are being simulated, they can be used to save the time and cost overhead which otherwise would be invested to perform the experiments. Moreover, there might be some attributes which would be tough to measure. For instance, currents in the order of nanoamperes would be difficult to measure with all the other second-order effects influencing it. The specifications are as follows:

- Structure specification: MESH, REGION, ELECTRODE, DOPING
- Numerical method specification: METHOD
- Solution specification: LOG, SOLVE
- Results specification: EXTRACT, TONYPLOT, OVERLAY
- Material model specification: MATERIAL, MODEL
- The following figure gives a brief overview of ATLAS input and output files.
- The ATLAS tool takes inputs from Athena (works in DeckBuild runtime environment) which creates structure files or from DevEdit which creates command files. The structure file can be input to TonyPlot and TonyPlot 3D for visualisation.
- ATLAS is called in the DeckBuild runtime environment to perform *device simulation* on the structure created by Athena/DevEdit.
- It provides runtime output (usually the parameters which are calculated using *Extract* command) and after simulation in the form of log files and solution files.
- The values stored in log files and solution files can be used for visualisation using TonyPlot tool.

The different approaches of the process of simulation are as given in Fig. 3.1. The approach adopted in this project is to:

1. Design a structure of TFT using DevEdit/Deckbuild.
2. Simulate the saved structure with the defined active layer in ATLAS device simulator.
3. View results of simulation in Tonyplot editor (Fig. 17.2).

The basic structure of a TFT was developed on the DevEdit tool as shown in Fig. 17.3. The same was saved as a structure to be invoked by the other tools later. The structure had the required active layer, source, gate and drain regions and their respective contacts, all in appropriate dimensions for the design.

The structure thus created is invoked in the ATLAS simulator. The required meshing is done. The active layer so defined is simulated for I_d v/s V_{gs} curve for a pre-defined range of values of V_{gs}. The active layer of the TFT is changed in the TFT structure in DevEdit, invoked in the ATLAS simulator and simulated. The output curves are observed and compared for different active materials [20].

Compounds that lack carbon and hydrogen atoms are organic compounds and are found in biological systems. Traditionally, inorganic compounds are viewed as being synthesised by the agency of geological systems. The materials used as active layers in the simulation are as follows:

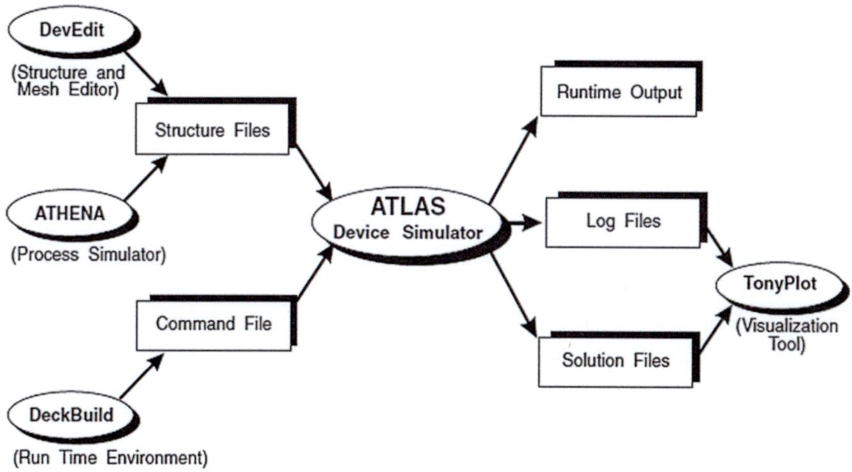

Fig. 17.2 ATLAS input and output environments [20]

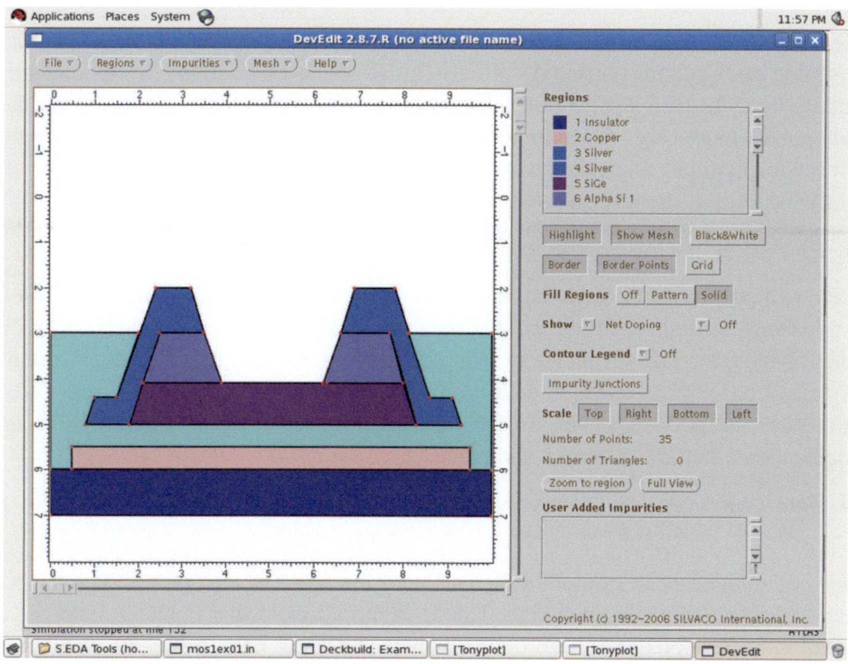

Fig. 17.3 Structure of TFT in DevEdit

1. SiGe

SiGe has an adjustable band gap. This makes it more versatile in terms of applications. This is also used for the manufacture of heterojunction bipolar transistors due to its properties. Such transistors have relatively higher forward gain than the conventional transistors.

2. Indium phosphide

Indium phosphide is a binary compound. It is a semiconducting material. Most of the group's III–V elements in the periodic table have a face-centred cubic crystal structure. This facilitates high carrier mobility. Therefore, this material is usually used in transistors which cater to high-frequency applications. This material has a direct band gap which is suitable for optoelectronic applications.

3. Mercury telluride (HgTe)

Mercury telluride is a binary compound of mercury and tellurium. This is not totally a semiconducting material. However, it is categorised under the II–VI category of semiconductor materials. HgTe shows metal-like qualities and properties.

4. Gallium arsenide phosphide ($GaAs_{1-x}P_x$)

Gallium arsenide phosphide is a semiconductor material, an alloy of gallium arsenide and gallium phosphide. This is an example of a ternary compound primarily used for LED applications. The proportion of the individual elements in the compound may vary, depending upon the resultant band gap required for light emission. Gallium arsenide phosphide is used in the manufacture of LEDs for emission of red, orange, and yellow light.

5. PPV (Poly phenylene vinylene)

PPV is a polymer which can be processed into a crystalline thin film. This is used exclusively for light-emitting and photovoltaic applications in the industry. It has a small band gap. The advantage of using this material is that its physical as well as chemical properties can be altered easily by introducing and replacing functional groups in the polymer chain. So far, this has not been used for chemical sensing applications, but is a potential candidate.

6. Pentacene

Among organic semiconductor materials, pentacene has been found to be having one of the highest carrier mobilities. This is by the virtue of its high packing density as compared to the other materials. Due to high intermolecular bonding, there are a lot of cohesive forces between the molecules. Different heat treatments were used based on thin-film studies and sought to optimise pentacene TFT performance on printed electronic substrates. Pentacene is a polycyclic aromatic hydrocarbon. It has five linearly fused benzene rings in its molecular structure. All the interatomic bonds in this compound are conjugate in nature. This compound is found to be an organic semiconductor. Since it has five benzene rings in its molecular structure, it is very rich in electrons. Pentacene is a linear acene in the family of acenes [21].

17.4 Results and Evaluation

The active layer is of primary concern for us. It should be so chosen that it provides a change in its electrical properties when it comes in contact with ions. The different parameters which decide the appropriate active layer are shown in Table 17.1.

Figures 17.4, 17.5, 17.6, 17.7, 17.8 and 17.9 show the simulation outputs corresponding to different materials.

The TFTs characteristics simulated with SiGe as active layer is shown in Fig. 17.4. This shows a typical transistor curve for the $I_d - V_{gs}$ characteristics.

The TFT characteristics simulated using indium phosphide as active layer as shown in Fig. 17.5 show a skewed curve for the $I_d - V_{gs}$ characteristics. This characteristic is not suitable as there are sudden variations in current with respect to specific voltages.

The TFT characteristics simulated using mercury telluride as active layer as shown in Fig. 17.6. Since this active material typically shows metallic properties, the current through the material reaches its maximum value and conducts in the same manner irrespective of the voltage applied.

The TFT characteristics are simulated using gallium arsenic phosphide as active layer as shown in Fig. 17.7. These $I_d - V_{gs}$ characteristics are similar to that of indium phosphide.

The TFT characteristics are simulated using PPV as active layer as shown in Fig. 17.8. The $I_d - V_{gs}$ characteristics are appreciable. However, the skewed nature of the characteristic might cause problems at the time of sensing.

Table 17.1 Electrical parameters of active layers

Eg	Energy band gap at room temperature
Nc	No. of majority carriers in conduction band per cc of volume of material at room temperature
Nv	No. of majority carriers in valence band per cc of volume of material at room temperature
Mup	Mobility of charge carriers in the p-region of transistor
Mun	Mobility of charge carriers in the n-region of transistor
NTA and NTD	Conduction and valence band edge intercept densities
WTA and WTD	Characteristic decay energy
EGA and EGD	Peak energy/peak distribution
SIGTAE and SIGGAE	Electron capture cross section for the acceptor tail and corresponding Gaussian states, respectively
SIGTAH and SIGGAH	Hole capture cross sections for the acceptor tail and corresponding Gaussian states, respectively

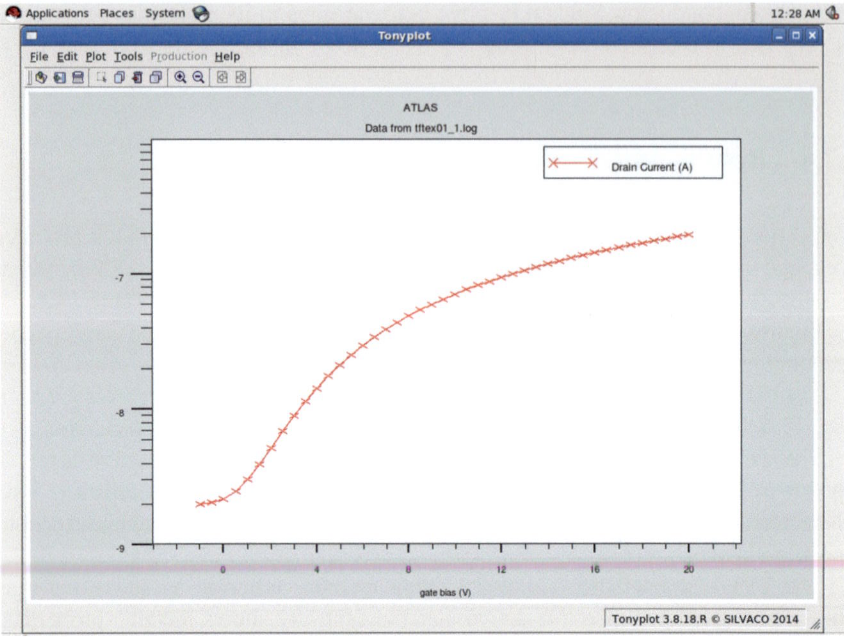

Fig. 17.4 ATLAS simulation output with SiGe as active layer

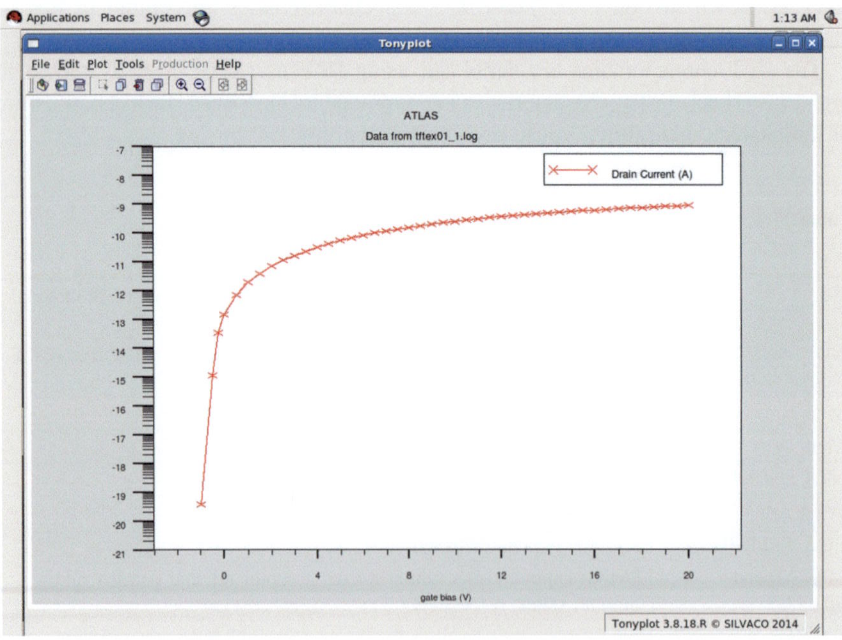

Fig. 17.5 ATLAS simulation output with InP as active layer

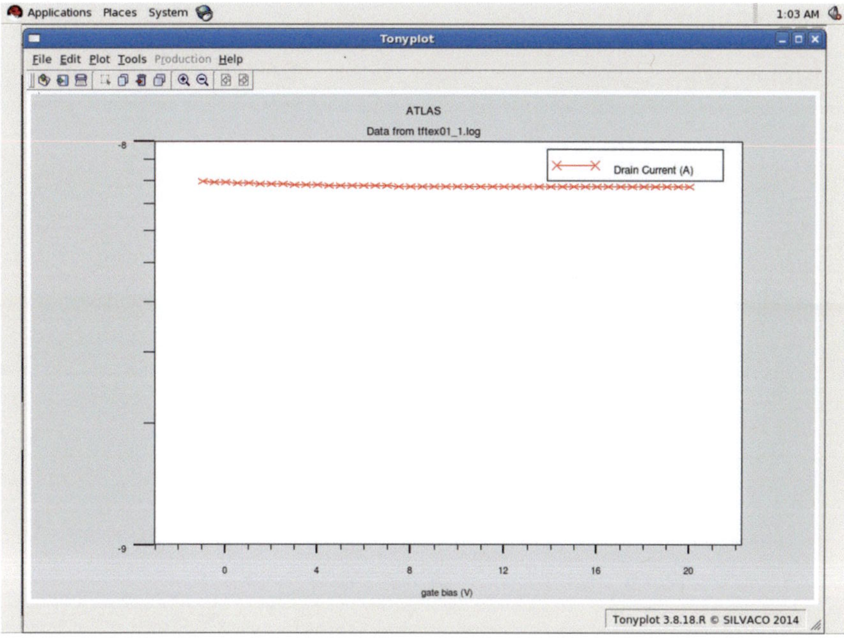

Fig. 17.6 ATLAS simulation output with HgTe as active layer

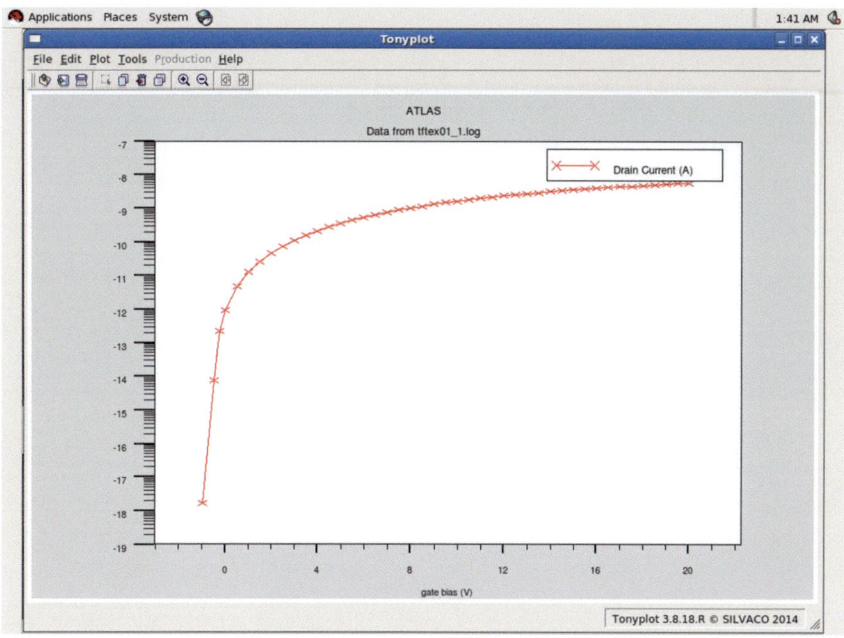

Fig. 17.7 ATLAS simulation output with GaAsP as active layer

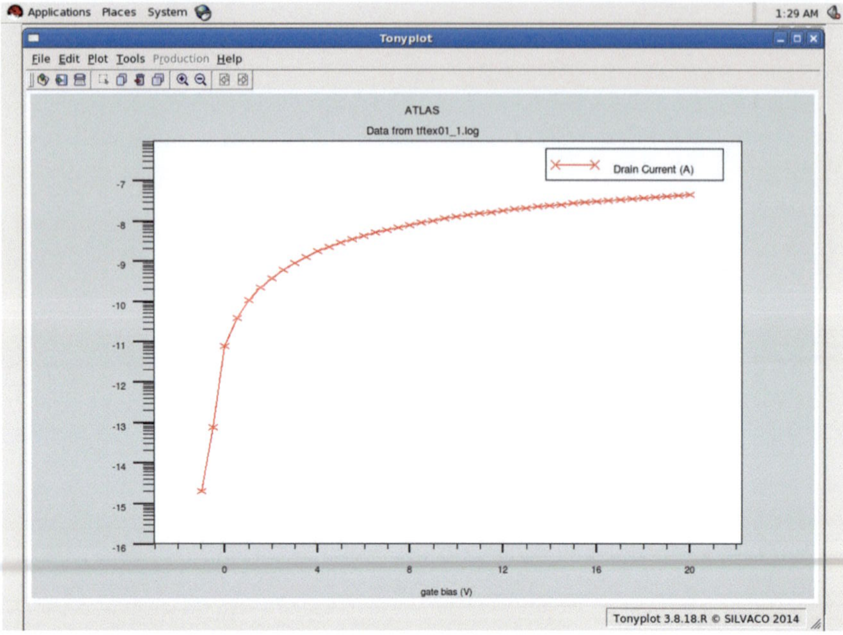

Fig. 17.8 ATLAS simulation output with PPV as active layer

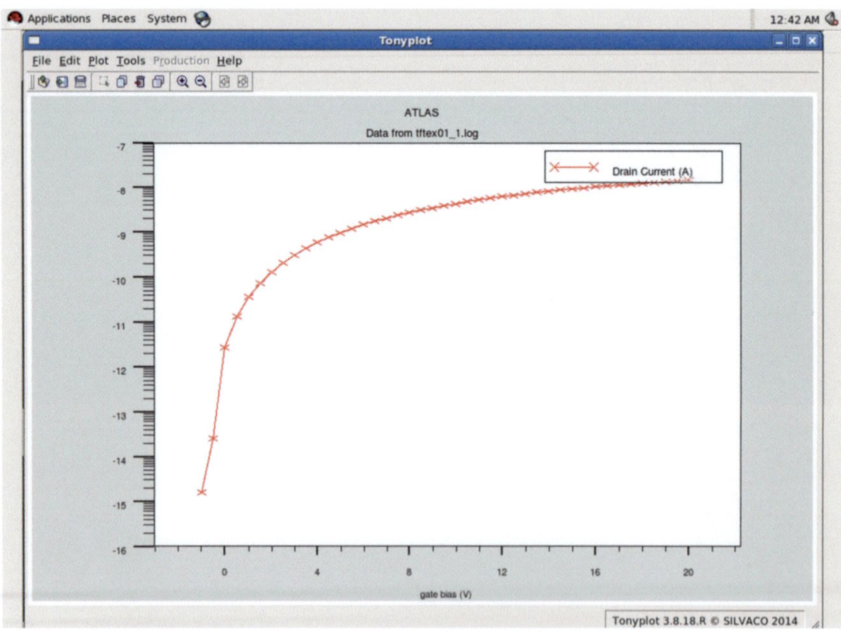

Fig. 17.9 ATLAS simulation output with pentacene as active layer

The TFT characteristics simulated using pentacene as active layer as shown in Fig. 17.9. For the voltage range of 0–8 V, this material shows a good linear response.

The chosen active layer must have a very high range of I_d for a given range of V_{gs} (0–20 V in this case). Moreover, the curve should be linear enough so that the sensitivity can be calibrated easily for corresponding values of V_{gs}.

17.5 Conclusion and Future Scope

It is observed that pentacene (organic) and SiGe (inorganic), as active layers, are a suitable choice for our applications. Pentacene shows good response for commonly used input voltage range in most of the applications. The corresponding graphs are linear for low applied voltages which increase their ability to select and sense the presence of ions. They also operate at low values of I_d which reduces power consumption. This behaviour of pentacene and SiGe make them a practical candidate for hand-held real-time systems where power consumption is a major constraint. Moreover, these materials are also available in amorphous form which makes it suitable for TFT fabrication processes.

The electrical parameters of pentacene and SiGe, like energy band gap, mobilities, doping concentration and electron, and hole trap parameters have admissible values for good sensing of the required metal ions. It can be concluded that since fabrication methods of pentacene and SiGe are well established, a single or an array of TFTs on flexible substrates with pentacene or SiGe as an active layer can be made to sense the ions in different media, that is, gaseous, liquid, or suspensions.

References

1. Venugopal SM, Allee DR (2010) Flexible electronics: what can it do? What should it do? In: International Reliability Physics Symposium (IRPS)
2. Wong WS, Salleo A (2009) Flexible electronics: materials and applications. Springer Publications, ISBN 978-0-387-74362-2
3. Kane MG, Campi J, Hammond MS, Cuomo FP, Greening B, Sheraw CD, Nichols JA, Gundlach DJ, Huang JR, Kuo CC, Jia L, Klauk H, Jackson TN (2000) Analog and digital circuits using organic thin-film transistors on polyester substrates. In: IEEE electron device letters, vol 21, No. 11
4. Shah A (2010) Thin film silicon solar cells, 1st edn. EPFL Press, ISBN 978-2-940222-36-0s
5. Rosan K (1989) Hydrogenated amorphous-silicon image sensors. IEEE Trans Electron Devices 36(12):2923–2927
6. Yamaguchi M, Kaneko Y, Tsutsui K (1993) Two-dimensional contact type image sensor using amorphous silicon photo-transistor. Jpn J Appl Phys 32(1B):458–461
7. Kaneko Y, Koike N, Tsutsui K, Tsukada T (1990) Amorphous silicon phototransistors. Appl Phys Letts 650(7):650–652
8. Stearns RG, Weisfield RL (1992) Two-dimensional amorphous-silicon photoconductor array for optical imaging. Appl Opt 31(32):6874–6881

9. Michel J, Liu J, Kimerling LC (2010) High-performance Ge-on-Si photodetectors. Nat Photon 4(8):527–534
10. Liu J, Cannon DD, Wada K, Ishikawa Y, Jongthammanurak S, Danielson DT, Michel J, Kimerling LC (2005) Tensile strained Ge p-i-n photodetectors on Si platform for C and L band telecommunications. Appl Phys Lett 87(1):011110-1–011110-3
11. Hack M, Lewis AG, Bruce RL, Lujan R (1991) Optically addressable input circuit for two-dimensional image sensing. In: Proceedings of Materials Research Society Symposium, pp 167–172
12. Nomura K, Ohta H, Takagi A, Kamiya T, Hirano M, Hosono H (2004) Room-temperature fabrication of transparent flexible thin-film transistors using morphous oxide semiconductors. Nature 432(7016):488–492
13. Varadan K (2003) Three dimensional polymer MEMS with functionalized carbon nanotubes and modified organic electronics, presented at Nanotechnology, 2003. In: 2003 Third IEEE Conference on IEEE-NANO 2003
14. Ativanichayaphong T, Tang S-J, Hsu L-C, Huang W-D, Seo Y-S, Tibbals HF, Spechler S, Chiao J-C (2010) An implantable batteryless wireless impedance sensor for gastroesophageal reflux diagnosis, presented at the International Microwave Symposium, Anaheim, CA, 23–28 May 2010
15. Shay S (2004) Esophageal impedance monitoring: the ups and downs of a new test. Am J Gastroenterol 99(6):1020–1022
16. Caboni A, Orgiu E, Barbaro M (2009) Flexible organic thin-film transistors for pH monitoring. IEEE Sens J 9(12)
17. Rogers JA, Bao Z, Dodabalapur A, Makhija A (2000) Organic smart pixels and complementary inverter circuits formed on plastic substrates by casting and rubber stamping. IEEE Electron Devices Lett 21(3):100–103
18. Vissenberg MCJM, Matters M (1998) Theory of the field-effect mobility in amorphous organic transistors. Phys Rev B 57(20):12967–12968
19. Choopun S, Hongsith N, Wongrat E Metal-oxide nanowires for gas sensors. URL: http://dx.doi.org/10.5772/54385
20. User manuals of SILVACO suite
21. URL: http://en.wikipedia.org

Chapter 18
Storage Optimization of Automated Storage and Retrieval Systems Using Breadth-First Search Algorithm

Amiya Sagar Das, Prashant Kumar Dwivedi, Amit Kumar Mondal, Roushan Kumar, R. Manohar Reddy and Adesh Kumar

18.1 Introduction

Automated storage and retrieval systems (ASRS) have been in practice in the industries since 1950s, particularly for warehouse management. Before the introduction of ASRS to the industry, the inventory control and management were done manually which led to accidents, slow execution time, and high labor cost. It was introduced to tackle the above-mentioned problem of storing and retrieving raw materials and saving the transportation time for the employees, thereby improving the efficiency of the entire system.

ASRS offers the benefits of enhanced inventory control and economical utilization of time, place, and equipment [1–3].

In recent times, the ASRS has been implemented in state-of-the-art applications like automated car parking system, automated library management system, and automated luggage storage system. The major parts which are constituted in an ASRS are the racks, stacker cranes, input/output points (I/O's), and aisles.

A.S. Das (✉) · P.K. Dwivedi · A.K. Mondal · R. Kumar · R.M. Reddy · A. Kumar
University of Petroleum and Energy Studies, Dehradun, Uttarakhand, India
e-mail: ron.amiya@gmail.com

P.K. Dwivedi
e-mail: dwivedi.kr.prashant@gmail.com

A.K. Mondal
e-mail: akmondal@ddn.upes.ac.in

R. Kumar
e-mail: rkumar@ddn.upes.ac.in

R.M. Reddy
e-mail: rmreddy@ddn.upes.ac.in

A. Kumar
e-mail: adeshkumar@ddn.upes.ac.in

V. Nath (ed.), *Proceedings of the International Conference on Nano-electronics, Circuits & Communication Systems*, Lecture Notes in Electrical Engineering 403,
DOI 10.1007/978-981-10-2999-8_18

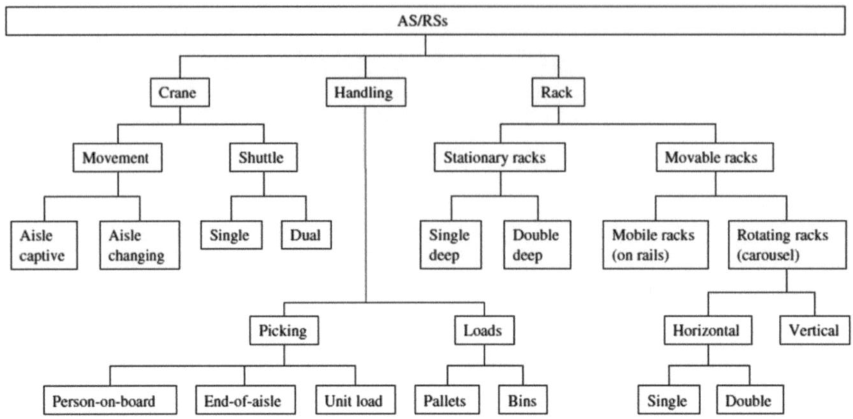

Fig. 18.1 Types of ASRS

The racks are majorly made up of steel or aluminum to store the material or load. The stacker cranes are automated machines which pick the material from the input/output point (I/O) and place the load or material in the storage cell. The vacant places among the racks where the stacker cranes can move are known as *aisles*. An I/O post is a place where incoming materials are picked up for storage and retrieved materials are dropped off. Pick points (if any) are places where single articles are removed from a retrieved load beforehand the load is sent back into the system by the human workers [4–6].

There are many types of ASRS that are used in the industry. The type depends upon the purpose. The major types are shown in Fig. 18.1. In this paper, *single-deep stationary rack* type of ASRS has been considered for analysis. The "single deep" term means that each rack has the storage capacity of only one load to be stored at a time. There are a few assumptions which were taken into consideration, they were as follows:

(1) All racks are empty initially, i.e., when the system is used for the very first time there is no load in the racks.
(2) All racks are equal in dimensions and are equidistance from each other, which reduces the complexity of the system.
(3) Interleaving is ignored. All the sequences are considered to start from the I/O and end at I/O. If a storage sequence is in run and a retrieval request is sent, then the system completes the storage sequence first, returns to the I/O, and then acknowledges its retrieval request.
(4) The system is independent of the retrieval time or the period of stay of the material in the rack.

This paper is divided into two major parts: firstly, the ASRS in which the algorithm is to be implemented is described along with the concept of virtual sensing. After this, the implementation of the algorithm in the system is elucidated.

18.2 Storage Assignment

There are many storage assignments which are being successfully implemented and used in ASRS. The majorly used storage assignment policies are as follows [7–9]:

- Dedicated storage assignment.
- Random storage assignment.
- Closet open location assignment.
- Full turnover-based storage assignment.
- Class-based storage assignment.

In Table 18.1, a brief description of all the storage assignments has been given [4]. In this paper, the storage assignment which is taken into consideration is random storage assignment.

18.3 Random Storage Assignment

In this storage assignment, the material is equally likely to be stored in any of the rack. In actual practice, it is often approximated with open closet location assignment. The open closet location rule states that, just before the storing of any material in the rack is initiated, the system scans for the list of all the empty racks and finds the closest one to the I/O. Then, the material is stored in the closest rack regardless of the turnover or priority [7]. It is to be noticed that according to this rule if a retrieval command and a storage command is done alternatively, then the storing will occur at thc samc location again and again. It should also to be kept in mind that any material stored is independent of the location it has been stored before. It is also independent of the material that has this rule if a retrieval command and a storage command is done alternatively, then the storing will occur at the same location again and again.. It is also independent of the material that has been stored in that same location. In real time, there are many applications to these types of the

Table 18.1 Types of storage assignments and their description

Type of storage assignment	Description
Dedicated storage assignment	Each product category is allotted to a fixed location. Every material has a specified storage rack
Random storage assignment	Every empty position in rack has equal probability of getting stored for an incoming material
Closet open location assignment	The first empty slot encountered is filled first
Full turnover-based storage assignment	Storage locations for loads are determined based on their demand frequency
Class-based storage assignment	Stores divide the materials on the basis of use and priority

assignments. One of the major applications that have been implemented till now is the automatic car parking system.

18.4 Virtual Sensor

Before going to the implementation of algorithm which will search the nearest empty rack and guide the stacker crane to store the material, it is important to find out the status (empty/filled) of the rack. Usually, array of sensors or limit switches are used to know the status of the rack [10]. The array of sensors and limit switches make it disordered and is less flexible, i.e., if the number of racks are increased, then it will need more sensors to get the status of the rack. In some applications like image processing, a sensor at the end effecter of the robot is attached to get the status [11] which makes them a costly affair. To solve such problems, there is a solution called as a virtual sensor.

The virtual sensors are not actual sensors but the status of the racks is stored in the memory of the controller. The concept is shown in Fig. 18.2. In the figure, "0" signifies that the rack is empty and vice versa. All the racks are marked empty initially. The basic concept behind it is that when the stacker crane visits a specific rack with a specific coordinate; supposedly (1, 1) whose status is "0" for storage, the status of that coordinate is changed and marked as "1" which means that the rack is visited once. The controller is associated with a counter which makes the count. When the stacker crane again visits the coordinate (1, 1) for retrieval, the controller resets the counter to "0," signifying that it is empty. While checking for status of the rack if the status is found to be "1," then the pointer moves to next neighbor to check the status if it is empty or filled and so on. It is to be noted that the change in status only takes place after the storage and retrieval sequence is completed and after the carne reaches back to the I/O. This technique can be used in any microcontroller, processor, or programmable logic controller (PLC) [12]. It is also important that the actuators for the stacker crane should be chosen wisely, i.e., the steps of the crane can be controlled or it should have feedback which will notify the controller about its position.

Fig. 18.2 Example of a rack with various status

(3, 1) 0	(3, 2) 1	(3, 3) 1
(2, 1) 1	(2, 2) 0	(2, 3) 0
(1, 1) 0	(1, 2) 1	(1, 3) 0

18.5 Breadth-First Search (BFS)

It is one of the basic and simple un-informed search techniques in graph theory. Breadth-first search (BFS) expatiates on all the states, which are at one step away from the beginning state, and then expatiates all states which are two steps from the beginning state, then three steps, etc., until a goal node is reached [13–15]. As it always expatiates on all nodes at a known depth beforehand expatiating any node at a larger depth, the first solution route found by BFS will be the one of the smallest spans. At certain times, BFS expatiates on all the nodes to a certain given depth to find the goal node. In general, the algorithm can be explained with reference to Fig. 18.3 as follows:

1. Form a queue with the root node at the initial position (a), check whether the root node is the goal node.
2. If the root node or the first element is the goal node, then do nothing and come out of the loop.
3. If the root node is not a goal node, then remove the root node from the queue, and if there are any children nodes then put them in the queue.
4. If the goal node is found, return the node and announce success, if not then return failure.

The general algorithm can be given as in Fig. 18.3:

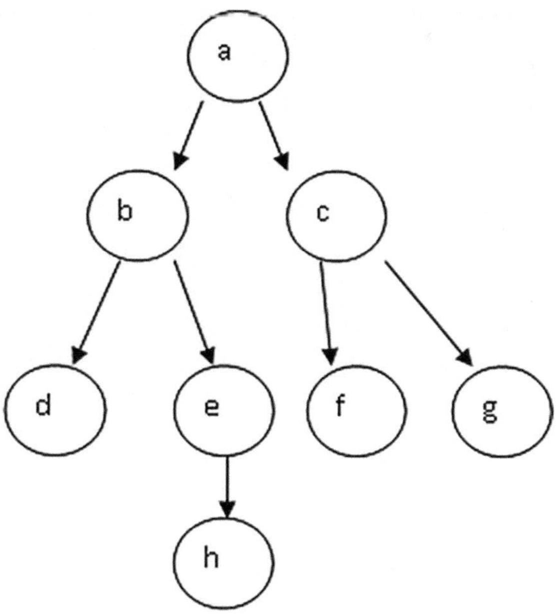

Fig. 18.3 Tree which denotes the state space to explain the BFS algorithm

18.6 Implementation of BFS in ASRS

The main reason to implement BFS in ASRS is to make it autonomous, so that it can make decision on its own to decide on the location of storage of the material. The proposed algorithm in this paper has been implemented using PLC and SCADA. The software that is used for the purpose of simulation is *Indraworks Engineering V7* which is actually based on the *Codesys v2.3*. The visualization tool in the software serves as SCADA which is helpful in monitoring the algorithm in progress.

There are many searching algorithms which have been implemented in association with ASRS which proposes to associate genetic algorithm (GA) with the system. It provides the intelligence to remember the pattern and store the object in the empty racks [16]. Although GA has its own advantages, BFS can also be implemented. The BFS searches and provides the nearest empty rack to the controller which then with the help of a crane stores the material. The BFS algorithm in Algorithm 1 can be modified and implemented to get the desired output from ASRS. The modified algorithm is given as Algorithm 2.

Algorithm 1: Generalized BFS

```
Algorithm BFS (graph, start, goal) // initialize the necessary
parameters
  {
       Queue (graph, start)
        While (queue (graph) != Empty) // get into the loop
            {
            Node= dequeue (graph)
              Print "node"
              If (node == goal) break  // check if the node
is goal node if true then break
              For ( I ← (n-1) to 0 ) do
              {
                If (edge (node,i))
                Enqueue (graph, i)
              }
            }
            Return 0
  }
```

The Algorithm 2 can be best explained through an example. Let us consider a case where the total number of racks is 16 and the matrix is an order of 4×4 as shown in Fig. 18.4. The I/O point is the starting position of the crane and as per the algorithm, and the controller will start looking for empty racks from the coordinate (1, 1). If the rack is empty, the searching will be terminated and the coordinates will be returned to the controller specifying the rack (1, 1) is empty. If the position is full, it will start searching the neighbor racks starting from (1, 2); if it is empty, then it will return the coordinates or else it will move to the next neighbor i.e. (2, 2); if this is also not empty, then it will move to the next neighbor, i.e. (2, 1). It will check for its vacancy and will act accordingly. After all the neighbors of (1, 1) are stored, the start node moves to (1, 2). As the (1, 2) is already filled, it will start checking its neighbors, i.e. (1, 3), then (2, 3), then (2, 2), and it goes on. It switches to next row when the boundary condition is reached.

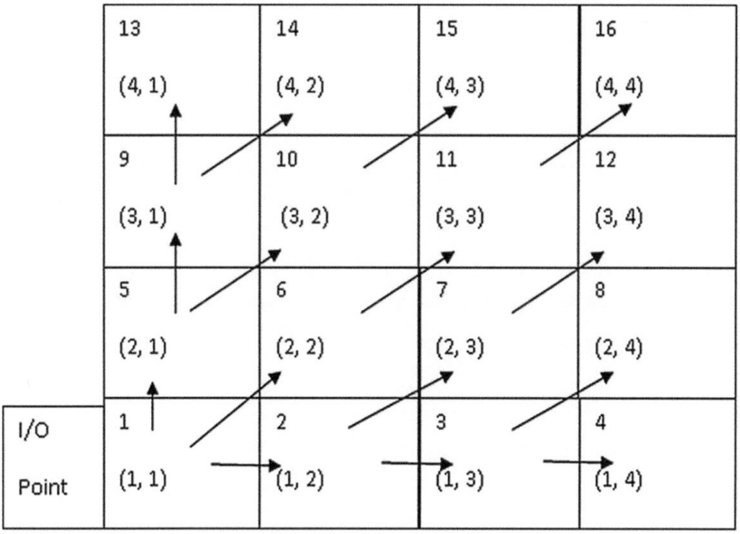

Fig. 18.4 Rack arrangement in order of 4 × 4 matrix

As discussed before, the initial condition of all the racks is to be considered empty. If suppose there is no retrieval request till all the racks are occupied, then the path in which the crane stores the material can be known by the path tree. For the above-mentioned example with 16 racks in total, the path tree can be derived as shown in Fig. 18.5. The tree also denotes the state space of the algorithm.

Algorithm 2: Modified BFS for storage optimization in random assignment of ASRS

```
If ( A [i , j] == 0 ) Then
 Return [ i , j];
          //check for the neighbor position in
same row and next column
     Else If ( A[i,j+1] == 0 )
          Return [i,j+1];
               //check for the neighbor position
in next row and same column
          Else If ( status [i+1,j] == 0 )
               Return [i+1,j];
          //check for the diagonal neighbor
       Else If ( A [i+1,j+1] == 0 )
          Return [i+1,j+1];
     End if
  End for
 End for
End
```

In [17] the author has compared the evolution program (GA) and simple heuristic searching techniques like lower tier first heuristic and adjacent vacant cell heuristic and has concluded that, although the GA has out run in many cases to the

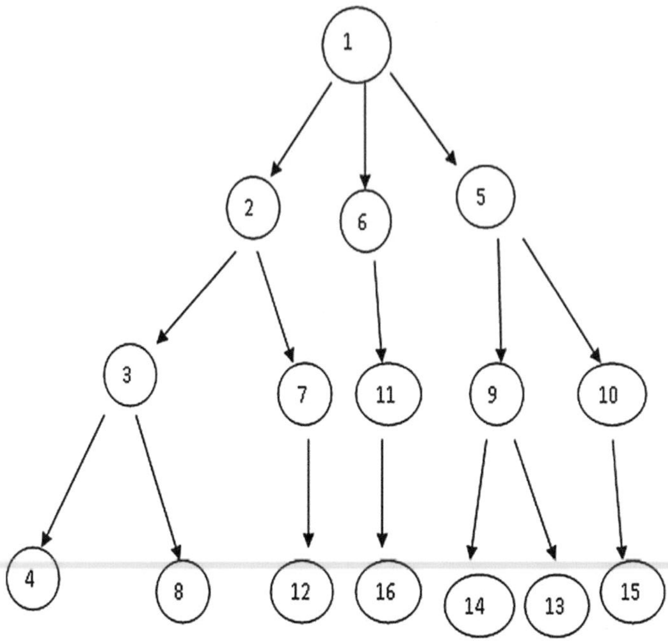

Fig. 18.5 Path of the storage of material according to algorithm

heuristic searching techniques but in case of adjacent vacant cell heuristics considering the latest vacant position gives results which are close to the evolution program. The paper also concludes that the evolution program takes more space in the memory of the controller, the reason being of its complexity. In contrast with it, the heuristic searching techniques are simpler in approach and utilize less space for searching. The BFS algorithm which is proposed in this paper takes care of both the adjacent vacant cell as well as lower tier first heuristic which is the gravity-centered priority too together, which makes its execution time less and faster.

A. *Time Complexity of the Algorithm*

The complexity of the algorithm decides the speed of its execution and scan time. The complexity of the algorithm is calculated through the number of iteration it does. The time complexity of the proposed algorithm, i.e. Algorithm 2 is the total time required to find the empty slot and place object in the empty slot.

Let us consider a time complexity notation for the system as $T(r, c)$, where the number of rows is r and columns is c. So the expression for it can be given as:

$$T(r, c) = t1 + t2 \tag{18.1}$$

where $t1$ is the time taken for the identification of the position and $t2$ is time taken to place the object in the position.

Equation (18.1) can also be written as:

$$\sum_{i=1}^{r}\sum_{j=1}^{c}1+\sum_{i=1}^{r}\sum_{j=1}^{c}1 \tag{18.2}$$

Solving Eq. (18.2) further, we get

$$2\left(\sum_{i=1}^{r}\sum_{j=1}^{c}1\right) \tag{18.3}$$

$$2\left(\sum_{i=1}^{r}(c-1+1)\right) \tag{18.4}$$

$$2c\left(\sum_{i=1}^{r}1\right) \tag{18.5}$$

By further solving Eq. (18.5), we get

$$2c(r-1+1) \tag{18.6}$$

Solving Eq. (18.6) finally, we get

$$2cr \tag{18.7}$$

The constant term can be excluded and the final time complexity notation can be expressed as:

$$O(c\cdot r) \tag{18.8}$$

18.7 Conclusion

As discussed, BFS searches the nearest neighbors and the returns the nearest empty rack from the I/O point available, thus making the storage faster and optimized. The BFS being a blind searching technique takes longer time to search, but for application in ASRS which is associated with faster processors can reduce the execution time. There are other heuristic searching algorithms which can also be implemented in storage optimization of ASRS. The concept of virtual sensors is very effective and can be implemented according to the need. It is very flexible and can be associated with any searching algorithm for searching the empty racks for ASRS. It can make the system cheaper and simple.

Acknowledgments The research paper is made possible through the inputs of Mr. Venkateshwaran PS, Senior Research scientist of UPES, and Mr. G.L. Prakash, assistant professor, CIT, UPES, Dehradun, for their support and assistance in revising the paper.

References

1. Hur S, Lee YH, Lim SY, Lee MH (2004) A performance estimation model for AS/RS by M/G/1 queuing system. Comput Ind Eng 46:233–241
2. Manzini R, Gamberi M, Regattieri A (2006) Design and control of an AS/RS. Int J Adv Manuf Technol 28:766–774
3. Van Den Berg JP, Gademann A (2000) Simulation study of an automated storage/retrieval system. Int J Prod Res 38:1339–1356
4. Roodbergen KJ, Vis IF (2009) A survey of literature on automated storage and retrieval systems. Eur J Oper Res 194:343–362
5. Gu J, Goetschalckx M, McGinnis LF (2007) Research on warehouse operation: a comprehensive review. Eur J Oper Res 177:1–21
6. Bargiotas D, Ktena A, Manasis C, Ladoukakis O (2009) A scalable low-cost automated storage & retrieval system. In: Systems, signals and image processing, 2009. IWSSIP 2009. 16th International Conference on, 2009, pp 1–4
7. Hausman WH, Schwarz LB, Graves SC (1976) Optimal storage assignment in automatic warehousing systems. Manage Sci 22:629–638
8. Graves SC, Hausman WH, Schwarz LB (1977) Storage-retrieval interleaving in automatic warehousing systems. Manage Sci 23:935–945
9. Hsieh S, Tsai K-C (2001) A BOM oriented class-based storage assignment in an automated storage/retrieval system. Int J Adv Manuf Technol 17:683–691
10. Rashid MM, Kasemi B, Rahman M (2011) New automated storage and retrieval system (ASRS) using wireless communications. In: Mechatronics (ICOM), 2011 4th international conference On, 2011, pp 1–7
11. Dawale AC, Sakhale CN. Design and fabrication of automatic storage and retrieval system. Int J Innov Technol Exploring Eng (IJITEE) 3075. ISSN: 2278
12. Zein El Din A (1996) High performance PLC controlled stepper motor in robot manipulator. In: Industrial Electronics, 1996. ISIE'96, Proceedings of the IEEE international symposium on, 1996, pp 974–978
13. Yoo A, Chow E, Henderson K, McLendon W, Hendrickson B, Catalyurek U (2005) A scalable distributed parallel breadth-first search algorithm on BlueGene/L. In: Supercomputing, 2005. Proceedings of the ACM/IEEE SC 2005 conference, pp 25–25
14. Korf RE (1985) Depth-first iterative-deepening: An optimal admissible tree search. Artif Intell 27:97–109
15. Tarjan R (1972) Depth-first search and linear graph algorithms. SIAM J Comput 1:146–160
16. Liu S-a, Wang Q, Sun J (2013) Integrated optimization of storage allocations in automated storage and retrieval system of bearings. In: Control and decision conference (CCDC), 2013 25th *Chinese*, pp 4267–4271
17. Jawahar N, Aravindan P, Ponnambalam S (1998) Optimal random storage allocation for an AS/RS in an FMS. Int J Adv Manuf Technol 14:116–132

Chapter 19
Design Analysis of Octagonal-Shaped Microstrip Patch Antenna at 5.70 and 8.00 GHz

Dinesh Kumar Cholkar and Abhishek Rawat

19.1 Introduction

The fast development in wireless communication systems requires the improvement in antenna design, to achieve light mass, low profile, and single-feed antennas. Some characteristics of microstrip antennas make it very attractive such as small profile, low weight, and easy production. Therefore, these antennas are widely used in many practical wireless applications. Basically, a microstrip antenna has a radiating patch on one side of the dielectric substrate. This substrate also has a ground plane on the opposite side.

The characteristics of microstrip antennas are closely related to the different physical parameters such as length, width, height, and type of substrate material. They can also be constructed to have many different shapes and dimensions [1]. These shapes closely affect the pattern and other radiation characteristics. In this paper, octagonal-shaped geometry is proposed for dual-band microstrip antennas. This design provides excellent results at the dual band at 5.7 and 8.00 GHz.

19.2 Microstrip Patch Antenna

Microstrip antenna came into existence in the year 1970s. A microstrip patch antenna (MPA) is made of a radiating patch on the dielectric substrate. This substrate has a ground plane on the opposite side. The patch is normally prepared by

A. Rawat (✉)
Institute of Infrastructure Technology Research and Management, Ahmedabad, Gujarat, India
e-mail: arawat@iitram.ac.in

D.K. Cholkar
Samrat Ashok Technological Institute, Vidisha, MP, India

V. Nath (ed.), *Proceedings of the International Conference on Nano-electronics, Circuits & Communication Systems*, Lecture Notes in Electrical Engineering 403,
DOI 10.1007/978-981-10-2999-8_19

the conducting materials such as copper and gold. The feed wires and emanate patch are typically photoetched on the dielectric substrate [2]. Nowadays, microstrip patch antenna (MPA) is extensively used in different applications of wireless communication systems, because of their very attractive properties such as low volume and thin profile characteristics [3, 4].

The size of MPA is closely related to its resonance length and width. Transmitting signals can insert by a variety of methods in microstrip patch antennas. These properties make it popularity for use in different wireless applications especially compatible for embedded antennas. Basic structure of rectangular microstrip patch antenna is shown in Fig. 19.1. We can feed the signal in microstrip patch antennas by different kinds of methods. These methods are contacting and non-contacting methods. In the contacting method, RF signal is fed instantly using a connecting element such as a microstrip line. On the other side in the non-contacting methods, electromagnetic field coupling is used. This coupling makes possible power transfer between the microstrip line and the radiating patch. Both methods have their different merits and drawbacks, so depending on the wireless system requirements, we can choose suitable feeding methods. Generally, there is a wide variety of feeding techniques used, which include microstrip line, coaxial probe, aperture coupler, and propinquity coupler.

In a coaxial probe feeding system, the inside conductor of the coaxial is connected to the radiating patch. At the same time, outer conductor is connected to the ground plane of an antenna. Simple fabricate, straightforward matching, and small spurious radiation are the key benefits of coaxial feed system. But narrow width and difficult to model specially for thick substrate are some drawbacks of this feeding method. In the proposed design, we are using coaxial probe feeding technique.

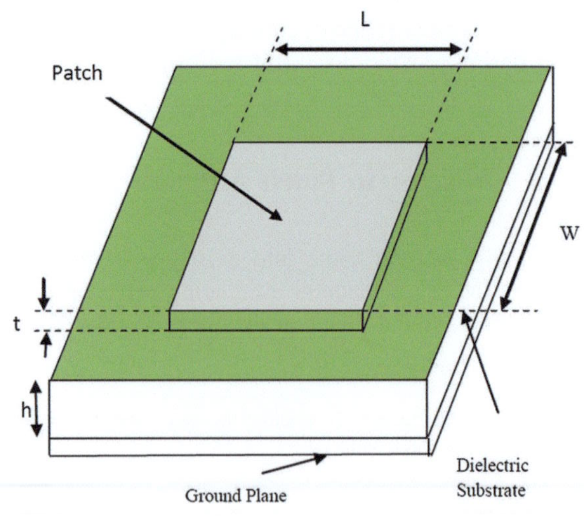

Fig. 19.1 Basic structure of rectangular microstrip patch antenna

19.3 Design of Octagonal-Shaped Microstrip Patch Antenna

The octagonal-shaped microstrip patch antenna design is based on the width and length of the patch which are measured by the permittivity and height of the dielectric material between the conductive microstrip patch and ground plane, called as substrate. In the proposed antenna design, the substrate material is GLASS_PTFE Reinf. The Ansoft HFSS simulator is used for the simulation and testing of the proposed design as shown in Fig. 19.2.

Different design parameters are derived from Eqs. (19.1) to (19.4). These standard derivations of microstrip antenna design are described below in short [5–8].

$$W = \frac{V_0}{2f_r}\sqrt{\frac{2}{\varepsilon_r + 1}} \tag{19.1}$$

$$L = \frac{V_0}{2f_r\sqrt{\varepsilon_{ef}}} - 2\Delta L \tag{19.2}$$

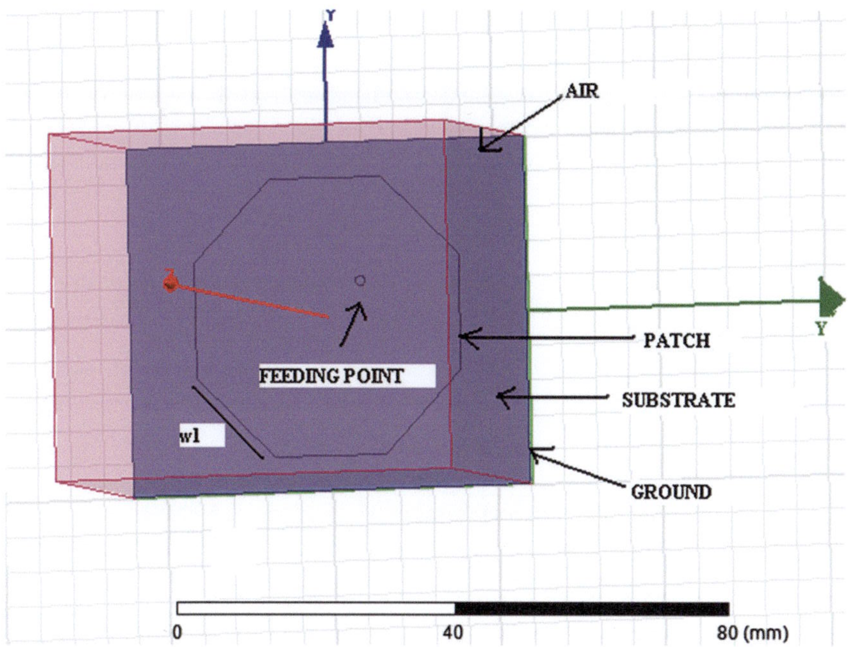

Fig. 19.2 The octagonal-shaped microstrip patch antenna

$$\Delta L = h \times 0.421 \left[\frac{\left(\varepsilon_{ef} + 0.3\right)\left(\frac{w}{h} + 0.264\right)}{\left(\varepsilon_{ef} - 0.258\right)\left(\frac{w}{h} + 0.8\right)} \right] \tag{19.3}$$

$$\varepsilon_{ef} = \frac{\varepsilon_r + 1}{2} + \frac{\varepsilon_r - 1}{2} \left[\frac{1}{\sqrt{1 + 12\frac{h}{W}}} \right] \tag{19.4}$$

where

V_0 speed of the light
ε_{ef} Effective dielectric constant
ε_r Substrate dielectric constant
f_r Fringing frequency
h Height of dielectric substrate
W Width of the patch
ΔL Extension of the length

The above-mentioned mathematical equations are used to calculate the appropriate dimensions of the proposed design of octagonal-shaped microstrip patch antenna which are given in Table 19.1. These parameters are used to develop the soft prototype of octagonal-shaped microstrip antenna at HFSS.

19.4 Simulation Results and Discursion

The proposed design is prepared and simulated in Ansoft HFSS, which is a high-performance full-wave electromagnetic (EM) field simulator. It consolidates simulator, resolution, strong modeling, and automation in the common simple

Table 19.1 Design specifications

Material	Glass-PTFE
Dimension (A) Height (B) Width (C) Length	1.6 mm 20 mm 8.2 mm
Dielectric constant [ε_r]	2.4
Operating frequency	5.2 GHz
Resonant frequency	5.70 GHz 8.00 GHz
Return loss	−16.17 dB −23.07 dB
Bandwidth	200 MHz 250 MHz
Material	Glass-PTFE

environment, where solution of 3D EM complication is rapidly and precisely obtained. Detailed simulation results of proposed design are listed below.

19.4.1 Return or Reflection Loss

If the load is mismatched, the total power will not transferred to the load and is a reflection of the power, that is, called loss, and this loss that is reflected is called the reflection loss. Greater reflection loss indicates immense power being radiated by the antennas which eventually increment the gain. Its unit is dB [9]. The reflection losses (RL), is the measurement of amount of power that is transferred to the load or the amount of power reflected back. Reflection loss of the proposed design is shown in Fig. 19.3. The proposed design has excellent return loss at 5.7 and 8 GHz.

19.4.2 Radiation Pattern

The radiation pattern of an octagonal-shaped microstrip patch antenna is the amount of power radiated/received by the antenna. Radiation pattern is disparate for different antennas and affected by the position of antenna with respect to ground. Its unit is volt per meter (in terms of field pattern) and power per unit solid angle (in terms of power pattern). The previous design [2] of hexagonal patch antenna at same substrate height and dielectric constant of 1.6 and 2.4 mm respectively,

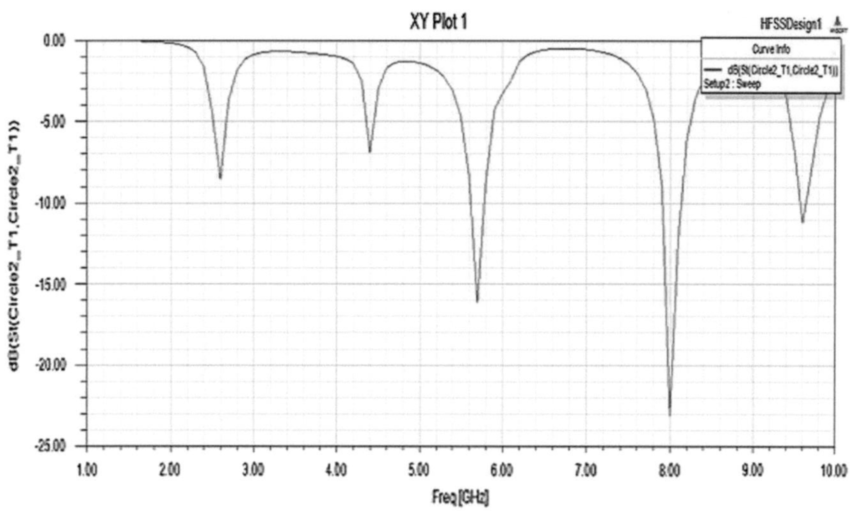

Fig. 19.3 Reflection loss of octagonal-shaped microstrip patch at 5.2 GHz and dielectric constant glass-PTFE

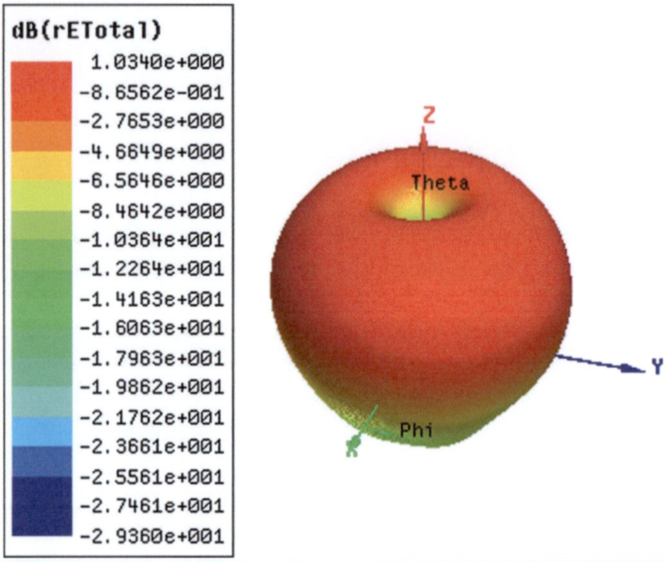

Fig. 19.4 Radiation pattern of octagonal-shaped microstrip patch antenna

provides only a single-band resonant frequency at 5 GHz. By modification of that design, we get dual-band resonant frequency of 5.70 and 8.00 GHz. The detailed radiation pattern plot is shown in Fig. 19.4.

19.4.3 VSWR

The voltage standing wave ratio (VSWR) is the function of reflection coefficient. It explains the amount of power reflected from antenna. It should be always a real and positive number for the antennas. The smaller the VSWR makes, the better the antenna matching to the transmission line. The simulation result of voltage standing wave ratio is shown in Fig. 19.5; at resonant frequencies 5.70 and 8.00, the VSWR value obtained is 2.72 and 1.22, respectively.

19.4.4 Gain

The gain of an antenna is the measure of actual power delivered to the antenna. It can also indicate the power radiated or dissipated within the antenna. A high-gain antenna radiates most of the power present at the input of the antenna, whereas in the lower-gain antennas, most of the power absorbed as losses or reflected away because of impedance mismatch. Generally, gain is quoted in terms of a percentage.

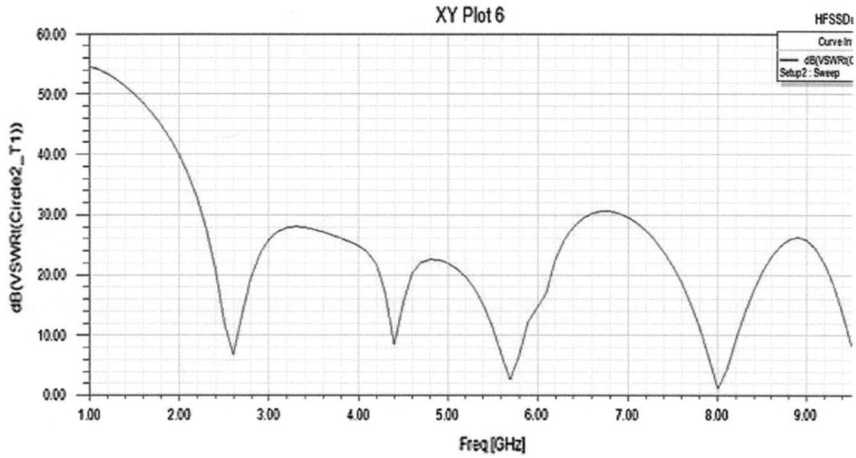

Fig. 19.5 Plot of VSWR for octagonal-shaped microstrip patch antenna

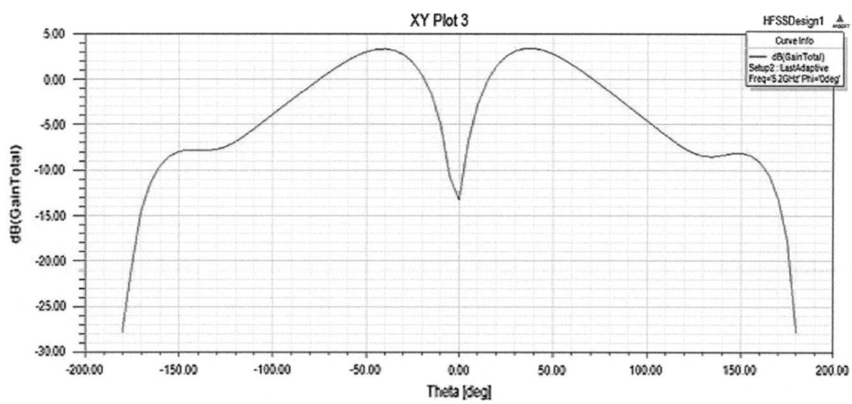

Fig. 19.6 Plot of gain for octagonal-shaped microstrip patch

Figure 19.6 shows the simulated result of gain of the proposed antenna. The achievable gain is −3.32 and 3.36 dB.

19.4.5 *Directivity*

The directivity is the directive gain (DG). It is the ratio of radiation intensity due to the test antenna to isotropic antenna (hypothetical antenna that radiates uniformly in all direction).

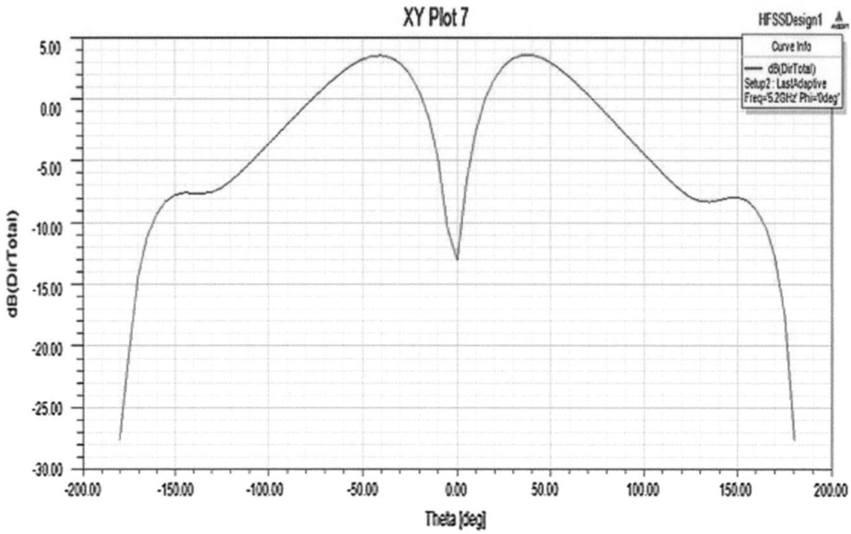

Fig. 19.7 Directivity plot for octagonal-shaped microstrip patch

$$DG = U/U_0 = 4\pi U/P_{rad}$$

where

U Radiation intensity of test antenna
U_0 Radiation intensity due to isotropic antenna
P_{rad} Total power radiated in watts

Figure 19.7 shows the total directivity plot for the proposed design.

19.5 Conclusion

The proposed design of the octagonal-shaped microstrip patch antenna can work on dual band (*c* and *x* band) at resonant frequency 5.70 and 8.00 GHz. Even for at least five bands, it provides excellent return loss performance. This design is particularly easy to fabricate, uses a low-cost substrate, and covers a less space; moreover, by this design good radiation properties are obtained. It can be used in wide range of satellite communications transmission, several Wi-fi devices, several cordless telephones, several weather radar system, terrestrial communication, and space communication.

References

1. Bancroft R (2009) Microstrip and printed antenna design, 2nd edn. SciTech Publishing
2. Mandal K, Pratim sarkar P (2013) A compact high gain micro strip antenna for wireless application. Int J Electron Comm (AEU) 67(12):1010–1014
3. James JR, Hall PS, Wood C (1981) Microstrip antenna theory and design. Peter Periginus, London
4. Singh I, Tripathi VS (2011) Microstrip patch antenna and its applications: a survey. Int J Comp Tech Appl 2(5):95–1599
5. Sanjeev Dwedi J, Rawat A (2013) Design of U-shape microstrip patch antenna for WIMAX application at 2.5 GHz. In: IEEE tenth international conference on wireless and optical communications networks WOCN 2013, Bhopal, India, 26–28 July 2013
6. Kuo JS, Wong KL (2001) A compact micro strip antenna with meandering slots in the ground plane. Microwave Opt Tech Lett 29:95–97
7. Ge Y, Esselle KP, Bird TS (20004) A broadband E-shaped patch antenna with a micro strip compatible feed. Microwave Opt Tech Lett 42(2):111–112
8. Kraus JD, Marhefka RJ (2002) Antenna for all applications, 3rd edn. McGraw-Hill, New York
9. Balanis CA (2005) Antenna theory analysis and design. Wiley, New York

Chapter 20
Blueprint of a CMOS Charge Pump for Phase-Locked Loop Synthesizers with High Efficiency

Adesh Kumar, Anurag Kumar Paliwal and Saurabh Sharma

20.1 Introduction

PLL is the most important part of the modern electronic equipment. A PLL is a circuit [1, 2] that makes a particular system to get synchronized with another system. PLL is a feedback system which found various applications in the areas such as clock generation, frequency multiplication, clock, and data recovery circuits. PLL is a circuit which synchronizes our output signal which is generated by an oscillator circuit with our input signal in terms of phase as well as frequency. In locked state, the phase difference and frequency difference in our two signals are zero and PLL is said to be in locked state, and if a phase error builds up, a control mechanism acts on the oscillator circuit in such a way that the phase error is again minimized to zero. Due to switching activity in digital supply, substrate noise is introduced. The basic block diagram of PLL system is shown in Fig. 20.1. The basic building block of phase-locked loop (PLL) system includes phase frequency detector (PFD), charge pump, and voltage-controlled oscillator (VCO). The operating principle of PLL system can be explained on the basis of angular [1, 3] frequency concept. If we assume the angular frequency of our input signal is equal to the angular frequency of the VCO circuit, then we get zero output (zero error signal) across the PFD circuit and similarly zero output across the charge pump

A.K. Paliwal (✉) · S. Sharma
Electronics & Communication Department, Meerut Institute of Technology, Meerut 250001, UP, India
e-mail: akpaliwal7@gmail.com

S. Sharma
e-mail: sharma.saurabh2006@gmail.com

A. Kumar
Instrumentation & Control Engineering Department, University of Petroleum and Energy Studies, Dehradun 248007, Uttarakhand, India
e-mail: adeshmanav@gmail.com

V. Nath (ed.), *Proceedings of the International Conference on Nano-electronics, Circuits & Communication Systems*, Lecture Notes in Electrical Engineering 403,
DOI 10.1007/978-981-10-2999-8_20

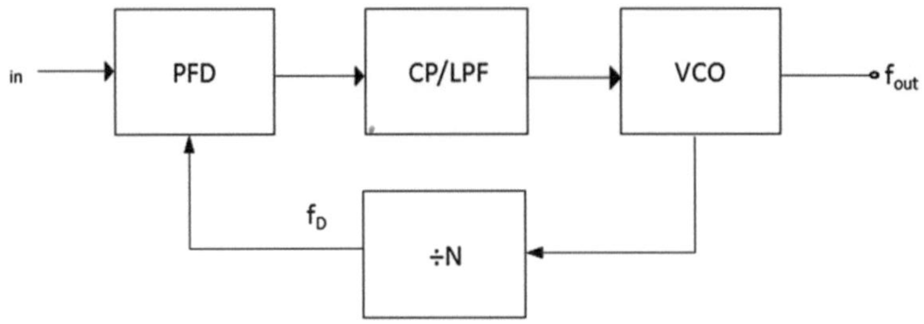

Fig. 20.1 Generalized structure of PLL system

circuit [4, 5] and this is the condition which allows PLL to operate at its center frequency value, but if the angular frequency of our input signal is greater than the angular frequency of the VCO circuit, then we get a nonzero output across the PFD as well as charge pump circuit and this [5, 6] condition will make the VCO circuit of PLL system to operate at higher angular frequency value than its previous angular frequency value in order to vanish the phase and frequency differences between the input signal and the output signal.

20.2 Basics of Charge Pump Circuit

Charge pump is one of the most important building blocks in the recent designs of PLL system, and that is why, it is sometimes also known to be the heart of the PLL system. It is placed in between the PFD and VCO. Charge pump is one of the most crucial elements that translate the phase and frequency information into voltage signal which tunes the frequency of the VCO circuitry. Charge pump is a kind of dc converter that uses capacitor as an energy storage element. The main purpose of charge pump is the convert the dc output voltage of PFD into an analog voltage signal as a input to the VCO circuitry, i.e., the output of PFD which are nothing but the UP and DOWN signals are converted to analog output voltage which further acts as a control signal to VCO. The output of our charge pump depends on the comparison between reference signal and feedback signal. Positive spike occurs at the output of charge pump when our reference signal is greater than the feedback signal, and it generates negative spike when feedback signal is greater than reference signal. The main aim of designing charge pump is that we can get wide voltage range with minimum power dissipation for the devices like PLL. Besides PLL, there are many other applications of charge pump in many wireless communication systems because of its tendency to operate at higher frequency range with larger system gain, and we also get nearly zero static phase error.

20.3 Implementation of Charge Pump Circuit

Charge pump circuits are implemented using MOS switches, but these MOS switches have non-ideal behavior when these circuits are practically implemented. The non-ideal behavior of the charge pump circuit can be observed based on charge pump circuit [7, 8]. These non-ideal effects include current mismatching, charge injection, clock feedthrough, and current sharing. These non-ideal effects disturb the working of VCO circuitry, and the total phase errors can be described as follows:

$$\begin{aligned}\Delta\emptyset_{\text{Total}} &= 2\pi\left(\Delta\varphi_{\text{leakage}} + \Delta\varphi_{\text{mismatch}} + \Delta\varphi_{\text{timing}}\right)\\ &= 2\pi\left(\frac{I_{\text{leakage}}}{I_{\text{cp}}} + \frac{\Delta i.t_{\text{on}}}{I_{\text{cp}}T_{\text{ref}}} + \frac{\Delta t_{\text{delay}}t_{\text{on}}}{T_{\text{ref}}^2}\right)\end{aligned} \tag{20.1}$$

where I_{cp} the rating is current, I_{leakage} is the leakage current, T_{ref} is the reference cycle time, c is the turn on time, and $\Delta i\,\&\,\Delta t_{\text{delay}}$ is the mismatching current and timing. Now, in all the above parameters mentioned, I_{leakage} parameter can be ignored as its value is very small (pA). The reference cycle time parameter is denoted as T_{ref} and is decided during the time of designing of circuit and is mainly a constant parameter. t_{on}—this parameter should be minimized to its least possible value in order to avoid the situation of the dead zone; hence, total phase errors $\Delta\emptyset_{\text{Total}}$ mainly depend on the parameter Δi.

The difference in the magnitude of charging current and discharging current leads to the phenomenon popularly known as current mismatch. The main factors that lead to the phenomenon of current mismatching in the circuits are as follows:

(1) There is no symmetry between the charging and discharging current sources of low pass filter.
(2) Random occurrence of UP and DOWN current pulses as switches is made on.

Leakage of charge across capacitor node even when switches are turned off across that node leads to the phenomenon which is commonly known as charge injection.

As shown in Fig. 20.2, when the switches are turned off, the charge remaining in the channel will enter into the charge pump causing spur at the output voltage. Figure 20.2 also represents the situation of clock feedthrough which arises when MOS switches couple the clock transition to the load capacitor C_{P} through its drain gate parasitic capacitor.

Charge sharing is the phenomenon that originates due to the fine capacitance at the current sources as shown in Fig. 20.3. If both switches S1 and S2 are off, then all the nodes are at different potentials, i.e., node a is charged to high value (V_{DD}), whereas node B is maintained at zero potential. And as the switches S1 and S2 are closed, then the charges on capacitor are distributed among the capacitances of node A and node B which have different values initially. It results that the ripple can be

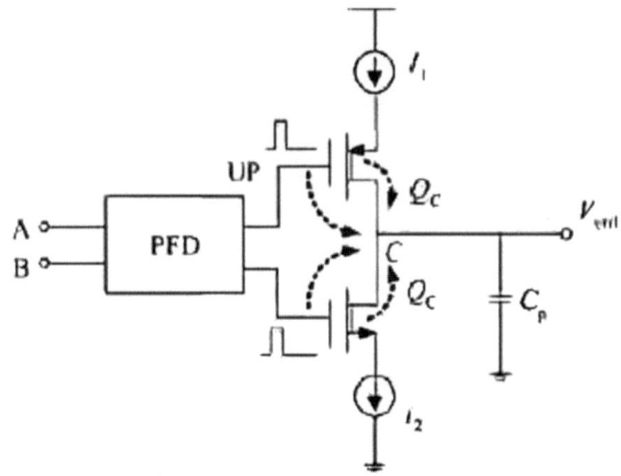

Fig. 20.2 Block diagram of the conventional charge pump having the problem of charge injection and clock feedthrough

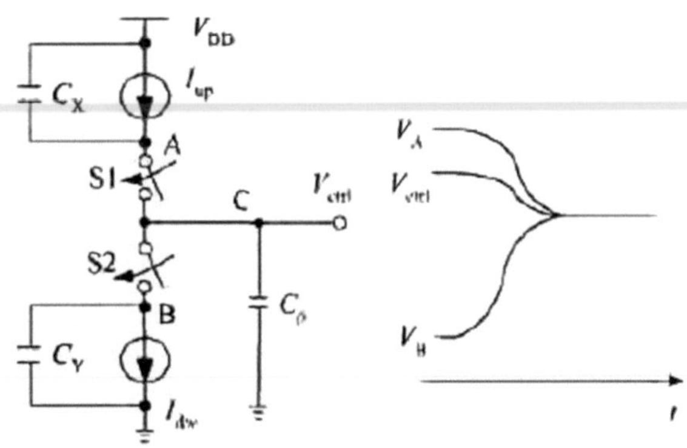

Fig. 20.3 Block diagram of the conventional charge pump having the problem of charge sharing

seen across the load output voltages which lead to the unwanted phenomenon's like jitters etc. It indirectly affects the working of VCO circuitry (VCO).

Figure 20.4 represents the modified charge pump design in which positions of the current sources and switches are interchanged in order to solve the problem of charge injection and clock feedthrough. This design is somewhat able to solve the problem of spur tones and jitters at the output of VCO, but this circuit is not the ultimate remedy for such problems.

20.4 Proposed Charge Pump Design

Hence, to solve the problem discussed above, the anticipated charge pump structure is shown in Fig. 20.5. Here, for proper current mismatch with large output voltage range, an error amplifier is used. Here, M1 is a switch of PMOS type placed in between V_{DD} and M2. Here, M1 is also known as charge switch. M2 and M4 are of

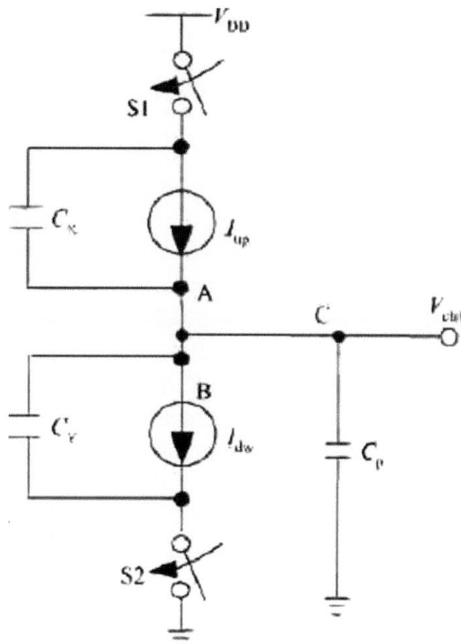

Fig. 20.4 Block diagram of conventional charge pump solving the problem of charge sharing

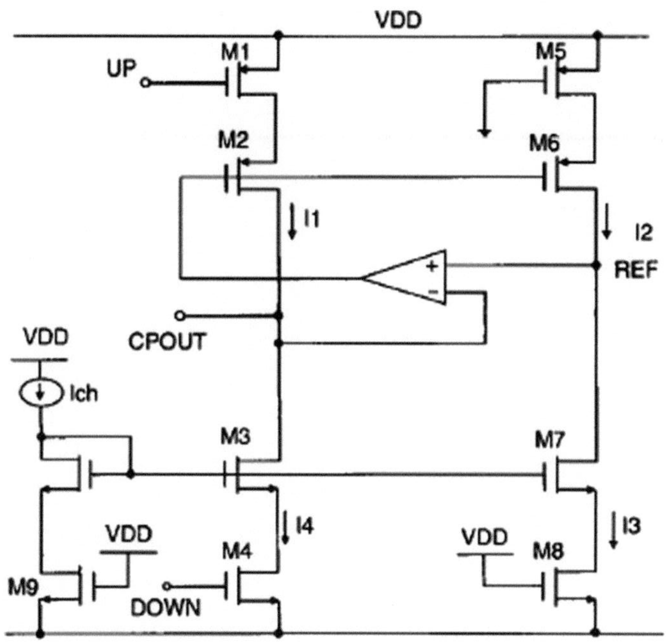

Fig. 20.5 Representing the anticipated charge pump design

NMOS types placed in between ground and M3. These switches are also known as discharge switch as shown in Fig. 20.5. Here, $V_{ref} = V_{cpout}$ until and unless when both the inputs are either high or low. If UP and DOWN both are low, then M1 and M4 are off and the currents I_3, I_2 and I_1 are same. If UP and DOWN both signals are high, M1 will be off and M4 will be on and we can say current I_4, I_3 and I_2 are same and hence we can say that we are able to minimize the current mismatch as source current I_1 and sinking current I_4 are nearly same and as a result phenomenon's like jitter will be reduced which indirectly affects the working of the VCO (Table 20.1).

Figure 20.6 represents the output waveform generated from the proposed charge pump design. It depicts that the spikes are generated at that points, where the magnitude of UP signal and DOWN signal is present. The constant voltage signal is obtained at the output of charge pump, when the magnitudes of UP signal and DOWN signals are at same level. The positive spike is generated at the output of the charge in case of the magnitude of UP signal is greater than the DOWN signal. In

Table 20.1 Performance summary of ideal charge pump

UP signal	DOWN signal	Capacitor status	Output current
1	0	Capacitor charges	Current increases
0	1	Capacitor discharges	Current decreases
0	0	Constant voltage across capacitor	Output current is equal to zero
1	1	Constant voltage across capacitor	Output current is not equal to zero

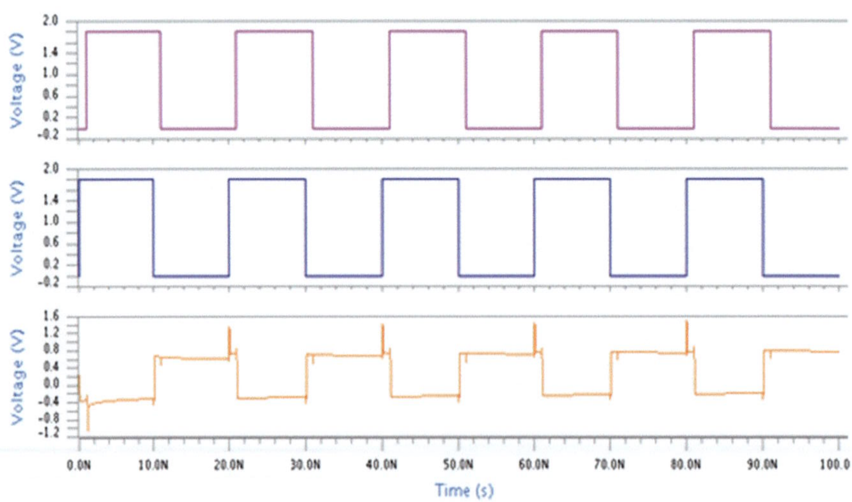

Fig. 20.6 Waveforms generated from the anticipated charge pump

the same way, a negative spike is obtained at the output of the charge pump, when the magnitude of the DOWN signal is greater than the UP signal. Hence the charge pump acts as switch with three positions. The positions are controlled by PFD. The charge pump delivers either positive or negative output voltage across its output, when switch is set in UP or DOWN position. In case of switch position in UP and DOWN direction and set as open condition, it isolates the output of the charge pump and PFD to loop filter.

20.5 Advantages of Proposed Charge Pump Design

Here, the charge pump design discussed above has number of advantages over traditional and conventional charge pump. The design of switches in the proposed charge pump design will reduce effects such as current mismatch, charge sharing, charge injection, and clock feedthrough as explained earlier in this paper. This proposed charge pump design also accelerates the switching speed as individual switches are only connected to individual transistor so there is less parasitic capacitance, Also the proposed charge pump design also consists of the current mirror as a result of which much steadier charging current and discharging current in the comparison to traditional charge pump circuits. Also, due to the presence of current mirror, the gain of the overall PLL system is increased. In the proposed design we are getting wide range of input voltages because NMOS and PMOS behave as a switch. In this way, PLL has better performance and overall efficiency of the PLL system increases. It can also prove the good results in wireless communication (Table 20.2).

20.6 Simulation Results

As shown in Fig. 20.7, in conventional charge pump, we can see that large amount of current mismatch occurs when output voltage is near to supply voltage or ground value and a little better result is observed when VCO is near to the center of the supply voltage, but in our proposed method, in charge pump, as source current and

Table 20.2 Performance summary of proposed charge pump design

Parameters	Results
Power supply (V_{DD})	1 V
Operating frequency	1 GHz
Pull-up and pull-down current ($I_1 \& I_2$)	3 μA
Technology	130-nm technology

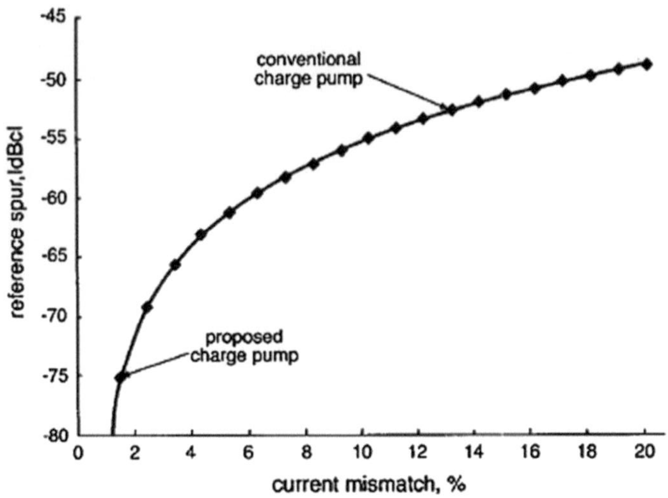

Fig. 20.7 Representing the comparison of spur against current mismatch in conventional charge pump design and proposed charge pump design

Table 20.3 Comparison of proposed charge pump design with traditional charge pump design

Parameters	Results of proposed charge pump design	Results reported in reference no. 1	Results reported in reference no. 5	Results reported in reference no. 4
Technology (nm)	130	180	500	180
Supply voltage (V_{DD}) (V)	1.0	1.0	3.0	1.8
Input frequency range	1 GHz	40–70 MHz	50–162 MHz	10 MHz
Power consumption (mW)	0.01013	0.38	4.1	19.8

sinking current are nearly the same, so we are getting wide amount of voltage range, and hence, our charge pump performs better near to supply voltage or ground.

The new charge pump circuit is designed at 130 nm technology simulated using T-spice (s edit, l edit, t edit) under a 1 V power supply. The pull up and pull down I_1 and I_2. both are set to 3 μA. The operating frequency is 1 GHz for the suggested design. Fig. 20.6 shows the waveforms of charging and discharging result of new charge pump proposed design. A summary of the new charge pump of proposed design is shown in Table 20.3.

20.7 Conclusion

In this paper, a high-frequency CMOS charge pump is designed using T-spice simulation tools, i.e., s edit, l edit, and t edit tools for PLL-based applications at 130-nm technology. The simple and symmetric structure of the circuit reduces serious problems such as spur tones and jitter noises which occur in traditional charge pump circuits and provides more stable operation under a 1-V power supply with the use of error amplifier. Pull-up and pull-down currents are kept to 3 μA and are operated at an operating frequency of 1 GHz.

References

1. Kasht S, Jaiswal S, Jain D, Verma K (2012) Designing of charge pump for fast locking and low power PLL. IJCTEE J 2(6):33–37
2. Raout PK, Panda BP, Achrya DP, Panda G (2011) Designing of a 1 Ghz PLL for fast phase and frequency acquisition. In: International conference of electronic system, 2011, pp 212–215
3. Ramesh J, Vanathi PT, Gunavathi K (2008) Fault classification in phase locked loops using back propagation neural networks. ETRI J 30:546–554
4. Chiu W-H, Huang Y-H, Lin T-H (2010) A dynamic phase error compensation technique for fast-locking phase-locked loops. IEEE J Solid State Circ 45:1137–1149
5. Chen K-H, Liao H-S, Tzou L-J (2008) A low-jitter and low-power phase -locked loop design. IEEE J 30:257–260
6. Mann A, Karalkar A, He L, Jones M (2010) The design of a low-power low-noise phase lock loop. IEEE J 528–531
7. Gupta J, Sangal A, Verma H (2011) High speed CMOS charge pump circuit for PLL applications using 90 nm CMOS technology. IEEE J 346–349
8. Razavi B (2002) Design of analog CMOS integrated circuits. Tata Mc-Graw Hill, pp 532–578

Chapter 21
An Efficient Algorithm for Removing RVIN in Grayscale Images

Neeti Singh, G. Vasugi and O. UmaMaheswari

21.1 Introduction

Digital images are often corrupted by impulse noise. Impulse noise is caused by malfunctioning pixels in camera sensors and faulty memory locations, and due to transmission in noisy channel [1–3]. Numerous algorithms based on median filter have been proposed by various researchers to remove impulse noise significantly [1–8]. All the algorithms work based on two stages: detection of noisy pixels and replacing the noisy pixel by applying filtering.

Center weighted median (CWM) filter [6], adaptive CWM (ACWM) filter [6], progressive switching median (PSM) filter [7], and noise adaptive fuzzy switching median filter (NAFSMF) [4] are few filters which remove RVIN significantly. In all these methods, the uncorrupted pixels are detected as corrupted pixels. In order to overcome this, two conditions with predefined threshold values are added in the detection stage for more investigation and validation of pixels to determine whether they are noisy or not [4].

Moreover, these methods produce visible noise patches at high noise density. Hence a new method, condition-based detection with fuzzy switching weighted median filter (CBD-FSWMF) and condition-based detection with fuzzy switching median filter (CBD-FSMF), is used to remove the noises and restore the images effectively. The CBD-FSMF employs the conditions to validate the pixels and uses a fuzzy switching median filter in the filtering stage [1, 4]. The CBD-FSWMF employs the conditions to validate the pixels and uses NAFSWMF in the filtering stage [1, 4].

N. Singh (✉) · G. Vasugi · O. UmaMaheswari
Department of Electronics and Communication Engineering,
CEG Anna University, Chennai 600025, India
e-mail: neetisingh83@gmail.com

V. Nath (ed.), *Proceedings of the International Conference on Nano-electronics, Circuits & Communication Systems*, Lecture Notes in Electrical Engineering 403,
DOI 10.1007/978-981-10-2999-8_21

21.2 Condition-Based Detection with Fuzzy Switching Weighted Median Filter

The RVIN model of equal probability is used in CBD-FSWMF. Simulations are performed with the images corrupted using the noise model [4].

The probability density function of $x_{(k,l)}$, i.e., $fx_{(k,l)}$ can be expressed as

$$f(x_{k,l}) = \begin{cases} p/2, & 0 \le x_{(k,l)}) < m \\ 1-p, & x_{(k,l)} = s_{(k,l)} \\ p/2, & (255-m) < x_{(k,l)} \le 255 \end{cases} \tag{21.1}$$

where $x_{(k,l)}$ is the (k, l)th pixel in the corrupted image, $s_{(k,l)}$ is the (k, l)th pixel in the original image, and p is the noise density. The dynamic range of an image intensity values is [0, M − 1], where M = 2^n and n is the number of bits per pixel. An 8-bit gray image is assumed, and hence, n = 8 and M = 256 (Fig. 21.1) [4, 8].

A. Detection of noisy pixels

In this stage, the two conditions are executed to distinguish between corrupted and uncorrupted pixels. Only the pixels that are found to be corrupted are processed in the filtering stage [4, 6].

B. Replacing noisy pixels by filtering

The filtering stage employs the fuzzy switching weighted median filter [1, 4]. The detected corrupted pixels are replaced by the fuzzy switching weighted median value of the uncorrupted neighborhood pixels.

21.3 Simulation Results and Discussions

Simulations are carried out in MATLAB R2011a on three standard test images (Lena, Boat, and Cameraman) after they are resized to 512 × 512 and corrupted with RVIN of equal probability. Also, noise densities varied from 40 to 90% are tested and compared using different filters. Denoised images are compared by the PSNR and MSE, and defined as

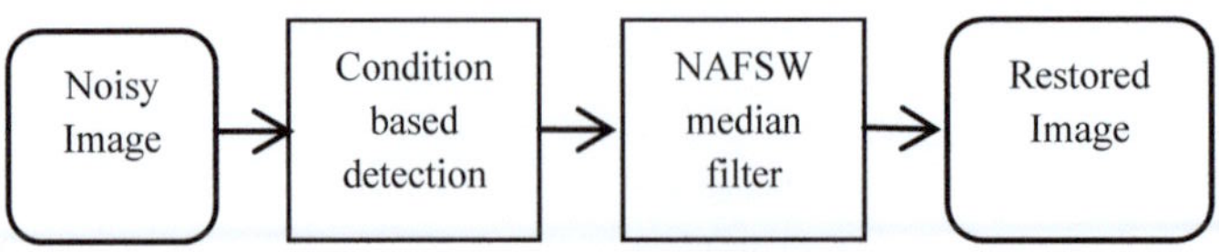

Fig. 21.1 Block diagram of restoration of image corrupted by RVIN

$$\text{MSE} = \frac{1}{\text{MN}}\left[\sum_{k=1}^{M}\sum_{l=1}^{N}(y(k,l) - s(k,l))^2\right] \tag{21.2}$$

$$\text{PSNR(db)} = 10\log_{10}\left(255^2/\text{MSE}\right) \tag{21.3}$$

Restoration results of “Boat” image corrupted with 90 and 95% RVIN, respectively, are shown in Figs. 21.2 and 21.3. The CBD-FSWMF has a better denoising ability when compared with other techniques.

From the visual perception of Figs. 21.2 and 21.3, the CBD-FSWMF is able to preserve the thin lines when compared with the other filters. The noise has been suppressed and the quality of the image has also maintained up to 95% of RVIN.

The median filter does not produce good results at high noise densities. The NAFSMF and NAFSWMF are able to suppress the noises till 60%. When the

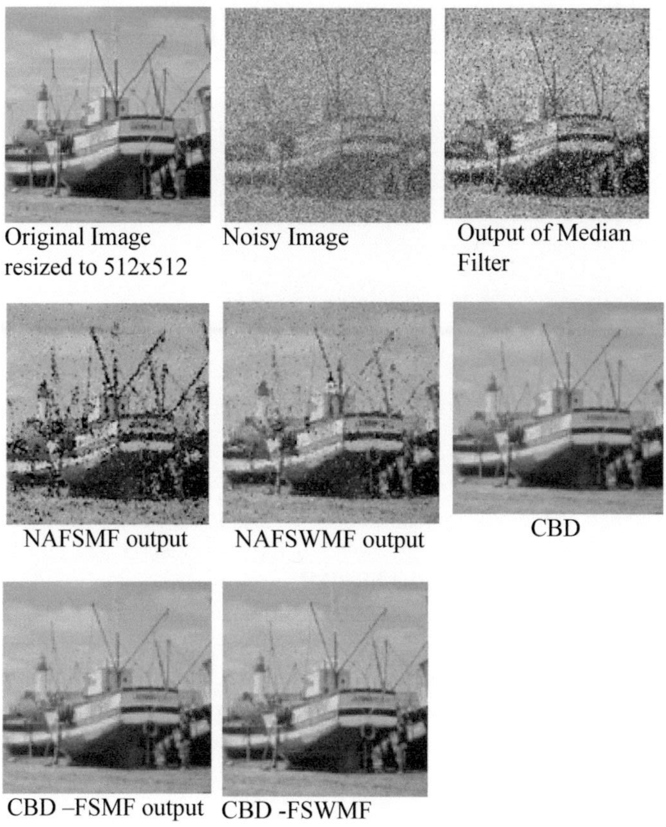

Fig. 21.2 Restoration performance of different filters for Boat image with 90% RVIN and m = 4

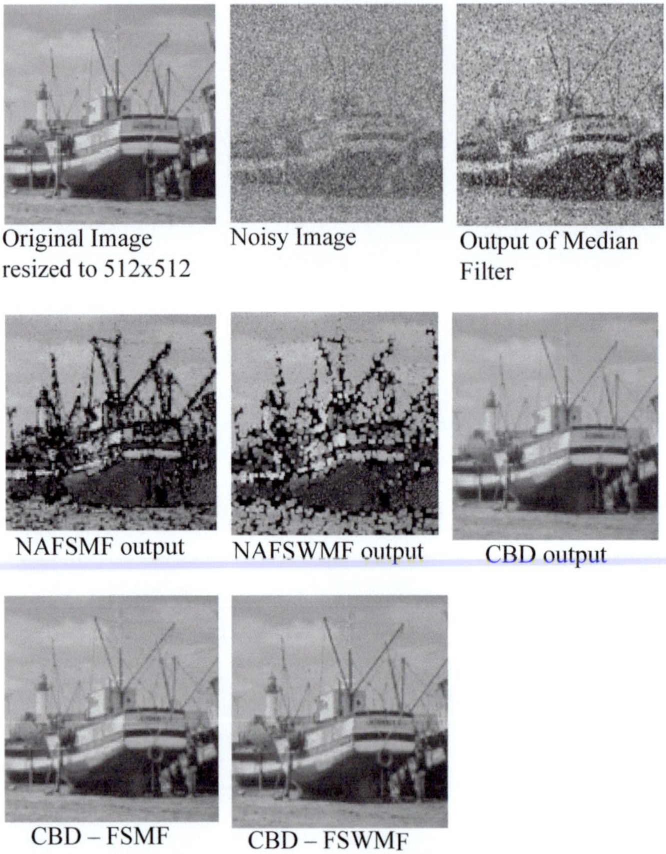

Fig. 21.3 Restoration performance of different filters for Boat image with 95% RVIN and m = 4

noise density is increased beyond 60%, the NAFSMF and NAFSWMF fail to detect the noises; hence, some noise patches and blurring of the image occur. Though the CBD method effectively detects the noise, it fails to preserve the thin lines when compared to CBD-FSMF and CBD-FSWMF.

Table 21.1 shows the MSE values for impulse noise for Lena, Boat, and Cameraman images. For performance comparison, the standard median filter, NAFSMF, NAFWSMF, CBD method, CBD-FSMF, and CBD-FSWMF have also been simulated [1, 4, 5]. The CBD-FSWMF achieves a significantly low MSE even in high noise density because of effective noise detection, and FSWM is used in filtering stage instead of standard median filter.

Table 21.2 lists the restoration results in PSNR (dB) of different methods for images “Lena,” “Boat,” and “Cameraman” corrupted by RVIN with various noise densities. It can be seen from the table that CBD-FSWMF provides the best results in PSNR at high noise density.

Table 21.1 Restoration results in MSE for "Lena," "Boat," and "Cameraman" images corrupted by RVIN

Noise density (%)	Images	Median filter	NAFSMF	NAFSWMF	CBD	CBD-FSMF	CBD-FSWMF
40	Lena	24.12	19.11	16.70	7.01	5.36	3.80
	Boat	23.13	19.62	16.83	8.87	7.5	4.17
	Cameraman	22.89	21.22	18.54	7.89	6.54	4.69
60	Lena	39.57	30.76	21.65	11.50	8.59	6.59
	Boat	40.12	33.14	22.23	14.19	11.75	7.6
	Cameraman	34.23	30.16	23.76	11.22	9.29	7.04
80	Lena	80.11	46.81	28.02	17.23	12.25	10.54
	Boat	82.09	49.93	30.25	18.06	14.53	11.27
	Cameraman	68.75	41.20	29.60	15.09	12.16	9.97
90	Lena	100.56	52.71	30.61	17.58	12.68	11.84
	Boat	103.78	55.73	33.01	22.16	17.72	14.45
	Cameraman	80.99	45.97	32.74	18.60	13.98	12.70
95	Lena	145.56	100.73	81.65	20.02	14.33	13.21
	Boat	147.33	104.40	81.97	22.9	19.29	15.29
	Cameraman	125.89	72.85	59.46	18.73	14.02	12.97

Table 21.2 Restoration results in PSNR (Db) for "Lena," "Boat," and "Cameraman" images corrupted by RVIN

Noise density (%)	Images	Median filter	NAFSMF	NAFSWMF	CBD	CBD-FSMF	CBD-FSWMF
40	Lena	34.30	35.31	35.90	39.67	40.83	42.32
	Boat	34.48	35.12	35.85	38.64	39.37	41.92
	Cameraman	34.53	34.86	35.44	39.15	39.96	41.41
60	Lena	32.15	33.25	34.77	37.52	38.78	39.94
	Boat	32.09	32.92	34.66	36.60	37.46	39.29
	Cameraman	32.78	33.33	34.37	37.60	38.38	39.65
80	Lena	29.09	31.42	33.65	35.74	37.24	37.89
	Boat	28.98	31.14	33.32	35.56	36.5	37.6
	Cameraman	29.75	31.98	33.41	36.89	37.27	38.28
90	Lena	28.10	30.91	33.27	35.51	37.29	37.39
	Boat	27.96	30.66	32.94	34.87	35.89	36.84
	Cameraman	29.04	31.50	32.97	35.43	36.67	37.09
95	Lena	26.50	28.09	29.01	35.11	36.56	36.91
	Boat	26.44	27.94	28.99	34.58	35.27	36.28
	Cameraman	27.13	29.50	30.38	35.40	36.66	37.00

The feasibility of the CBD-FSWMF is compared with the other state-of-the-art impulse noise filters based on their simulation results. It is seen that the performance of the CBD-FSWMF is definitely better than the other filters even when the noise density is higher than 80%. The simple median filter suppresses very little noise and hence results in low PSNR value. The NAFSMF and NAFWSMF perform much better than the median filter, but some noises remain and the quality of the image also degrades. The CBD-FSWMF replaces noisy pixels without affecting uncorrupted pixels. And also, CBD-FSWMF preserves image details very well.

21.4 Conclusion

In this paper, random-valued impulse noise is suppressed by using "condition-based detection with fuzzy switching weighted median filter." Simulation results show that this filter could remove RVIN while preserving image details and yield better results than the existing methods. The CBD-FSWMF is able to recover noisy pixels till 95% of noise density. The results prove that the CBD-FSWMF achieves excellent performance than the other state-of-art filters.

References

1. Suganya C, UmaMaheswari O (2011) Image restoration using noise adaptive fuzzy switching weighted median filter for removal of fixed-valued and random-valued impulse noise. In: Defence science research conference and expo (DSR)
2. Luo W (2006) An efficient detail-preserving approach for removing impulse noise in images. IEEE Signal Process Lett 13(7):413–417
3. Wang Z, Zhang D (1999) Progressive switching median filter for the removal of impulse noise from highly corrupted images. IEEE Trans Circuits Syst II Analog Digital Signal Process 46 (1):78–80
4. Ramadan Zayed M (2014) A new method for impulse noise elimination and edge preservation. Can J Electr Comput Eng Winter 37(1):2–10
5. Toh KKV, Isa NAM (2010) Noise adaptive fuzzy switching median filter for salt-and-pepper noise reduction. IEEE signal process lett 17(3)
6. Chen T, Wu HR (2001) Adaptive impulse detection using center weighted median filters. IEEE Signal Process Lett 8(1):1–3
7. Srinivasan KS, Ebenezer D (2007) A new fast and efficient decision based algorithm for removal of high-density impulse noises. IEEE Signal Process Lett 14(3):189–192
8. Ibrahim H, Kong NSP, Foo TF (2008) Simple adaptive median filter for the removal of impulse noise from highly corrupted images. IEEE Trans Consum Electron 54(4):1920–1927

Chapter 22
Reliability Analysis and Energy Benefit Analysis of Distribution System Incorporating Wind Turbine Generator

Mohini Yadav, Prem Prakash and R.C. Jha

22.1 Introduction

Distribution systems are designed to deliver electrical energy to customers. Much of the work has been done regarding the integration of distributed generators at generation end and transmission end, but little work is done regarding DG integration at distribution end. The numbers of techniques have been developed to evaluate generation system reliability [1–7], and simulation techniques are being developed for evaluating distribution side reliability [8]. Electrical transient analyzer program (ETAP) is software used to simulate the six-bus RBTS system integrating WTG to evaluate the system reliability. The reliability analysis of the distribution system is done separately, wherein the indices from HLII evaluation or transmission sides are inputs to the reliability evaluation of the distribution system [9]. With the advancements of technology, wind turbine generators are inserted at the distribution end to make the supply available to each and every customer point, even when the main generation is not in service. In the presence of both main grid supply and DG, DG will supply power to the customer end while reducing the pressure on the grid side in terms of power and also in terms of total cost. The various contexts regarding purpose, locations, rating, technology, environmental impacts, power delivery area, etc., are considered here for acquiring a general definition and classification of DGs [10].

Integrating both conventional and non-conventional sources of energy at the distribution level increases reliability at the increased level of capital cost. In any system, two main costs are considered, i.e., maintenance cost and fuel cost, and in case of unavailability of supply, when DG is integrated there is no fuel cost so it will return the benefit in terms of total cost reduction.

M. Yadav (✉) · P. Prakash · R.C. Jha
EEE Department, BIT Mesra, Ranchi, India
e-mail: mohini.yadav18@yahoo.com

V. Nath (ed.), *Proceedings of the International Conference on Nano-electronics, Circuits & Communication Systems*, Lecture Notes in Electrical Engineering 403,
DOI 10.1007/978-981-10-2999-8_22

This work is intended to make a way to utilize the assets at a level so that maximum benefit can be achieved from the technology used. This will necessitate the use of two most important criteria: One is cost of system, and another one is energy received from the system. On the basis of these two costs and energy benefit criteria, this work will provide an idea about the location of DG at system bus that gives more economical results.

22.2 Description of Distribution Network

It is a six-bus RBTS system having 4 load buses, 2 generator buses, and 6 generating units. Peak load is 185 MW, and the installed capacity is 240 mW; voltage levels preferred are 230, 11, 33, and 138 kV (Fig. 22.1) [11].

In this system, Bus 1 is a PV bus without any external loads having three different generations of 80, 20, and 10 mW. Bus 2 is a PV bus with external loads connected having three different generations of 40, 80, and 10 mW [12]. Bus 3, Bus

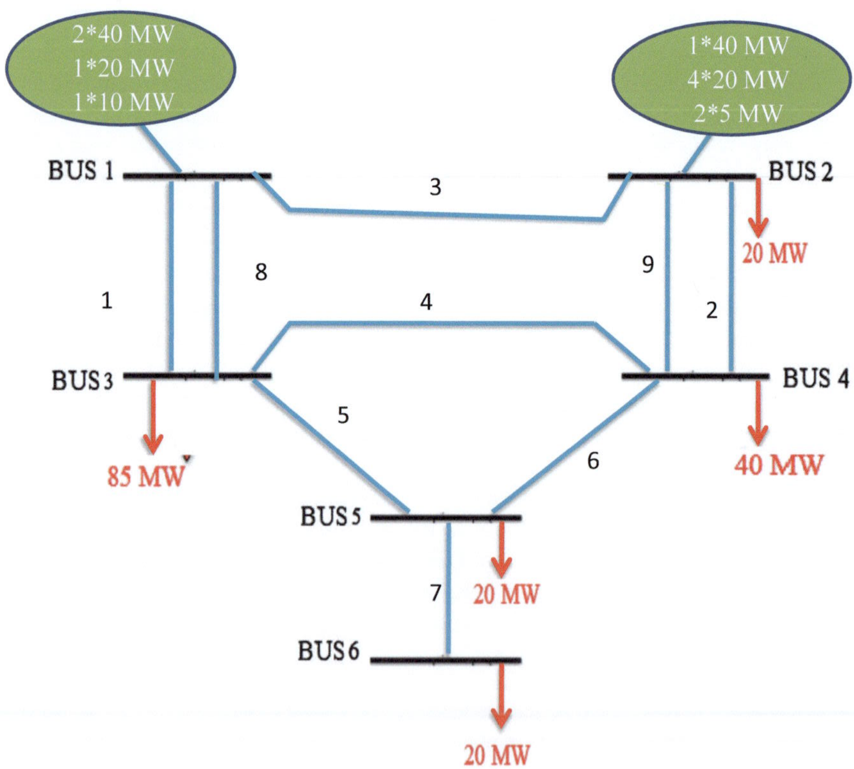

Fig. 22.1 Single-line diagram of six-bus RBTS system

Table 22.1 General details of six-bus RBTS distribution system

No. of feeders	27	
No. of buses	126	
No. of CBs	194	
No. of transformers	170	
No. of transmission line	272	
No. of generators	6	
No. of load points	170	
Peak load	185 mW	
Average load	112.125 mW	
Loads (mW)	R	62.8904
	C	18.539
	LU	55.5
	SU	22.85
	G/I	11.05
	F	7.4
	SI	3.05
	Office	3.69
	Total	184.9694 (mW)

Table 22.2 General information regarding individual buses

General information	Bus 2	Bus 3	Bus 4	Bus 5	Bus 6
Total peak load (mW)	20	85	40	20	20
Average load (mW)	12.91	52.63	24.58	11.29	10.72
No. of customers	1908	5805	4779	2858	2938
No. of feeders	4	8	7	4	4
No. of load points	22	44	38	26	40

Where *R* residential loads, *C* commercial loads, *SU* small user loads, *LU* large user loads, *SI* small industries, *G/I* Government institutions, *F* Farms

4, and Bus 5 are completely loaded buses or PQ buses; no generations are available on these buses.

The complete information of all the components used and the total load on the system is shown in Table 22.1.

To evaluate reliability indices, the general information regarding different buses such as total peak load, average load, number of customers connected, and loads at individual buses is shown in Table 22.2 [12].

22.3 Wind Turbine Generator Specifications

The wind turbine generator used here is of 3.6 mW. Its availability is 0.959634 and unavailability is 0.040365; repair rate is 0.3/day, and its failure rate is 0.0126/day.

22.4 Economy and Energy Benefit Indices

In distribution system reliability evaluation, two basic load point and system reliability cost indices used are the expected energy not supplied (EENS) and expected interruption cost (ECOST). These two indices are helpful in evaluating economy criteria of the system.

The reliability worth in terms of energy gain from six-bus distribution system can be represented by an index called wind generation interrupted energy benefit (WGIEB).

$$\text{WGIEB} = \frac{\text{EENS}(\text{Without WTG}) - \text{EENS}(\text{With WTG})}{\text{Incremental WTG Capacity}} \tag{22.1}$$

The reliability worth in terms of cost is represented by an index called wind generation interrupted cost benefit (WGICB).

$$\text{WGICB} = \frac{\text{ECOST}(\text{Without WTG}) - \text{ECOST}(\text{With WTG})}{\text{Incremental WTG Capacity}} \tag{22.2}$$

22.5 Reliability Analysis and Energy Benefit Analysis

1. Develop an ETAP model for different load buses using ETAP software used for simulation.
2. Find the various load points (i) in an ETAP model and then determine the failure rate (λi), annual outage time (Ui), and number of customers connected to particular load point (Ni).
3. Evaluate expected energy not supplied (EENS) and expected interruption cost (ECOST) using following equations,

$$\mathrm{EENS} = \sum_{i=1}^{\mathrm{Ni}} \mathrm{La}(i)\mathrm{Ui} \quad (22.3)$$

$$\mathrm{ECOST} = \sum_{i=1}^{\mathrm{Ni}} C(i)\mathrm{La}(i) \quad (22.4)$$

where

La (i) Average load connected to load point i
C (i) per unit cost calculated using composite customer damage function (CCDF)

4. Evaluate SAIFI (System Average Interruption Frequency Index) that is defined as the average number of interruptions that a customer experiences during the time period or the year calculated as:

$$\mathrm{SAIFI} = \frac{\sum \lambda i\,\mathrm{Ni}}{\sum \mathrm{Ni}} \quad (22.5)$$

5. Evaluate SAIDI (System Average Interruption Duration Index) that measures interruption hours for an average customer during a year:

$$\mathrm{SAIDI} = \frac{\sum \mathrm{Ui}\,\mathrm{Ni}}{\sum \mathrm{Ni}} \quad (22.6)$$

6. Evaluate ASAI (Average Service Availability Index) which is the ratio of the total number of customers hour's service availability during a year to the total customer hours demanded,
7. Evaluate ASUI (Average Service Unavailability Index) which is equal to (1-ASAI), and
8. The above procedure is repeated for different load buses in a distribution system.

22.6 Results

Three cases have been considered: Case 1 is done without DG and without feeder's consideration, Case 2 is done without DG but with feeder's consideration, and Case 3 is done with DG.

Case 1: The analysis is performed on individual buses without involving DG and shows different values of cost benefits and energy benefits at the time of interruptions. The result is shown in Table 22.3, in which Bus 6 gives higher value of these indices compared to another.

Table 22.3 The analysis is performed on individual buses without involving DG

Buses	Indices	
	WGICB ($/year mW)	WGIEB (h/year)
Bus 2	416553.02	43.229
Bus 3	60184.44	20.6319
Bus 4	314,045	16.024
Bus 5	89853.61	12.3783
Bus 6	742,550	87.74583

Table 22.4 The analysis and its result of Bus 2 and Bus 3

Buses	Indices	
	WGICB ($/year mW)	WGIEB (h/year)
Bus 2		
F1	58370.361	12.3283
F2	11.65	3.563
F3	26980.5	7.4941
F4	46291.027	11.54305
Bus 3		
F1	34512.88	11.477
F2	0	1.8697
F3	50498.19	7.4497
F4	22697.80	10.889
F5	83314.72	14.7177
F6	74224	12.226
F7	0	0.0321
F8	0	0.0261

Case 2: Then, the analysis is done on different buses while taking into consideration different feeders within the system.

(a) Bus 2 and Bus 3
The analysis is performed firstly on Bus 2 and Bus 3, and the result is shown in Table 22.4, which shows that feeder with maximum values of WGICB and WGIEB is more economical.

(b) Bus 4, Bus 5, and Bus 6
In this case, the analysis is done on different feeders of Bus 4, Bus 5, and Bus 6 which is shown in Table 22.5.

Case 3: The final analysis is done with DG on entire six-bus distribution system, and this shows that in entire loop system, Bus 2 has higher values of WGICB and WGIEB. That means this particular bus is more economical, and higher value of energy benefits is received (Table 22.6).

Table 22.5 The analysis is done on different feeders of Bus 4, Bus 5, and Bus 6

Buses	Indices	
	WGICB ($/year mW)	WGIEB (h/year)
Bus 4		
F1	23290.66	8.14
F2	0	1.3019
F3	25194.80	8.335
F4	27201.416	10.5364
F5	0	0.8325
F6	0	0.5842
F7	18486.88	7.372
Bus 5		
F1	28572.5	5.8521
F2	33888.80	7.4827
F3	57573.527	9.04361
F4	79871.472	12.562
Bus 6		
F1	26739.5	16.1005
F2	31847.016	18.8122
F3	13890.33	2.9755
F4	61851.694	87.6988

Table 22.6 The final analysis is done with DG on entire six-bus distribution system

Buses	Indices	
	WGICB ($/year mW)	WGIEB (h/year)
Bus 1	546344.7	238.7119
Bus 2	546354.7	248.648
Bus 3	142485.8	170.339
Bus 4	287285.83	204.626
Bus 5	287285.8	204.626
Bus 6	290829.16	173.0436

22.7 Reliability Analysis

This analysis shows the impact of WTG on the reliability of six-bus distribution system. The reliability results are shown in Tables 22.7 and 22.8 without WTG and with WTG, respectively.

(a) Without WTG.
(b) With WTG.

Table 22.7 This analysis shows the impact of WTG on the reliability of six-bus distribution system

Indices	Bus 2	Bus 3	Bus 4	Bus 5	Bus 6
SAIFI	2.1525	2.5622	2.2570	2.1742	7.9075
SAIDI	18.5371	19.2395	16.3535	18.3562	51.0564
ASAI	0.9967	0.9978	0.9981	0.9979	0.9941
ASUI	0.00240	0.00220	0.00187	0.00210	0.00588

Table 22.8 The reliability results are shown in Tables 22.7 and 22.8 without WTG and with WTG, respectively

Indices	Bus 2	Bus 3	Bus 4	Bus 5	Bus 6
SAIFI	1.1164	1.4567	1.3173	1.2621	6.1081
SAIDI	15.0420	14.1911	12.4305	16.234	43.4133
ASAI	0.9983	0.9984	0.9981	0.9981	0.9950
ASUI	0.00172	0.00162	0.00185	0.0018	0.00496

22.8 Conclusion

From the above, it may be observed that for the system to be reliable in terms of energy and cost benefits, higher value of WGIEB and WGICB should be obtained and the values of these two indices are improved when the WTGs are connected. From economical point of view, Bus 2 is more reliable bus in the RBTS six-bus distribution system. Moreover, while considering different reliability indices, the low value of SAIFI and SAIDI is obtained in case of Bus 2 that shows customer experiences a less number of interruptions and duration of interruptions at this particular bus as compared to other buses and has a higher value of availability at the time of interruptions.

This study will help the investment planners to take decision regarding the system reliability of the distribution system.

References

1. Hunt VD (1981) Wind power—a handbook on wind energy conversion systems. Van Nostrand Reinhold Company, New York
2. Giorsetto P, Utsurogi KF (1983) Development of a new procedure for reliability modeling of wind turbine generators. IEEE Trans Power Apparatus Syst 102(1):134–143
3. Wang X, Dai H, Thomas RJ (1984) Reliability modeling of large wind farms and electric utility interface system. IEEE Trans Power Apparatus Syst 103(3):569–575
4. Singh C, Lago-Gonzalez A (1985) Reliability modeling of generation systems including unconventional energy sources. IEEE Trans Power Apparatus Syst 104(5):1049–1056

5. Billinton R, Chowdhury AA (1992) Incorporation of wind energy conversion systems in conventional generating capacity adequacy assessment. IEE Proc C 139(1):46–56
6. Billinton R, Chen H, Ghajar R (1996) A sequential simulation technique for adequacy evaluation of generating systems including wind energy. IEEE Trans Energy Convers 11 (4):728–734
7. Desrochers G, Blanchard M (1986) A Monte Carlo simulation for the economical assessment of the contribution of wind energy to power systems. IEEE Trans Energy Convers 1(4):50–56
8. Allan RN, Sayas FC (1997) Reliability modeling and evaluation of distribution systems containing wind farm. In Proceedings of the probability methods applied to power systems conference, Vancouver, BC, Canada, Sept 1997
9. Ackermann T, Andersson G, Söder L (2001) Distributed generation: a definition. Electr Power Syst Res 57:195–204
10. Billinton R (1970) Power system reliability evaluation, Taylor & Francis
11. Billinton R, Jonnavithula S (1996) A test system for teaching overall power system reliability assessment. IEEE Trans Power Syst 11(4), Nov 1996
12. Allan RN, Billinton R, Sjarief I, Goel L, So KS (1991) A reliability test system for educational purposes–basic distribution system data and results. IEEE Trans Power Systems 6 (2), May 1991

Chapter 23
Development of Smart Grid System in India: A Survey

A.K.P. Kovendan and D. Sridharan

23.1 Introduction

The recent development in the information and communication technology has lead to a sophisticated life by providing various types of tiny-sized gadgets performing *n* number of task by reducing the efforts to be made by man power. This advancement has also reached the existing power grid technology. The first alternating current power grid system was installed in 1886. This system has been facing a number of challenges like aging of equipment, cultural value, and outdated engineering. This in turn will reduce the reliability of these power grids. Apart from reliability issues, the annual maintenance cost spent on these power grids for maintaining its infrastructure is very high. To solve these issues, smart grid technology has been emerged as advancement in the power electronic technologies [1]. Smart grid is nothing but the automation of existing power grid by improving its efficiency, reliability, and sustainability. Apart from these issues was the carbon emission which causes a major environmental problem [2].

Smart Grid being a next-generation power gridsupports all parts of the nation [3]. Though the traditional electrical grid and smart grid perform the task of delivering generated power from generating end to consumer end, reliability and online information are the key factors of smart grid technology. This can be achieved by integrating fields like information technology, communication technology, and electrical grid. Information technology will hold the database of the entire system. The role of information technology is to hold the database of generation, demand, and consumption. This probably requires cloud storage for ease of

A.K.P. Kovendan (✉) · D. Sridharan
Department of Electronics and Communication Engineering,
CEG Campus, Anna University, Chennai, India
e-mail: kpkvendan@gmail.com

D. Sridharan
e-mail: sridhar@annauniv.edu

V. Nath (ed.), *Proceedings of the International Conference on Nano-electronics, Circuits & Communication Systems*, Lecture Notes in Electrical Engineering 403,
DOI 10.1007/978-981-10-2999-8_23

access. The information technology will gather information from the electrical grid through monitoring system. Electrical grids are being maintained and monitored and even diagnosed by wired communications. Reference [4] states that the regular maintenance of such wired systems will results in loss of economy. In addition to the economical impact, the wired systems need a separate infrastructure that will further be costlier. Smart grid technology will maximize the usage of renewable energy resources (RES) by domestic costumers. This will further reduce the need for electricity and emission of CO_2 into atmosphere. There are three major divisions in power grid technology, namely generation side, transmission side, and distribution side. The next technology to be integrated is the communication technology. The communication should be efficient and should be able to fit into any kind of infrastructure. The optimal candidate for this application is the wireless sensor networks. The development of IEEE 802.15.4 standard has enabled wireless sensor network technology. This technology has been emerged as advancement in the wireless communication technologies. Wireless sensor networks are a collection of sensor nodes that are capable of sensing a parameter (e.g., voltage, current, and power), and sensed parameter will be transmitted to base station commonly called as a sink. There are three substations in WSN, namely sensing substation, processing substation, and transceiver substation. WSN cloud will be an ideal candidate for smart grid applications. Integration of cloud computing for computing, storing, and networking will result in the formation of WSN cloud [5]. The final technology is the existing power grid. This grid has to support the two-way communication. This survey has been organized into various sections for briefing the overview of smart grids followed by the need for smart grids in India. This paper has been concluded by summarizing the role of information and communications technology (ICT) in smart grids by briefing the challenges in it. Possible research directions and issues are also presented.

23.2 Overview of Smart Grids

Smart grids will be the next-generation's power grid. This combines information technology, integrated communication technology, sensing equipments, automation for increasing the reliability, efficiency and allows bidirectional flow of electricity and information. Existing power grids have served the nation for a long time, and now it became less efficient due to transmission and distribution losses. Since the transmission lines are old, renovation or replacement may be necessary for those lines. Instead of random replacement and renovation, modernizing of the existing grids by enabling monitoring features will be an ideal solution. This can be achieved by integrating advanced communication infrastructure to the existing grid. The flowchart shown in Fig. 23.1 explains the framework adopted in smart grids [6]. This shows the overall components of the system.

Electricity is the fastest growing component of the total global energy demand [7]. Smart grid is not only for maintaining and monitoring of power grid, but also

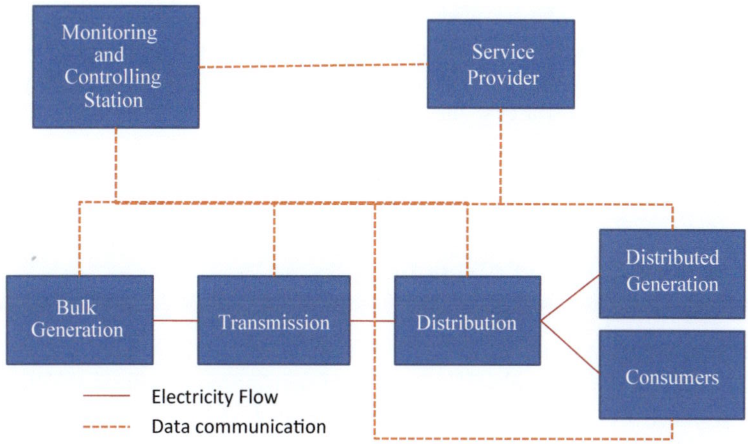

Fig. 23.1 Smart grid framework [7]

for ensuring the clean energy. As the population and technology has increased rapidly, pollutions and need for clean and sustainable technology has also increased. It has been stated that the average age of power line is around 60 years [8]. To meet out the problem with power evacuation, bidirectional power and information lines in the existing grid become mandatory [9]. References [10, 11] states there are seven key benefits for the consumers, business, utilities, and the nation.

A. Real-time interaction of customers: Interaction with customers/consumers will be facilitating the real-time usage monitoring, smart appliances control, building automation, and real-time pricing. This will facilitate the change of tariff from peak and non-peak hours.
B. Distributed generation and storage: Smart grid enables the local generation of power with scalable quantity by RES like solar, wind, and biomass. This also facilitates the storage by various means.
C. Reliability: Reliability can improved by enabling real-time power quality monitoring, asset management, optimal utilizations, and distributed automation and protection.
D. Quality of service: The quality of service can be improved by enabling features like self-healing nature of the system, frequency monitoring, and control in the T & D network, load forecasting and determining the anticipation of disturbances.
E. Prevention from attacks and power theft: Smart grids will provide added security services in the accessing mechanism for ensuring the theft prevention and secured operation.

F. Enabling markets: Smart grids will support both investing and innovating concepts by supporting energy markets for a dependable operation.
G. Economical: Smart grids will reduce building of new infrastructure by utilizing all the available resources; this will in turn reduce the time and needs less investments on operating and maintaining the grid.

23.3 Need for Smart Grids in India

Current technology adopted in India for Transmission and Distribution (T & D) of Electrical energy in electrical grid, a system of transmission medium which carries electrical energy from generation end to consumer end. In 1947, installed capacity of electricity (alternate current) was 16.3 kWh, and it has reached 255,681 kWh by December 2014 [3, 4, 12]. This capacity is not sufficient to meet out the requirements of Indian population and technologies. This rapid growth will not sustain at this level and will reach about 900 GWh. The availability of non-renewable resources like coal will not be enough to meet out those requirements. This states that only renewable sources can help India meet the rising demand for electricity. The modern tradition has changed the size and capacity of generation plant into a highly scalable one. Conversion of electrical grids needs no alteration in the existing infrastructure, instead the advancement in technologies like controlling systems, sensors, advanced metering infrastructure (AMI), and synchrophasors has to be added with the existing infrastructure. As a result, in 2014, 12.4% of the generated power has been drawn from RES [4]. Despite of enormous generating capacity, still India is a power deficit. Transmission capacity bottleneck has caused around 1.93 billion units of generated power to be wasted [6]. Hence, power evacuation has become a bigger problem than power generation for the country. The Northern (Northern, North-Eastern, Eastern, Western) region is surplus of about 2.3% during peak hours [6, 13]. But, southern region is at 26% shortage during the peak hours. Though there are surplus generations in some parts, the evacuation problem has resulted in loss of generated power. The resource-rich states like Chhattisgarh are also unable to evacuate the excess power. The current 12th year plan will result in 30,000 MW generation capacity where the total need for Chhattisgarh is about 3300 MW alone [14]. The transmission capacity of the same is around 7000 MW. Though there is a lot development in the industrial sector, surplus amount will be there in such resource-rich areas. Power evacuation capacity has to be added to address these issues. Besides evacuation, there exist several other problems with the existing grids, namely theft of power, the connector system, and service connection losses which leads to premature failure of capital equipments like transformers, average transmission, and distribution losses exceeding 30%. This causes the system more incompetent and fly-by-night. For the development of existing power grid, smart grid will be an optimal solution. Smart grid is the needy for the development of the nation's gross domestic product [15–17]. India being a developing nation suffers a lot from T & D losses, and smart grid will be an optimal candidate. These issues can be addressed by the smart grid concept. In order to reduce the maintenance cost and to increase the reliability, a monitoring system is needed. Wireless communication-based controlling and monitoring system is an ideal candidate for this application.

23.4 Related Works

In India, several initiatives have been taken for the remodeling of electric T & D network by the convergence of integrated communication systems with information technology that enhance the security, reliability, and efficiency to utilities [18]. Accelerated Power Development and Reforms Programme (APDRP) proposed by Indian Electricity Act 2003 for the reduction of distribution losses and to increase the reliability has been restructured to Restructured Accelerated Power Development and Reforms Programme (R-APDRP) for the integration of RES as a part of distributed generation [19, 20]. About 30–40% of the total power consumption must be from RES for the successful implementation of smart grids [21]. Rural electrification will enhance the economic growth of India [22]. The climate group [23] on behalf of Global e-Sustainability Initiative [24] has started a vision for reducing the carbon emission in 2008, India's Ministry of Power being a participant in SMART2020 has taken some initiatives for the transformation of Indian power grid to smart grid [25, 26]. The initiatives taken in India by government, private bodies, and allies are as follows: Power Grid Corporation of India Limited (PGCIL) has proposed a system for the implementation of PMU with GPS system, smart load control, online condition monitoring, data communication using optical fiber communication in the northern region [27, 28], and in the western region intelligent monitoring and controlling of interconnected electric power grid using wide area monitoring [29]. In Delhi, several initiatives on integrated SCADA solution, smart metering solution, SCADA-controlled grid station, AMI, hi-tech automation control, monitoring by Crompton Greaves Limited (CGL) for Govt. of India [30, 31], and North Delhi Power Limited (NDPL) for North and West Delhi [32]. In Karnataka, BESCOM (Bangalore Electricity Supply Company) has taken initiatives on reducing T & D losses and also for increasing the reliability and power quality in 8 districts of Karnataka [33]. The successful completion of these initiatives has not been achieved in the specified time but still the way forward has been drilled by the India Smart Grid Forum [34]. Few projects like the one in Pondicherry have been completed with 1400 smart meters [35]. The integration of green and sustainable energy into the grid for reduced emission of carbon has been facilitated by MNRE through prime minister's JNN mission launched on January 2010 [36]. This will further reduce the dependency on grid.

23.5 Information and Communication Technologies

Existing power grid when combined with integrated communication technologies (two-way power communication, information, and control signal flow) will become as smart grid [37]. The basis of power system communication is the reliable communication infrastructure. Figure 23.1 structures the players involved in the operation and maintenance of smart grids. It is evident that ICT will be the enabling technology for smart grid communications [38]. The communication system in smart grid will correspond to all the three substations of electrical grid and maintenance station and service provider. The operation of interconnected elements are being protected, monitored, and optimized automatically which covers traditional central generator and distributed renewable generator through communication network [39]. The main characteristics of smart grid were the bidirectional communication (power and information) [40]. Structures that demand side management and balance the available resources were the main tasks [41]. These can be achieved only when a reliable ICT infrastructure has been incorporated. The reliability in ICT will not stop at this level of providing a greater power management scheme. It has to ensure the integrity of data, availability of the network, and so on. The role and responsibility of ICT will be discussed here as follows.

23.5.1 Role

Electrical grids are "smart" grids when there is a flow of data (information for control and monitoring) along with bidirectional flow of electricity in the grid [42]. References [43, 44] explains the most critical and complex nature of the electrical generation and distribution networks in the country. Gungor et al. [45] states that ICT is a leading role player for smart grids. Reliable ICT in smart grids will increase the efficiency, reliability, and reduced carbon emission. Reliability can be achieved by overcoming the challenges in the communication technologies [46]. The role of ICT includes security [47], reliability [48], scalability [49], availability, and quality of service [50]. ICT should ensure confidentiality, integrity, authentication, and availability for the secured access to the network [51]. The robustness of the electrical grids infrastructure and its deployed area will be a source of deployment, and installation challenges for ICT infrastructure in it. The network in smart grid may be home area network (HAN), local area network (LAN), neighborhood area network (NAN), and wide area network (WAN), but it should support the applications like automatic metering infrastructure (AMI) for the collecting history of consumed energy [52–56] demand response for the variation in tariff for peak and non-peak times, situational awareness for the monitoring of grid condition, distributed generation [40], T & D losses, and grid management [57].

23.5.2 Communication Media

ICT will integrate the telecommunication and computers for the information transfer. As discussed earlier, the smart grid communication networks like HAN, NAN, WAN, LAN can use the medium as wired or wireless. Though there are so many communication technologies for support, the main medium is either wired or wireless [45, 58]. In wired communication, the prone to attack/failure is more common in the case of environmental disasters, and the infrastructure need to be developed and maintained. The ease of connection is available in wireless compared with wired even for remote location. Wired system has a drawback of high economical investment but intern supports higher data speed. In the case of wireless, the technologies like Zigbee [59–61] WIMAX, GSM, GPRS, 3G support applications like AMI, demand response, and real-time control and monitoring. Zigbee operates in low power with lesser data rate, coverage, less complex, and easier to design [43]. The other cellular systems will facilitate the higher data rate with increased complexity and computational power. The smart grid infrastructure is vulnerable to challenging environments like harsh environmental conditions, geographical environments, and mobile communication [62]. The possible solutions can be achieved through reliable wireless networks like wireless sensor networks or cognitive radio network [63–65].

23.5.3 Obligatory Technological Advancements

India is a third largest emitter of carbon dioxide resulted from thermal power plants used for incineration and so many other sources [66, 67]. Apart from carbon dioxide emission, water pollution and water wastage are also more common because of power plants. The major energy consumer and a cause for pollution is the ICT infrastructure in the country [68]. It has been observed that the initiatives taken for the reduction of emission and for the efficient usage of available resources has been developed. Green communication will facilitate the efficiency in power usage and reduced pollution for the better future of generation [69]. References [70, 71] proposed the model for green cellular networks and green communication. Green communication like green telecom towers and load centers has to be regulated like peak time billing in the interest of preserving the nature and ecosystem of the society.

23.5.4 Open Research Issues

Smart grid is a “saucer full of secrets” [72]. The domain which attracted most of the academicians, industry, and government for research is smart grids. The possible fruitful research areas of smart grids can be broadly classified into two major components: First is the communication part, and the next one is electrical grid. In communication part, areas like integrity, availability, maximum utilization of resources, scalability, security, and link reliability have to be still improved. Improvement in reliability and availability of grid can be achieved only by a proper communication technology which provides high level of security, scalability, and adaptability to heterogeneous networks and devices. The heterogeneity is of prior importance in this case as the network will be incorporating various devices for communication and controlling operations [73]. The electrical grid has to communicate with different platforms for information sharing, which has to be managed (SCADA) [74]. In grid, renewable energy integration [75] will have to be monitored for the synchronization of phase angle, frequency, and voltage. Demand-based power production, Security issues like power theft, global id for online billing through online metering via communication medium are yet to be developed. Development of test bed for the learning of performance metrics like efficiency, accuracy, and so on will be a potential research topic.

23.6 Conclusion

Smart grid is a new and wide area for research which attracts the attention of academicians, industry and the government. In this study, we presented an ample survey on the present status of Indian power grid and the need for smart grid. We also summarize the role and responsibilities of ICT architecture in the smart grid. In addition to that, the present context and the possible enhancements have been discussed. The inference of this survey is "major components of smart grid like integrated communications system are still under construction level for achieving high reliability." The other components like integration of renewable energy and reliability in electrical grid will be a challenging and fruitful research topic in the future.

References

1. Gao J et al (2012) A survey of communication/networking in smart grids. Future Gen Comput Syst 28.2:391–404
2. Yang JI, Qian AI, Da XIE (2010) Clean energy grid-connected technology based on smart grid. Low Voltage Apparatus 4:005
3. Hou H et al (2011) A brief analysis on differences of risk assessment between smart grid and traditional power grid. In: Knowledge Acquisition and Modeling (KAM), 2011 Fourth International Symposium on. IEEE, 2011
4. U.S. Department of Energy (2004) Assessment study on sensors and automation in the industries of the future. Office of Energy and Renewable Energy

5. Aslam MS, Rea S, Pesch D (2012) Service provisioning for the wsn cloud. In: Cloud Computing (CLOUD), 2012 IEEE 5th International Conference on Cloud Computing. IEEE, 2012
6. National Institute of Science and Technology, Smart Grid—Beginners guide (online). http://www.nist.gov/smartgrid/beginnersguide.cfm. Accessed 10 Jan 2015
7. International Energy Agency, Technology Roadmaps, Smart Grids (online) http://www.iea.org/publications/freepublications/publication/smartgrids_roadmap.pdf accesses 28–12-2014
8. Yang Y et al (2006) Power line sensornet-a new concept for power grid monitoring. In: Power Engineering Society General Meeting, 2006. IEEE. IEEE, 2006
9. Federation of Indian Chambers of commerce and Industry, "Power Transmission the Real Bottleneck," September 2013 (online). http://www.ficci.com/spdocument/20311/power-transmission-report_270913.pdf. Accessed 17 Sep 2014
10. IEEE emerging technology portal, 2006–2012 (online) http://www.ieee.org/about/technologies/emerging/emerging_tech_smart_grids.pdf. Accessed 17 Sep 2014
11. Ministry of Power, Government of India, "India Smart Grid knowledge Portal," (online) http://indiasmartgrid.org/en/smartgridgyan/Pages/sgg2.aspx. Accessed 17 Sep 2014
12. Omer A et al (2013) Indian power system: issues and opportunities. Int J Adv Res Electr, Electron Instrum Eng 2(3):1089–1094
13. Central Electricity Authority, Ministry of Power, Government of India, "Load Generation Balance Report," (online) http://www.cea.nic.in/reports/yearly/lgbr_report.pdf. Accessed 13 Jan 2015
14. Akyildiz IF et al (2002) A survey on sensor networks. Commun Mag, IEEE 40.8:102–114
15. Wissner M (2011) The smart grid—a saucerful of secrets? Appl Energy 88.7:2509–2518
16. Simoes MG et al (2012) A comparison of smart grid technologies and progresses in Europe and the US. Ind Appl, IEEE Trans 48.4:1154–1162
17. Depuru SSSR, Wang L, Devabhaktuni V (2011) Smart meters for power grid: Challenges, issues, advantages and status. Renew Sust Energy Rev 15.6:2736–2742
18. Datta A, Mohanty P, Gujar M (2014) Accelerated deployment of Smart Grid technologies in India-Present scenario, challenges and way forward. In: Innovative Smart Grid Technologies Conference (ISGT), 2014 IEEE PES. IEEE, 2014
19. Initiatives of Ministry of Power, Ministry of Power, Government of India, "Restructured Accelerated Power Development and Reforms Programme" (online) http://www.apdrp.gov.in/Forms/Know_More. Accessed 25 Jan 2015
20. Arthapedia, Indian Economic Service, "RAPDRP" (online) http://www.arthapedia.in/index.php?title=Restructured_Accelerated_Power_Development_and_Reforms_Programme(R-APDRP). Accessed 25 Jan 2015
21. Jadhav GN, Dharme AA (2012) Technical challenges for development of smart grid in India. In: Advances in Engineering, Science and Management (ICAESM), 2012 International Conference on. IEEE, 2012
22. Modi V (2005) Improving electricity services in rural India. Center on Globalization and Sustainable Development Working Paper Series: The Earth Institute at Columbia University
23. The Climate Group, Non Profit Organization, London from 2004 (online) http://www.theclimategroup.org/. Accessed 09 Jan 2014
24. Global e-Sustainability Initiative (2002) Industry as a partner for sustainable development: information and communications technology. United Kingdom
25. Ministry of Power, Government of India, "India Smart Grid Knowledge Portal" (online) http://indiasmartgrid.org/en/Pages/Projects.aspx. Accessed 13 Jan 2015
26. Ministry of Power, Government of India, "India Smart Grid Knowledge Portal" (online). http://indiasmartgrid.org/en/Pages/smrtcities.aspx. Accessed 25 Jan 2015
27. Power Grid Corporation of India Limited (PGCIL) (online). http://www.powergridindia.com/_layouts/PowerGrid/User/ContentPage.aspx?PId=154&LangID=English. Accessed 09 Nov 2014
28. Northern Region Load Despatch Centre, A wholly owned subsidiary of PGCIL (online). http://www.nrldc.org/docs.aspx. Accessed 26 Jan 2015

29. Western Region Load Despatch Centre, A wholly owned subsidiary of PGCIL (online). http://www.wrldc.com/docs.aspx. Accessed 26 Jan 2015
30. Balijepalli VSKM et al (2010) SmartGrid initiatives and power market in India. Power and Energy Society General Meeting, 2010 IEEE. IEEE, 2010
31. Crompton Greaves Limited (CGL) (online). http://www.cgglobal.com/frontend/news_detail.aspx?str=vrsXCPqmMRw=&prd=8x1dMSBwHT8. Accessed 26 Jan 2015
32. North Delhi Power Limited (NDPL) and Tata Power—DDL as a joint venture (online) http://www.ndpl.com. Accessed 28 Jan 2015
33. BESCOM (Bangalore Electricity Supply Company) (online). http://bescom.org/en/smart-grid-vision-for-india/. Accessed 12 Feb 2015
34. India Smart Grid Forum, Ministry of Power (online). http://indiasmartgrid.org/en/Lists/SmartGrid_Project/Attachments/9/14%20Smart%20Grid%20Pilots%20-%20Updates.pdf. Accessed 10 Dec 2014
35. Power Grid Corporation of India Limited, Government of Puducherry (online). http://apps.powergridindia.com/smartgrid/functionalities.aspx#. accessed 12 Feb 2015
36. Ministry of New and Renewable Energy, Government of India (online). http://www.mnre.gov.in/solar-mission/jnnsm/introduction-2/. Accessed 12 Feb 2015
37. Amin SM, Wollenberg BF (2005) Toward a smart grid: power delivery for the 21st century. Power and Energy Mag, IEEE 3.5:34–41
38. Fan J, Borlase S (2009) The evolution of distribution. Power Energy Mag, IEEE 7(2):63–68
39. Rahimi F, Ipakchi A (2010) Demand response as a market resource under the smart grid paradigm. Smart Grid, IEEE Trans 1(1):82–88
40. Zaballos A, Vallejo A, Selga JM (2011) Heterogeneous communication architecture for the smart grid. Network, IEEE 25(5):30–37
41. Li F et al (2010) Smart transmission grid: vision and framework. Smart Grid, IEEE Trans 1.2:168–177
42. Sood VK et al (2009) Developing a communication infrastructure for the smart grid. Electr Power Energy Conf (EPEC), 2009 IEEE. IEEE, 2009
43. Wang W, Yi X, Khanna M (2011) A survey on the communication architectures in smart grid. Comput Netw 55(15):3604–3629
44. Aggarwal A, Kunta S, Verma PK (2010) A proposed communications infrastructure for the smart grid. Innov Smart Grid Technol (ISGT), 2010. IEEE, 2010
45. Gungor VC, et al (2011) Smart grid technologies: communication technologies and standards. Ind Inform, IEEE Trans 7.4:529–539
46. Sauter T, Lobashov M (2011) End-to-end communication architecture for smart grids. Ind Electron, IEEE Trans 58(4):1218–1228
47. Yang Q, Barria JA, Green Tim C (2011) Communication infrastructures for distributed control of power distribution networks. Ind Inform, IEEE Trans 7(2):316–327
48. Moslehi K, Kumar R (2010) Smart grid-a reliability perspective. Innov Smart Grid Technol (ISGT), 2010. IEEE, 2010
49. Gungor VC, Hancke GP (2009) Industrial wireless sensor networks: challenges, design principles, and technical approaches. Ind Electron, IEEE Trans 56(10):4258–4265
50. Li H, Zhang W (2010) QoS routing in smart grid. Global Telecommun Conf (GLOBECOM 2010), 2010 IEEE. IEEE, 2010
51. Kuzlu M, Pipattanasomporn M, Rahman S (2014) Communication network requirements for major smart grid applications in HAN, NAN and WAN. Comput Netw 67:74–88
52. Saputro Nico, Akkaya Kemal, Uludag Suleyman (2012) A survey of routing protocols for smart grid communications. Comput Netw 56(11):2742–2771
53. Hart DG (2008) Using AMI to realize the Smart Grid. In: Power and energy society general meeting-conversion and delivery of electrical energy in the 21st century, 2008 IEEE. IEEE, 2008
54. Siano Pierluigi (2014) Demand response and smart grids—A survey. Renew Sustain Energy Rev 30:461–478

55. Gellings CW (2009) The smart grid: enabling energy efficiency and demand response. The Fairmont Press, Inc.
56. Law YW et al (2014) Control and communication techniques for the smart grid: An energy efficiency perspective. World Congr 19(1)
57. Ranganathan V (2005) Determining T & D losses in India: their impact on distribution privatisation and regulation. Econ Polit Weekly (2005):657–668
58. Fadlullah ZM et al (2011) Toward intelligent machine-to-machine communications in smart grid. Commun Mag, IEEE 49.4:60–65
59. Lee J-S, Su Y-W, Shen C-C (2007) A comparative study of wireless protocols: Bluetooth, UWB, ZigBee, and Wi-Fi. In: Industrial Electronics Society, 2007. IECON 2007. 33rd Annual Conference of the IEEE. IEEE, 2007
60. Luan S-W et al (2009) Development of a smart power meter for AMI based on ZigBee communication. Power Electronics and Drive Systems, 2009. PEDS 2009. International Conference on. IEEE, 2009
61. Bauer M et al (2009) Packet-oriented communication protocols for smart grid services over low-speed PLC. Power Line Communications and Its Applications, 2009. ISPLC 2009. IEEE International Symposium on. IEEE, 2009
62. Yan Ye et al (2013) A survey on smart grid communication infrastructures: motivations, requirements and challenges. Commun Surv Tutorials, IEEE 15.1:5–20
63. Kaur S (2013) Intelligence in wireless networks with cognitive radio networks! Pushing frontiers with the first lady of emerging technologies. IETE Tech Rev 30(1):6–11
64. Yu R et al (2011) Cognitive radio based hierarchical communications infrastructure for smart grid. Network, IEEE 25.5:6–14
65. Ranganathan R et al (2011) Cognitive radio for smart grid: theory, algorithms, and security. Int J Digital Multimedia Broadcast 2011 (2011)
66. The Statistics Portal (online). http://www.statista.com/statistics/271748/the-largest-emitters-of-co2-in-the-world/. Accessed 20 Feb 2015
67. Mishra UC (2004) Environmental impact of coal industry and thermal power plants in India. J Environ Radioact 72(1):35–40
68. Fettweis G, Zimmermann E (2008) ICT energy consumption-trends and challenges. In: Proceedings of the 11th International Symposium on Wireless Personal Multimedia Communications, vol 2(4)
69. Kakemizu Mitsuaki, Chugo Akira (2009) Approaches to green networks. Fujitsu Sci Tech J 45(4):398–403
70. Hasan Ziaul, Boostanimehr Hamidreza, Bhargava Vijay K (2011) Green cellular networks: a survey, some research issues and challenges. Commun Surv Tutorials, IEEE 13(4):524–540
71. Erol Kantarci M, Mouftah HT Energy-efficient information and communication infrastructures in the smart grid: a survey on interactions and open issues
72. Hossain MR, Oo AMT (2010) Evolution of smart grid and some pertinent issues. In: Universities Power Engineering Conference (AUPEC), 2010 20th Australasian. IEEE, 2010
73. Fan Z et al (2010) The new frontier of communications research: smart grid and smart metering. In: Proceedings of the 1st international conference on energy-efficient computing and networking. ACM, 2010
74. Boyer SA (2009) SCADA: supervisory control and data acquisition. International Society of Automation.
75. Mehta Gitanjali, Singh Sajjan Pal (2013) Power quality improvement through grid integration of renewable energy sources. IETE J Res 59(3):210–218

Chapter 24
System-Level Performance Analysis of Embedded Systems for GSM Applications

M. Rajendra Prasad and D. Krishna Reddy

24.1 Introduction

Embedded computing systems are characterized by the execution of the processors, and optimized applications are running continuously on dedicated hardware systems [1]. As computing system complexity increases as per the requirement of industry, the processor performance has to be increased to meet the constraints of the specifications; hence, embedded computing system design methodology for application-specific system is becoming more challenging. These make many researchers to rethink and make efforts in the design and analysis of customization of hardware system and optimization of application code in the area of engineering applications [1, 2]. In this paper, a detailed procedure is described to compare the low-level performances and analysis of three different processor boards based on MPC 8548E PowerPC based board, S3C2440 board, and Raspberry Pi board. MPC 8548E processor board consists of MPC 8548E PowerPC processor 1.3 GHz from Free Scale with DDR, two flash banks and SDRAM for storage, and two 10/100 M Ethernet RJ-45 (DM9000) for communication interfaces [3]. ARM processor board comprises Samsung ARM920t processor with speed of 405 MHz, 3.5 inch touch screen LCD, and 1G nand flash. Raspberry Pi is an ARM11-based processor with speed of 700 MHz and consists of onboard storage called SD/MMC/SDIO card slot. This SD card consists of images of boot loader, linux, and rootfs.

Lmbench is a system-level performance evaluation tool and used to evaluate and analyze the hardware system for Unix/Linux/POSIX operating system. It evaluates two key features: latency and bandwidth of data movement among processor,

M. Rajendra Prasad (✉)
Department of ECE, Vidya Jyothi Institute of Technology, Hyderabad, India
e-mail: rajendraprasad@vjit.ac.in

D. Krishna Reddy
Department of ECE, Chaitanya Bharathi Institute of Technology, Hyderabad, India
e-mail: dkreddy@cbit.ac.in

V. Nath (ed.), *Proceedings of the International Conference on Nano-electronics, Circuits & Communication Systems*, Lecture Notes in Electrical Engineering 403,
DOI 10.1007/978-981-10-2999-8_24

memory, and network which are commonly used in optimized application code. The latency benchmarks include memory read latency, context switching, networking, process creation, signal handling, and system call overhead. The bandwidth benchmarks include memory read, memory write, cached file read, TCP, and Pipe [2, 4]. It not only evaluates these benchmarks but also evaluates other miscellaneous benchmarks such as processor clock rate calculation, cache line size, i/o timing measurements, how much memory size we have to use, menu sleep which gives information on sleep, parallelism in the memory hierarchy, John McClain's stream measurements, translation lookaside buffer used to determine the size of pages, and disk specifications.

24.2 Related Work

Benchmarking and system-level performance analysis are extremely important stage for telecom application developers who are working on dedicated platforms such as embedded systems. There are different types of approaches to benchmarking, which are greatly differ from lmbench benchmarks. L. McVoy discussed an overview of lmbench and its benchmarks. This paper also describes the bandwidth and latency benchmarks, disk, and system call benchmarks. It also compared the shown benchmarks of different systems [5]. J.D McCalpin developed STREAM benchmark description method. It also elaborates the procedure to measure the benchmark of memory bandwidth for a large number of customized hardware systems. It gives the sustainable memory bandwidth and latency. It also presented broad survey of memory bandwidth for a large variety of current computers [6]. Mogul [4] presented an outline of various benchmarking approaches, current benchmarks, and their shortcomings. Gal-On and Levy [7] described multicore benchmarking techniques and their performance. In this paper, the detailed procedure to evaluate and analyze the system-level performance is described using PowerPC-based MPC 8548E processor, ARM920T processor, and ARM1176JZF-S processor-based custom boards. Using lmbench system-level performance analysis, tool bandwidth, latency, and other benchmarks measurements are analyzed which are greatly helpful for the telecom application developers at early stages of their product development [8].

24.3 Experiment Setup

An experiment setup is designed to analyze the performance of three processor boards used as MS, BTS, BSC, and MSC for GSM application development. The technical specifications of customized processor boards are shown in Table 24.1.

Nowadays, RTLinux is an emerging operating system because of open source, powerful features, and very suitable for embedded telecom-based applications. In

Table 24.1 Technical specification of processor boards

	MPC8548E	ARM920T	Raspberry Pi
CPU	Freescale MPC8548E PowerPC Processor 1.3 GHz	Samsung S3C2440 ARM920t 405 MHz with 3.5 inch touch screen LCD	700 MHz ARM1176JZ F-S core (ARM11 family, ARMv6 instruction set)
Memory	DDR2-256 MR BootFlash-32 MB UserFlash-32 MB 2 serial EEPROMS	SDRAM-64 MB, 1 GB of NAND flash and 2 MB of NOR flash with BIOS, one SD card socket	SDRAM-256 MB or 512 MB (shared with GPU), onboard memory Storage-SD/MMC slot
Ethernet	Two 10/100 Ethernet ports	10/100 M Ethernet RJ-45 (DM9000)	10/100 Mbit/s Ethernet (8P8C) USB adapter on the third port of the USB hub
Operating system	Linux-2.6.10	Linux-2.6.32.2	Linux-3.10.25

Fig. 24.1 Top side view of MPC8548E board

three boards, we have chosen RTLinux as an operating system [8]. The top side view of MPC 8548E processor-based board is shown in Fig. 24.1 [9].

The top view of ARM920T and ARM1176JZF-S processor-based boards is shown in Figs. 24.2 and 24.3.

Fig. 24.2 Top side view of ARM920T board

24.4 Building RTLinux Kernel Image for ARM Board

In this paper, a procedure to port Linux kernel on ARM processor is described, and the method of transplanting and booting of Linux kernel on MPC8548E board is discussed [10] Before starting the performance analysis of this board, here is the procedure to ARM-based board that brings up with Linux kernel image. For this development process, arm-linux-gcc-4.4.3 cross-compiler is used.

24.4.1 Install Cross-Compiler

After downloading cross-compiler arm-linux-gcc-4.4.3.tgz, it is untared and installed by executing the following commands.

Fig. 24.3 Top side view of Raspberry Pi board

```
# sudo tar xfvz arm-linux-gcc-4.4.3.tgz
# cd opt/ARM/toolschain/4.4.3/bin
# export PATH = $PWD:$PATH
```

To test whether it is installed or not, use the following command.

```
# arm-linux-gcc –v
```

This command will get the version of cross-compiler and information about whether it is correctly installed or not. Then, download Linux kernel source code from Kernel.org. This processor board supports Linux 2.6 version.

24.4.2 Configuring the Kernel for ARM 9mini2440

Configure kernel for mini2440 using menuconfig.

To set configuration for the target board and build the kernel image, use the commands as shown in Fig. 24.4:

```
# cd linux-2.6.32.2
# cp config_mini2440_x35.config
# make ARCH = arm CC = arm-linux- menuconfig
```

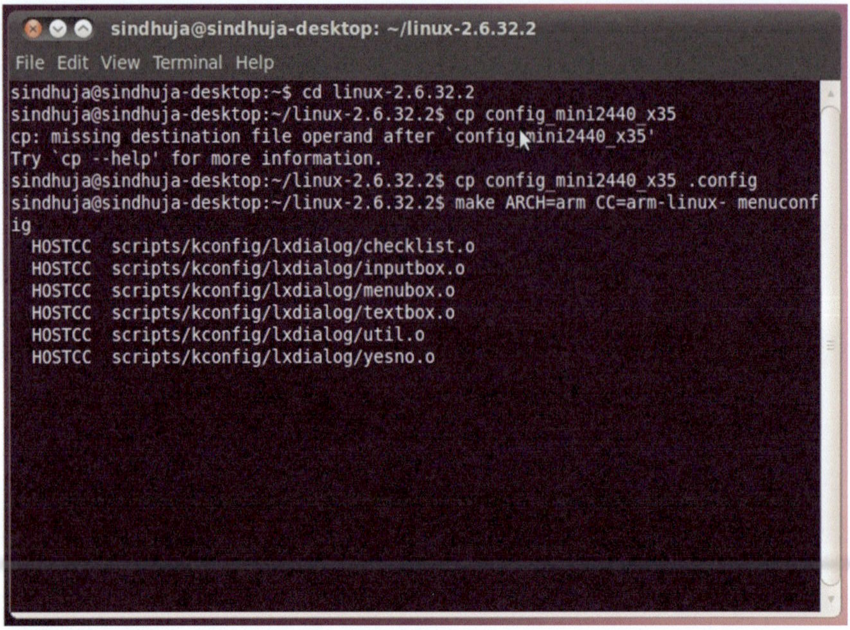

Fig. 24.4 Configuration of Kernel for ARM mini2440

24.4.3 Cross-Compiling Kernel

The Linux kernel source code is compatible for many CPU architectures. In this code, each architectural CPU has its system-related files/codes that are available in respective directory—for example, for ARM processor, arch/arm; for PowerPC, arch/ppc. To build the Linux kernel image for respective processor board, the following command is used, and for ARM-based processor boards, the following command is used.

make ARCH = arm CC-arm-linux-

Now required cross-compiled Linux kernel image supportive to boot on ARM processer hardware system is available in arch/arm/boot directory.

24.4.4 Flashing Kernel Image on to Processor Board

To flash this zImage onto board using windows operating system, a ARM USB Download Driver and DNW tool are used. Then, PC is connected to the processor board using RS232 female-to-female cable and USB cable for transferring the image from the host system to target board. The processor board is configured in

NOR mode; then, the power is turned ON. The serial port settings are shown in the DNW title bar when port is opened successfully (COM1, 115200 bps).

After flashing kernel, power off the processor board and switch on to the NAND mode. The booting messages on hyperterminal in windows operating system are viewed. Reboot the processor board with the new kernel image. Similar procedure is used to port Linux kernel image to MPC 8548E processor board and Raspberry Pi processor board [11].

24.5 Transplanting Lmbench on Arm Processor Board

IP-BTS is a telecom application that implements the different functionalities such as multiplexing, channel encoding, call establishment, speech communication through full rate speech channel, and transmission of signaling over TCP/IP on Abis interface on Linux operating system transplanted on different processor boards [7, 12]. For optimized implementation of IPBTS software, telecom application developer should characterize the early system-level performance details of hardware and operating system of the processor board [13]. To evaluate these results at early stages, the latest version of lmbench can be downloaded from http://sourceforge.net/projects/lmbench/ and its version is lmench-3.0-a9.

24.5.1 Cross-Compiling Lmbench

Uncompress or untar the downloaded file using the following commands to cross-compile.

```
# tar –xvf lmbench-3.0-a9.tgz
# cd lmbench-3.0-a9/src
# make CC = arm-none-linux-gnueabi-gcc AR = arm-none-linux-gnueabi-ar RANLIB = arm-none-linux-gnueabi-ranlib OS = arm-none-linux-gnueabi
```

After doing this, “bin” folder is created in home directory of lmbench.

24.5.2 Porting Lmbench on ARM Processor Board

Power on the ARM processor board, then go to directory of lmbench, and run the following commands.

```
# cd lmbench/src
# OS = arm-none-linux-gnueabi../scripts/config-run
```

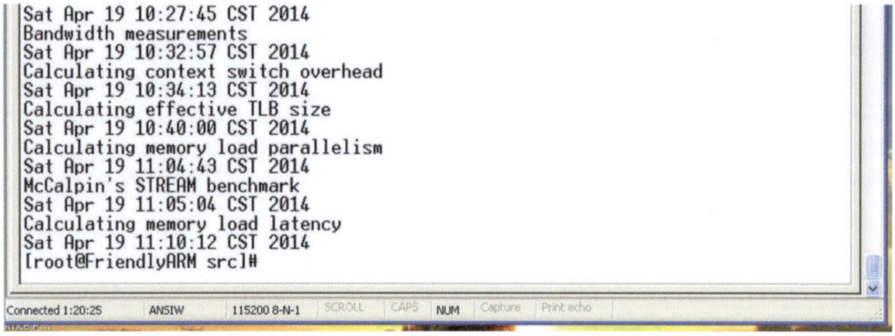

Fig. 24.5 Lmbench results on ARM mini2440 board

To evaluate the bandwidth and latency measurements of board for any application at early stage, the following command is used.

OS = arm-none-linux-gnueabi../scripts/results

After running lmbench on ARM, the following messages are displayed as shown in Fig. 24.5.

Technical report of this board is generated at lmbench/results/arm-none-linux-gnueabi directory. A file called ARM.0 is created which gives full report of the board.

24.5.3 Lmbench on Raspberry Pi

Similar procedure is used to transplant and run the lmbench on Raspberry Pi board. After configuring the lmbench for Raspberry Pi ARM processor board, the results will be displayed as shown in Fig. 24.6.

24.5.4 Lmbench on MPC 8548E

Lmbench running on MPC 8548E processor board is shown in Fig. 24.7 to evaluate latency, networking, and bandwidth measurements of PowerPC processor board.

24.6 Results and Discussions

Lmbench is an open-source micro-benchmark tool to evaluate and analyze the low-level system-related feature measurements for operating system and customized hardware system at early stages of telecom application development

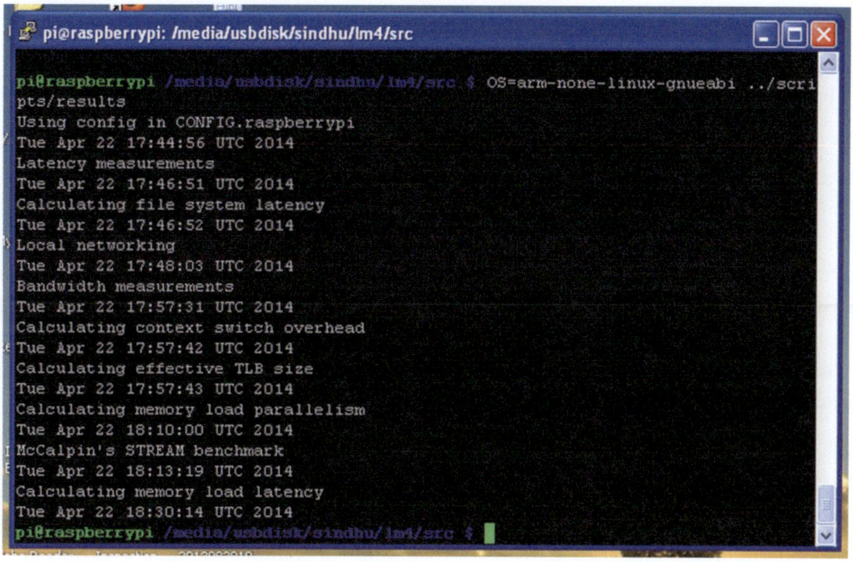

Fig. 24.6 Lmbench results on Raspberry Pi board

Fig. 24.7 Lmbench on MPC 8548 processor board

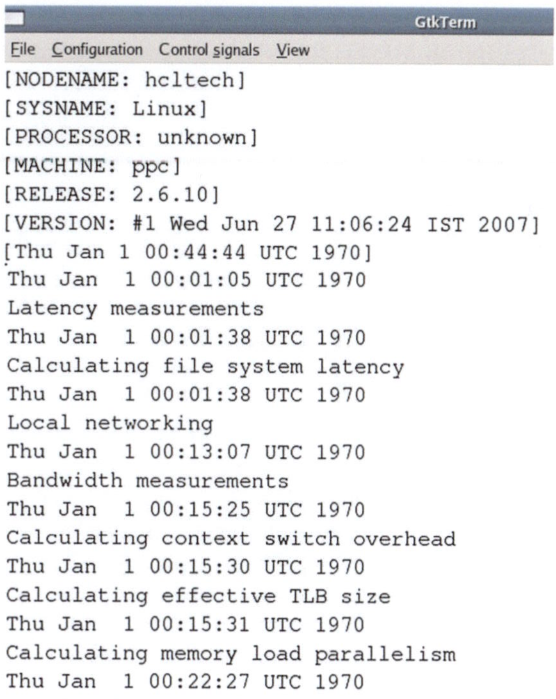

process. These are classified into three categories: bandwidth, latency, and measurements such as IP-network measurements [14].

(1) **Bandwidth Benchmarks**:

Bandwidth is defined as rate of data transfer, bit rate, or throughput, and is evaluated in bits per second which is fundamental in most embedded systems [3]. Lmbench comprises different facilities to measure the data movement features such as library bcopy, hand-unrolled bcopy, direct read and write, pipes, the read interface, TCP sockets, and the mmap interface which are very much useful for GSM telecom application developer at very early stage. Bandwidth benchmark is mainly divided into two types: operating system overhead and memory speeds [15]. The comparison of different bandwidth benchmarks in three processor boards is discussed in this paper.

(i) **Memory Bandwidth**

It is used to evaluate the ability to copy, read, and write data over different sizes of data processing on customized hardware system. It measures bandwidth of different memory block sizes that are evaluated in megabytes and corresponding memory bandwidths in megabytes per second. The comparison of bandwidth memory read in three processor boards is given in Table 24.2.

From the above analysis, read bandwidth of 1.05 MB memory size, MPC 8548E processor board takes 353.06 MB/s, ARM takes 103.09 MB/s, and Raspberry Pi takes 160.38 MB/s [15]. By observing above table, MPC 8548E is faster than other two boards and ARM is the slowest one. Any telecom engineer developing an application may involve more in the memory bandwidth which is a critical factor for network performance. Hence, PC 8548E processer board is suitable [16].

(ii) **Pipe Bandwidth(bw_pipe)**:

This benchmark evaluates the data movement through pipes, which is used in BTS application as shown in Table 24.3. Bw_pipe creates a Unix pipe between two processes and moves 50 MB through the pipe in 64 KB chunks. From this table, MPC 8548E is faster than other two boards.

(iii) **bw_tcp**:

It evaluates data movement over TCP/IP sockets. TCP bandwidth is identical to pipe bandwidth except the transport mechanism. Socket is an endpoint of an inter-process communication flow across a computer network. Socket bandwidth using localhost is shown in Table 24.4, and socket bandwidth using localhost is shown in Fig. 24.8. In IPBTS application, Layer 1 communicates with FPGA through raw socket over Ethernet (SPIF). The raw socket is created during the Layer 1 startup. This system level performance investigation is much helpful for a Layer 1 telecom developer in GSM application development while TCP/IP sockets are created [12, 17].

Table 24.2 Memory read bandwidth benchmark

MPC8548E		ARM920T		Raspberry Pi(ARM11)	
Memory read		Memory read		Memory read	
Bandwidth		Bandwidth		Bandwidth	
0.000512	2874.97	0.000512	529.85	0.000512	675.73
0.001024	2851.07	0.001024	530.90	0.001024	675.80
0.002048	2895.11	0.002048	531.63	0.002048	681.55
0.004096	2917.85	0.004096	532.05	0.004096	681.93
0.008192	2929.22	0.008192	531.89	0.008192	682.42
0.016384	2934.98	0.016384	333.79	0.016384	654.54
0.032768	2834.31	0.032768	103.96	0.032768	339.85
0.065536	1059.53	0.065536	103.94	0.065536	305.19
0.131072	1059.69	0.131072	103.99	0.131072	223.84
0.262144	1059.75	0.262144	103.02	0.262144	161.05
0.524288	532.63	0.524288	103.09	0.524288	160.21
1.05	353.06	1.05	103.05	1.05	160.38
2.10	345.84	2.10	103.09	2.10	160.84
4.19	342.84	4.19	103.07	4.19	160.98
8.39	342.34	8.39	103.05	8.39	161.11
16.78	342.35	16.78	103.08	16.78	161.00
33.55	342.38			32.55	160.96
67.11	342.38			67.11	160.94
134.22	342.39			134.22	160.78
				268.44	160.84

Table 24.3 Pipe bandwidth benchmark

MPC8548E	ARM920T	Raspberry Pi(ARM11)
Pipe bandwidth: 396.29 MB/sec	Pipe bandwidth: 21.25 MB/sec	Pipe bandwidth: 120.50 MB/sec

(2) **Latency Benchmarks**:

On most embedded systems, control messages are also fundamental for the system-level performance analysis. The latency benchmarks are to evaluate how fast a system can be told to do some operation [18].

Latency values are evaluated to measure timing constraints for raw socket transmission and shown here in microseconds per operation.

(i) **lat_ops**:

It is used to evaluate the latency of basic CPU operations such as integer, uint64 bits, integer add, float and double add, integer mul, maximum parallelism for integer, uint64, float, double XOR, ADD, MUL, DIV, and MOD operations [16]. The comparisons of lat_ops measurement in three processor boards are shown in Table 24.4.

Table 24.4 lat_ops benchmark

MPC8548E	ARM920T	Raspberry Pi(ARM11)
integer bit: 1.00 ns	integer bit: 3.86 ns	integer bit: 1.46 ns
integer add: 1.00 ns	integer add: 3.65 ns	integer add: 1.53 ns
integer mul: 4.00 ns	integer mul: 2.05 ns	integer mul: 0.72 ns
integer div: 35.02 ns	integer div: 554.32 ns	integer div: 187.09 ns
integer mod: 16.01 ns	integer mod: 135.72 ns	integer mod: 50.31 ns
int64 bit: 1.01 ns	int64 bit: 7.29 ns	int64 bit: 2.96 ns
int64 add: 1.26 ns	unit64 add: 10.95 ns	uint64 add: 3.14 ns
int64 mul: 8.05 ns	int64 mul: 6.35 ns	int64 mul: 1.89 ns
int64 div: 137.60 ns	int64 div: 2075.69 ns	int64 div: 667.38 ns
int64 mod: 105.11 ns	int64 mod: 1565.47 ns	int64 mod: 432.19 ns
float add: 4.00 ns	float add: 143.04 ns	float add: 44.79 ns
float mul: 4.00 ns	float mul: 113.47 ns	float mul: 39.58 ns
float div: 29.02 ns	float div: 528.40 ns	float div: 179.16 ns
double add: 6.00 ns	double add: 211.29 ns	double add: 63.20 ns
double mul: 6.01 ns	double mul: 195.27 ns	double mul: 52.81 ns
double div: 32.03 ns	double div: 2075.61 ns	double div: 842.27 ns
float bogomflops: 28.04 ns	float bogomflops: 1135.07 ns	float bogomflops: 407.86 ns
double bogomflops: 32.05 ns	double bogomflops: 3885.94 ns	double bogomflops: 1179.80 ns

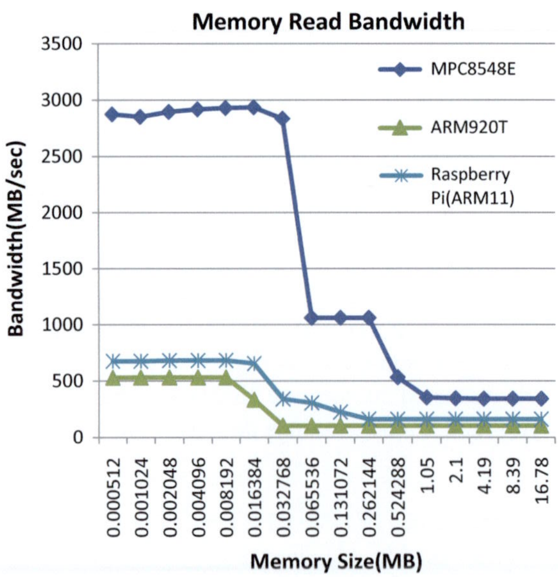

Fig. 24.8 Memory read bandwidth benchmark

By observing all the above values, MPC 8548E is faster and ARM is slower.

(ii) **Context Switching Measurement**:

Context switch time is referred as the time taken to store and restore the state of a context of process running customized hardware system [19]. The format of context switch time is multiline, and each subsequent line is a pair of numbers which indicates the numbers of processes running on the hardware system with benchmark timings as shown in Table 24.5.

Table 24.5 Context switching benchmark

MPC8548E		ARM920T		Raspberry Pi (ARM11)	
"size = 0 k ovr = 1.67		"size = 0 k ovr = 12.04		"size = 0 k ovr = 7.50	
2	0.63	2	162.83	2	15.11
4	0.90	4	186.18	4	16.32
8	2.29	8	188.04	8	19.31
16	2.58	16	192.10	16	23.75
24	2.68	24	188.66	24	27.03
32	2.84	32	195.54	32	27.55
64	4.06	64	191.94	64	29.57
96	4.06	96	190.45	96	30.21
"size = 4 k ovr = 3.08		"size = 4 k ovr = 22.34		"size = 4 k ovr = 14.73	
2	0.63	2	205.01	2	20.44
4	1.16	4	233.49	4	30.50
8	5.04	8	245.13	8	34.94
16	5.51	16	243.58	16	46.15
24	5.79	24	242.58	24	50.35
32	5.85	32	239.14	32	53.83
64	9.93	64	234.38	64	55.77
96	13.57	96	233.14	96	57.03
"size = 8 k ovr = 4.48		"size = 8 k ovr = 34.83		"size = 8 k ovr = 22.59	
2	0.71	2	235.00	2	30.14
4	4.05	4	259.48	4	41.41
8	8.12	8	261.16	8	50.03
16	8.15	16	261.06	16	69.24
24	8.98	24	271.73	24	73.7
32	11.31	32	266.90	32	74.38
64	24.05	64	263.85	64	76.82
96	28.18	96	262.43	96	77.24

Table 24.6 Stream benchmarks

MPC8548E	ARM920T	Raspberry Pi(ARM11)
STREAM copy latency: 26.31 ns	STREAM copy latency: 146.80 ns	STREAM copy latency: 61.33 ns
STREAM copy bandwidth: 608.18 MB/sec	STREAM copy bandwidth: 108.99 MB/sec	STREAM copy bandwidth: 260.88 MB/sec
STREAM scale latency: 26.68 ns	STREAM scale latency: 245.28 ns	STREAM scale latency: 95.54 ns
STREAM scale bandwidth: 599.76 MB/sec	STREAM scale bandwidth: 65.23 MB/sec	STREAM scale bandwidth: 169.25 MB/sec
STREAM add latency: 25.65 ns	STREAM add latency: 410.12 ns	STREAM add latency: 201.93 ns
STREAM add bandwidth: 935.58 MB/sec	STREAM add bandwidth: 58.52 MB/sec	STREAM add bandwidth: 118.85 MB/sec
STREAM triad latency: 52.73 ns	STREAM triad latency: 609.18 ns	STREAM triad latency: 202.65 ns
STREAM triad bandwidth: 455.19 MB/sec	STREAM triad bandwidth: 39.40 MB/sec	STREAM triad bandwidth: 118.43 MB/sec
STREAM2 fill latency: 12.17 ns	STREAM2 fill latency: 47.11 ns	STREAM2 fill latency: 10.55 ns
STREAM2 fill bandwidth: 657.12 MB/sec	STREAM2 fill bandwidth: 169.80 MB/sec	STREAM2 fill bandwidth: 758.38 MB/sec
STREAM2 copy latency: 26.31 ns	STREAM2 copy latency: 146.80 ns	STREAM2 copy latency: 61.35 ns
STREAM2 copy bandwidth: 608.14 MB/sec	STREAM2 copy bandwidth: 108.99 MB/sec	STREAM2 copy bandwidth: 260.81 MB/sec
STREAM2 daxpy latency: 39.28 ns	STREAM2 daxpy latency: 529.45 ns	STREAM2 daxpy latency: 180.14 ns
STREAM2 daxpy bandwidth: 611.06 MB/sec	STREAM2 daxpy bandwidth: 45.33 MB/sec	STREAM2 daxpy bandwidth: 133.23 MB/sec
STREAM2 sum latency: 25.65 ns	STREAM2 sum latency: 239.46 ns	STREAM2 sum latency: 90.46 ns
STREAM2 sum bandwidth: 311.86 MB/sec	STREAM2 sum bandwidth: 33.41 MB/sec	STREAM2 sum bandwidth: 88.44 MB/sec

(3) **Stream Benchmarks**:

The STREAM benchmark was developed by McCalpin while working as a faculty at the University of Delaware [20–22]. It is used for measuring sustainable memory bandwidth and latency, and the corresponding computation rate for simple vector kernels is shown in Table 24.6.

(4) **tlb**:

Translation lookaside buffer (TLB) is a cache that memory management hardware used to improve the virtual address translation speed [23, 24]. It determines the size in the pages of tlb. TLB of three processor boards is given in Table 24.7.

Table 24.7 tlb Benchmark

MPC8548E	ARM920T	Raspberry Pi(ARM11)
tlb: 32 pages	tlb: 64 pages	tlb: 10 pages

24.7 Conclusion

In this paper, lmbench is transplanted into three processor boards with different specifications and analyzed their system-level performances. From the results of system-level performance analysis of these three processor boards, we observed that MPC 8548E is best among three processor boards. For memory bandwidth and latency benchmarks, TCP benchmarks, context switching benchmarks MPC 8548E processor board is faster. So, these boards are not only used as BTS/BSC/MSC also used for any telecommunication signaling application platforms running on different protocol stacks. In simple system call functions, there is not much difference. In memory load parallelism measurements, ARM is best. In ARM and Raspberry Pi lmbench results, there is not much difference. We can use these two boards for mobile stack applications.

Acknowledgements We would like to thank Correspondent and Director of Vidya Jyothi Institute of Technology, Hyderabad, for their encouragement to publish this paper.

References

1. En-Nouaary A, Khendek F, Dssouli R (2000) Testing embedded real-time systems. In: Seventh international conference on real-time systems and applications, pp 417–424
2. Rajendra Prasad M, Krisha Reddy D (2013) Design methodology for IP secured tunel based embedded platform for AAA server. Int J Mob Netw Commun Telematics (IJMNCT) 3(6)
3. Chen H, Qin Z (2010) Reliability demonstration testing method for embedded operating systems. In: 2nd International conference on software engineering and data mining (SEDM), pp 272–275
4. Mogul JC (1999) Brittle metrics in operating systems research. In: 7th IEEE workshop on hot topics in operating systems(HotOS-VII), Rio Rico, AZ, pp 90–95
5. McVoy L, Staelin C (1996) lmbench: Portable tools for performance analysis. In: Processings of Winter 1996 UNSENIX, pp 279–284, Jan 1996
6. McCalpin JD (1995) Memory bandwidth and machine balance in current high performance computers. IEEE Tech Committee Comput Archit Newsl, pp 1–10
7. Gal-On S, Levy M (2008) Measuring multicore performance. Computer 41(11):99–102
8. Saeed Abou Trab M, Counsell S, Hierons RM (2011) GeTeX: a tool for testing real-time embedded systems using CAN applications. In: 18th IEEE international conference and workshops on engineering of computer-based systems, pp 61–70
9. Liu Y, Li Y, Guo L, Sun X (2005) An electrical machine testing system on embedded system. In: Eighth international conference on electrical machines and systems, vol 01, pp 57–59
10. Rajendra Prasad M, Ramasubba Reddy S, Sridhar V (2011) Framework to port linux kernel on powerPC based embedded system used for telecom application—IPBTS. Int J Softw Eng Appl (IJSEA) 2(4):127–139
11. Vudatha CP, Jammalamadaka S, Nalliboena S, Duvvuri, BKK (2011) Automated generation of test cases from output domain and critical regions of embedded systems using genetic algorithms. In: Second national conference on emerging trends and applications in computer science (NCETACS), pp 1–6

12. Jamal MH, Mustafa G, Waheed A, Mahmood W (2009) An extensible infrastructure for benchmarking multi-core processors based systems. Perform Eval Comput Telecommun Syst 41:13–20
13. Rajendra Prasad M, Sharath V, Nagendra G, Damodara Reddy K (2012) Research method to optimizes logger for telecom application running on embedded system. Int J Eng Res Dev 1(6):01–09
14. Wang H, Ding Z, Zhong Y (2008) Static analysis test platform construction for embedded systems. In: International conference on audio, language and image processing (ICALIP), pp 808–812
15. Rajendra Prasad M, Nagendra G, Mathey R, Sridhar V (2012) Research method for transmission of speech through RTP packets on abis interface for IPBTS. Int J Comput Inf Syst 3(6)
16. Iyenghar P, Pulvermueller E, Westeramp C, Wuebbelmann J (2011) Integrated model-based approach and test framework for embedded systems. In: Forum on specification and design languages (FDL), pp 1–8
17. Karmore SP, Mahajan AR (2013) Universal methodology for embedded system testing. In: The 8th international conference on computer science and eduction (ICCSE), pp 567–572
18. Mitsching R, Weise C, Fiedler F, Kowalewski S, Bohnenkamp H (2011) TripleT: improving test responsiveness for high performance embedded systems. In: IEEE Fourth international conference on software testing, verification and validation workshops, pp 67–74
19. Kuo F-C, Chen TY, Tam WK (2011) Testing embedded software by metamorphic testing: a wireless metering system case study. In: IEEE 36th conference on local computer networks, pp 291–294
20. Krishnan P, Venkatesh R, Bokil P, Muske T, Suman V (2012) Effectiveness of random testing of embedded systems. In: 45th Hawaii international conference on system sciences, pp 5556–5563
21. Wahler M, Ferranti E, Steiger R, Jain R, Nagy K (2012) CAST: Automating software tests for embedded systems. In: IEEE Fifth international conference on software testing, verification and validation, pp 457–466
22. Keranen JS, Raty TD (2012) Model-based testing of embedded systems in hardware in the loop environment. Softw IET 06(4):364–376
23. Wehrmeister MA, Ceron LM, da Silva JL (2012) early verification of embedded systems: testing automation for UML models. In: Brazilian symposium on computing system engineering (SBESC), pp 119–124
24. Karmore SP, Mahajan AR (2013) Universal methodology for embedded system testing. In: The 8th international conference on computer science and eduction (ICCSE), pp 567– 572

Chapter 25
IC Packaging: 3D IC Technology and Methods

Adesh Kumar, Gaurav Verma, Vijay Nath and Sushabhan Choudhury

25.1 Introduction

The technological development in the field of IC packaging [1, 2] is involved day by day to miniaturize the chip size, and industries are trying to integrate more functionality in the same area. To meet the current functional requirement and cost-effective solutions, Integrated chip package system (ICPS) has been proved for flexible solutions and more dimension of integration than Moore's Law. The future of IC integration includes MEMS, photonics devices, and optical solutions into subsystems. ICPS is helpful in the development of System on a Chip (SoC) [3, 4] and for solving the problems of memory distribution for larger systems, customizing the product for several applications, and integrating multiple devices into single chip package. In the manufacturing of integrated circuits (ICs), the final stage used in semiconductor ICs in the fabrication process is packaging. The process that provides the electrical contacts to connect the device with board is known as 'packages.' In the semiconductor industry, it is called 'packaging' or simply 'assembly.' It is also referred as seal or encapsulation or 'semiconductor device

A. Kumar (✉)
Department of Electronics, Instrumentation and Control Engineering,
University of Petroleum and Energy Studies (UPES), Dehradun, India
e-mail: ADESHKUMAR@ddn.upes.ac.in

G. Verma
Department of Electronics & Communication Engineering,
Jaypee Institute of Information Technology (JIIT), Noida, India

V. Nath
VLSI Design Group, Department of Electronics & Communication Engineering,
Birla Institute of Technology (BIT), Ranchi, India

S. Choudhury
Department of Electronics, Instrumentation and Control Engineering,
University of Petroleum and Energy Studies, Dehradun, India

V. Nath (ed.), *Proceedings of the International Conference on Nano-electronics, Circuits & Communication Systems*, Lecture Notes in Electrical Engineering 403,
DOI 10.1007/978-981-10-2999-8_25

Fig. 25.1 IC packaging

System
Printed Circuit Board (PCB)
Package
Chip

assembly.' IC testing is the part of IC packaging and provides large number of I/O interconnections on a bare chip to shrink its size. The main purpose of IC packaging is to prevent physical damage and corrosion of tiny block of semiconductor [5] materials. It is helpful to provide connections such as input signal, ground, and power. Effective surface area is also increased with aids of heat dissipation into PC board; protection of physical part is against the IC breakage; and environmental protection is against corrosion or moisture. The level of IC packaging is shown in Fig. 25.1. IC packaging is the intermediate process between the final process of chip fabrication and printed circuit board (PCB) design.

Integrated circuits were packed ceramic flat in the earliest beginning of ICs and used by the military for their small size and reliability. The packing technique was scaled very fast to dual in-line package (DIP) with plastic flat. In the 1980s, DIP packaging technique was replaced with Pin Grid Array (PGA) in which packages are rectangular and square arranged (2.54 mm, 0.1″ apart) underside of package and Leadless Chip Carrier (LCC) has rounded pins on the edge of molded plastics or ceramics. The reason of the DIP is because VLSI pin exceeds the practical limit for DIP of 48–64 pins and low performance. In the late 1980s, Surface Mounting Technology (SMT) [6] was used which occupies area 30% less in comparison with DIP. The SMT package is having 'gull wing' leads or J-Lead, protrude in two larger sides and one more lead spaced 0.050 in. A greater number of interconnections are increased with the help of area array package. The first chip package of such system is plastic Ball Grid Array (BGA) or ceramic Pin Grid Array (PGA). Flip Chip Ball Grid Array (FCBGA) [6, 7] allows higher number of pin counts. In the FCBGA, die is connected to upward and downward help to connect the substrate and package balls which is similar to PCB rather than wires. FCBGA permits I/O to distribute over entire die instead of die periphery only. Recent developments and research is carried out in stacking many dies in a single package, called System in Package (SIP) or Three-Dimensional integrated circuits (3D) ICs packaging technique. In SIP method, several dies are grouped on a small substrate, generally [8] called often Multi-Chip Module (MCM) ceramic. Moreover, the boundary areas between a small printed circuit board and big MCM are blurry.

3D IC packaging allows multiple devices to be integrated in as single die with larger space. For example, 3G cell phones require higher level of integration, and in such technology, 3D IC integration is used. ICPS processes follow the power integrity and signal integrity. Flip chip packaging yields the solution for higher reliability and throughput for larger I/O. Flip chip provides interdensity in small area and scalable. The technology of flip chip provides low power materials, larger die size, and ultra-low-k dielectrics (Figs. 25.2 and 25.3).

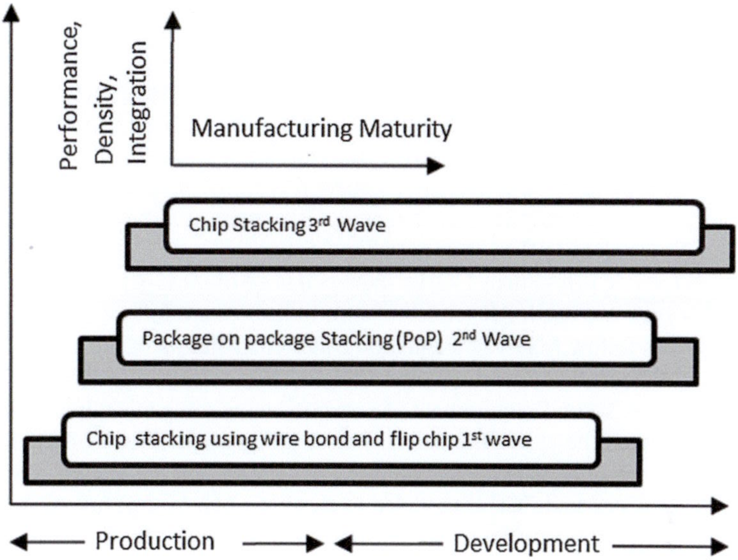

Fig. 25.2 IC packaging technology [9]

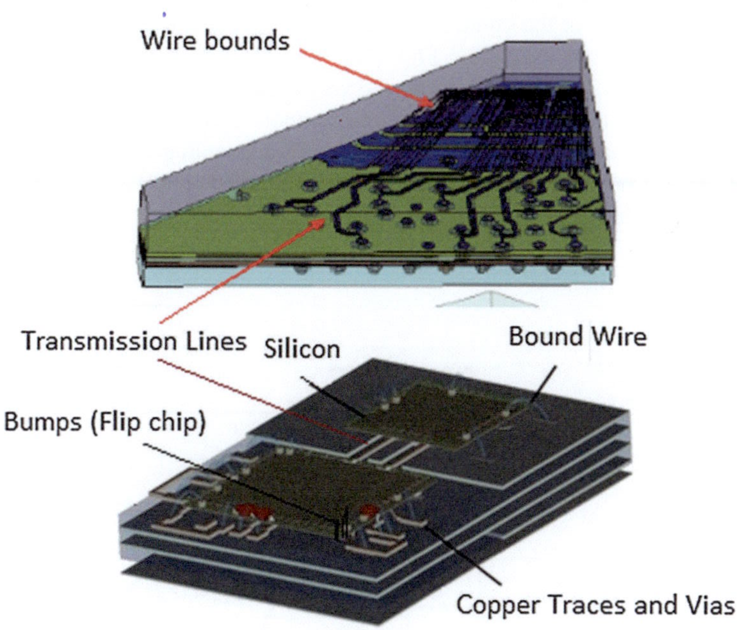

Fig. 25.3 3D EM models [9]

25.2 IC Packaging Design Flow

IC packaging design is a flow, followed by the IC packaging semiconductor [10] companies for final product. The essential part for the packaging that should be known very first is the frequency of the final product. This helps the engineers to select the appropriate tool. It can be understood with the help of example—if the frequency of final product is greater than 1 GHz, there is a need of 3D electromagnetic (EM) solvers [9] as shown in Fig. 25.4.

If the frequency is less than 1 GHz, there is a need of quasi-static tools in which 2.5D EM solvers are used. All semiconductor industries are following EM solvers. 3D EM solvers follow Maxwell's equation code. The solution of these equations is provided by the vendors Ansoft [9] and CST [9]. The selection of right packaging technology is the second requirement of semiconductor companies. In the practical scenario, engineers have the limitation with requirement and cost of technology. After the selection of IC packing technology, package models are integrated with the chip in the third step and PCB is modeled to check the behavior of full effect [9, 11] in the device. Selections of the right tools can help the engineers in different steps. Fourth step is to measure the bare chip package alone. The packager models are correlated with the measurement results of chip.

The radiation of EM [7] affects the each packaging component. In digital circuits packaging induction is the common problem. To reduce the packaging emission, several techniques are used. The Flip Chip Ball Grid Array (FCBGA) has less parasitic effect [7, 12] and signal counts are very high. In high-speed digital circuits, a packaging induction reduction is a very common technique. The radiation and noise depression approach are advantageous with decoupling capacitors. With the help of ground bounce, it is possible to suppress radiated emissions, developing short vias in multilayer.

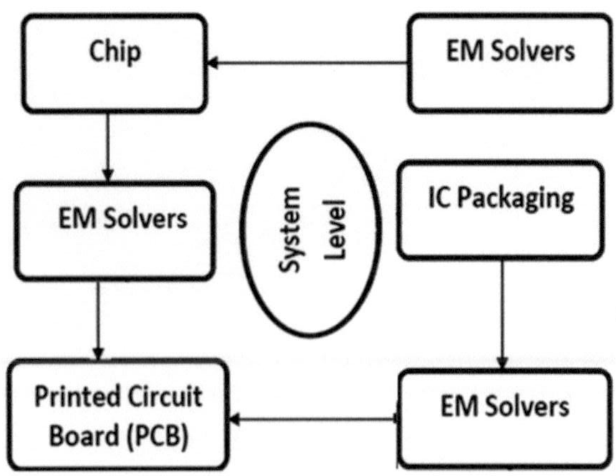

Fig. 25.4 EM model integration of whole device [9]

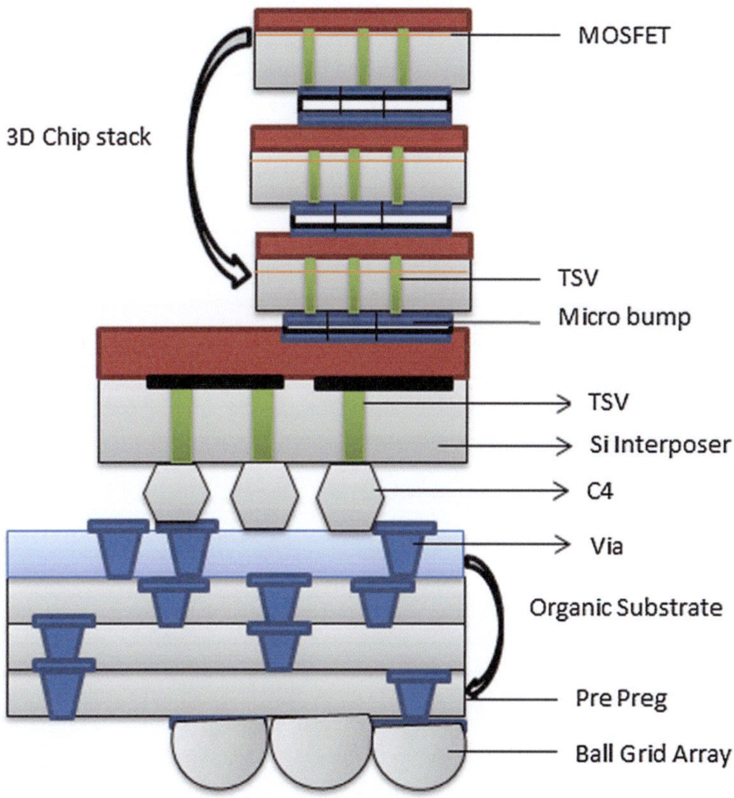

Fig. 25.5 Schematic of 3D packaging [13]

25.3 Chip-Joining Technology

The mechanical stress and strain can damage the chip during the chip packaging. A differential heating/cooling (H/C) chip-joining technique [13] is studied for Pb-free chip packaging based on ultra-low-k (ULK) technology applied on substrate to prevent the damage occurred due to Chip–Package–Interaction (CPI). The substrate and bonder head is mounted using a chip on baseplate. In the assembly process, the temperature difference between organic and Si chip matches the thermal expansion and eliminates the stress caused by the mismatch of coefficient of thermal expansion (CTE).

In case of larger chip size, there exist large CTC mismatch between organic substrate and silicon chip with fine chip interconnects. The mismatch affects the back-end-of-line (BEOL) chip structures due to concentrations of the mechanical strain and stress at interconnects. It can be understood with example, the value of CTE of an organic substrate and silicon chip is 17 and 2.6 ppm/°C [13], respectively. Thus, in flip chip assembly process, organic substrates are expanding and

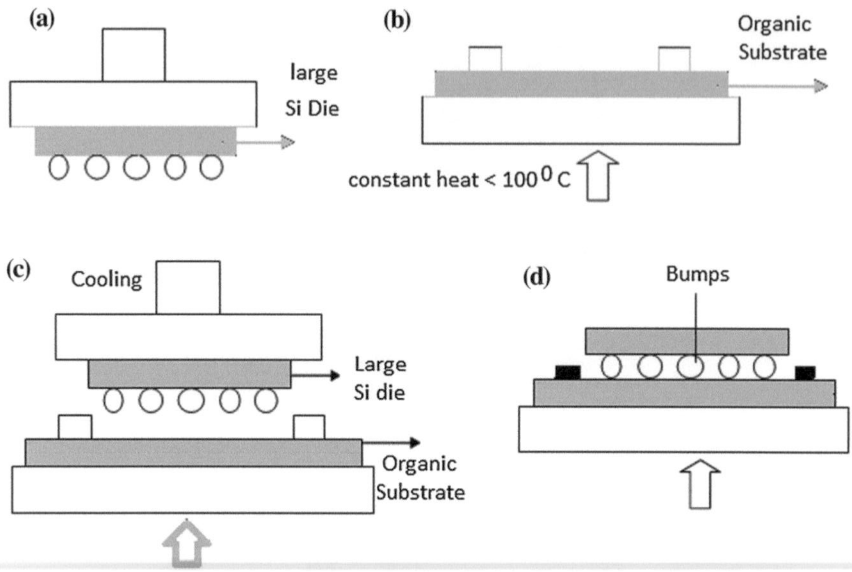

Fig. 25.6 Chip joining using differential heating and cooling [13]

contracting more than silicon chip which cause high level of stress and strain. In 3D integration process, silicon dies are connected in stack as shown in Fig. 25.5. The warpage change of organic substrate and chip decreases the assembly yield and it is big issues during solder reflow. Subsequently, there are also the chances of fracture within the ULK layers before underfelt process and solder reflow. The ULK layers are having low amount of dielectric constant in BEOL chip structures that is also the cause of fracture.

The differential heating/cooling (H/C) chip-joining technique is used to prevent the damage occurred during chip joining using Chip–Package–Interaction (CPI). The ULK semiconductor chips are having CPI as reliability issue to provide Pb-free chip packaging. The differential H/C technique [13] is understood with the description of Fig. 25.6a–d. The bonder head vacuum is placed by a chip, and baseplate vacuum is covered by substrate. Bonder head is used to hold the chip after the alignment between the substrate and chip. It descends until the solder bumps touch the substrate underneath the chip. Bonding process of chip and substrate follows for several elevated temperatures. After the completion of chip alignment and chip placement on substrate, the solder is heated than its melting temperature due to the heat from the chip. The technique matches thermal expansion between the chip and substrate.

The temperature of solder bumps is increased so much by heating of chip that it can melt. The bonder head is disconnected from the chip, as the solder joints are getting their desired height and flip chip is cooled until solder joints are solidify. Throughout the cooling step, optimal distance is kept between substrate and

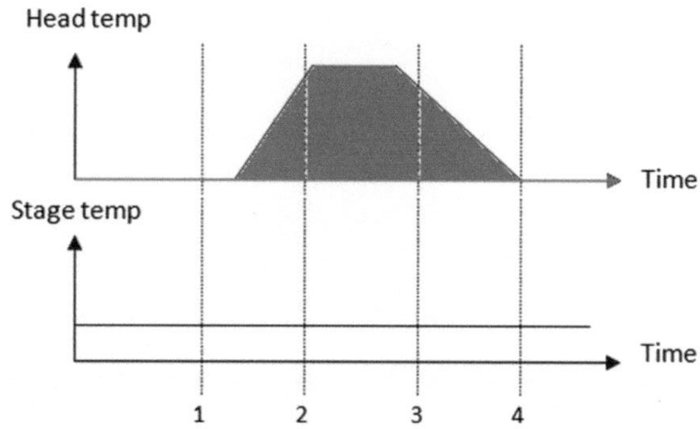

Fig. 25.7 Bonding profile for chip joining [13]

chip. The bonded head releases the vacuum after the bonding is completed. As a result, the discussed joining method eliminates the CPI issue such as solder opening, solder bridging, cracking in ULK interfaces, and delimitation in low k. The cooling rate of solder joints is also controlled by bonder head. The vacuum pressure is always maintained to affix the chip on the bonding head during entire bonding cycle. Differential H/C technique mitigates the CPI issue [13] and improves assembly yield. The bonding temperature profiles for various differential H/C chip-joining technique are shown in Fig. 25.7. The temperature expansion between the substrate and chip is matched to many elevated temperatures of chip. The substrate is kept on the baseplate by strong vacuum that eliminates warpage change in the substrate. The differential heating and cooling method H/C technique is different from conventional reflow. In conventional reflow, there is the need of replacing the chip after ashing and fluxing. The changes in assembly process include the change in equipment based on differential H/C technique instead of convention technique of chip reflow.

25.4 TSV Packaging Technology

The system integration and packaging with 3D IC is much cost-effective solution that can help the industries to provide flexibility in comparison with die solution, in which multifunctional integration does not require scale equivalently in terms of system performance and cost by node. The integration of 3D chip use through-silicon via (TSV) technique that dives 3D IC integration includes Package on Package (PoP) and 3D IC die stacks followed wire bonding and flip chip technique. Semiconductor companies are actively doing research on TSV in IC packaging. The technology needs commitment in terms of cost investment in tools and equipment.

3D IC integration follows low-temperature TSV fabrication technique is used for Benzocyclobutene (BCB) technology for hybrid bonding. TSV has the reliability of adhesive bonding and stacking approach is used, called Wafer-on-Wafer (WoW).

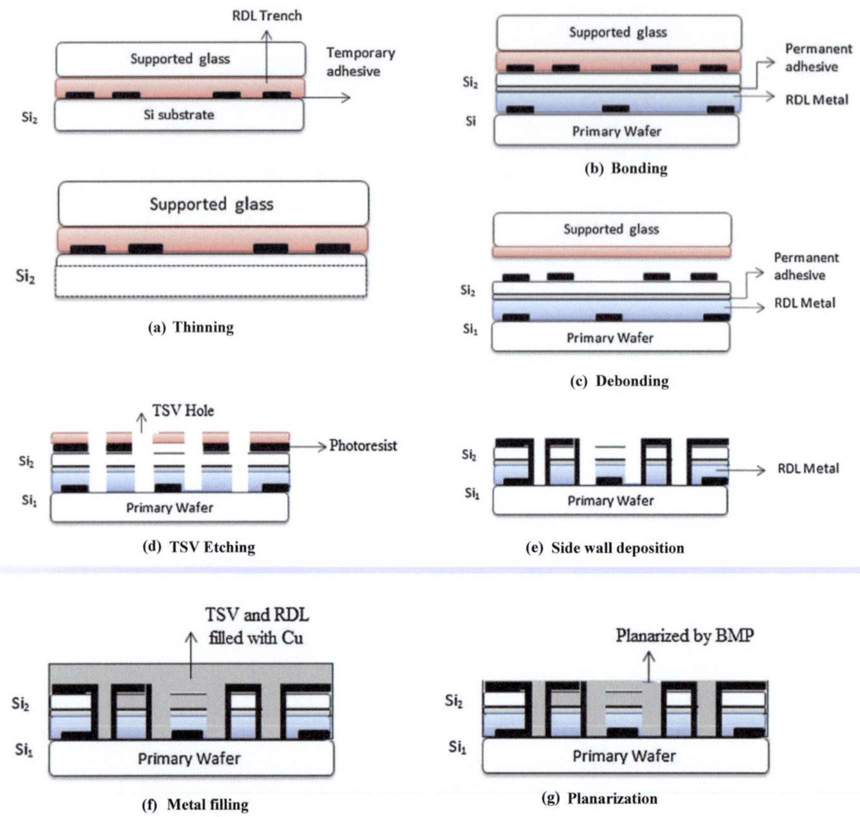

Fig. 25.8 WoW technique for stacked buffers [10]

It is reported by alliance following BCB adhesive bonding. The flow and process of WoW stacking technique is shown in Fig. 25.8. Interconnect metallization and the secondary wafers (Si 2) are bonded temporarily before thinning down to less than 20 μm with transistors to support glass. The same wafer is fixed and bonded to primary wafer (Si 1) and BCB adhesive. The support glass is removed from water stack, and Cu dual-damascene is used to fabricate TSV and RDL subsequently. The photographs of multiple TSV and stacked wafers on BEOL are shown in Fig. 25.8, which shows the image view of FIB–SEM (Focused Ion Beam)–(Scanning Electron Microscope). TSV process is applied at 150 °C low temperature to decrease the pumping stress and increases the reliability on Cu-TSV (Fig. 25.9).

The variable thermal stress based on pumping height in TSV is calculated. The corresponding results are shown in Fig. 25.10. As the process temperature is changing, the pumping stress is decreasing and mismatch with the dielectric film. Dielectric layers are analyzed with the help of BS-SIMS [10] and Cu diffusion profile is also checked in the diffusion at 400 °C, for thermal aging. As a result,

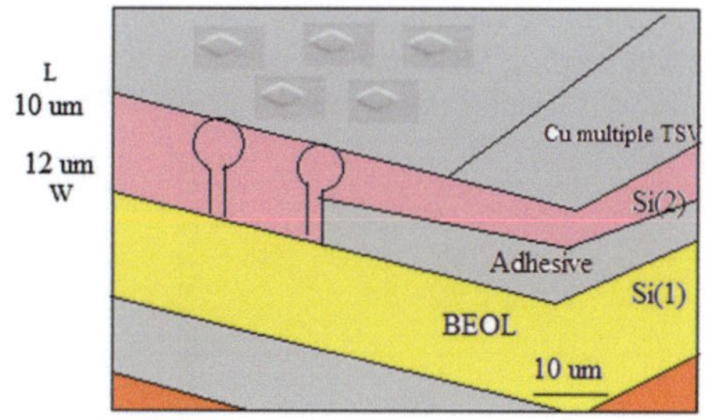

Fig. 25.9 Stacked buffer with motile TSVs [10]

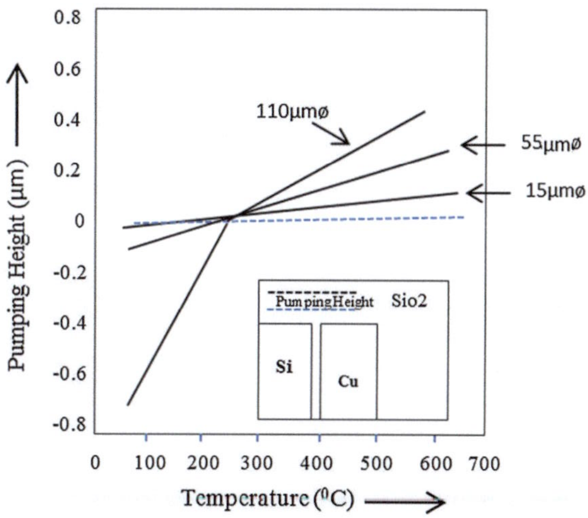

Fig. 25.10 Variation of pumping height with temperature [2, 10]

the deposition temperature is decreasing with Cu diffusion increment. The corresponding graph is shown in Fig. 25.11.

In TSV process, barrier developed at 150 °C is enough against Cu diffusion for SiON film. The film is deposited using Plasma-Enhanced Chemical Vapor Deposition (PECVD) method. Cu diffusion is a function of film density. The values of Cu diffusions are calculated with the Cu depth profiles and relative amount of bulk Si_3N_4 and SiO_2. At low temperature, the diffusion rate is higher at low density film. If the film density found to be less than 50%, the thickness of dielectric layer more than 1000 nm suppressed the Cu diffusion into Si substrate. The thermal stress of Cu diffusion TSV and BEOL is calculated with temperature −50 to 130 °C. Table 25.1 shows the change in resistance of Cu interconnects and it is analyzed that there is no significant change in the resistance of with TSV process tested at 150 °C low-temperature PECVD dielectric film after 1000 cycles. In the overall process, induced stress and temperature are lowered than low-temperature through-silicon via (LT-TSV) method and are suitable very much for TSV reliable structure and applicable 3D ICs stacking method for wafer level.

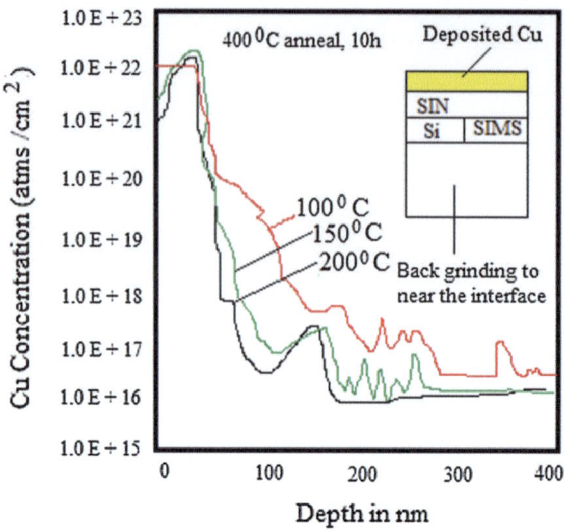

Fig. 25.11 Variation in SIMS depth in SiON barrier layer of Cu diffusion in TSV technique [3, 10]

Table 25.1 Variation in resistance in Cu-TSV and BEOL interconnects after TC testing [3, 10]

	Before TSV method	After TSV method	After TC 1010 cycles
0.5 μm dense chain (Ω)	1.08E+09	1.02E+08	1.05E+08
5 μm single line (Ω)	14.06	14.09	10.55

25.5 Reliability of TSV Interposer

The chip designers are having the limitations that the number of pins is less in the package. The limitation prevents to test the design for individual functional module within the chip. The technological advancements have changed the demand to drive the expansion of functional module on a single chip at low power and low cost per transistor. Interposers [10, 11] permits massive larger parallel busses between memory and logic cells, improving speed and significantly reducing power consumption. The design which is already in 'system on a chip' can integrate multiprocessor chip with heterogeneous die designed and optimized for a particular technology and the same is connected by means of an interposer. A 3D system in package (SIP) contains two or more integrated chips, stacked vertically so that it behaves as a single device, occupies less space, and has greater connectivity. Another type of 3D IC package is provided by IBM, where multiple ICs are not stacked sequentially but a substrate containing TSVs is used to form multiple packages with the help of interconnection of multiple ICs. Due to the interfacing of multiple IC, the length of edge wiring and width of package is increased slightly.

The solution of this is the use of extra 'interposer' between two chips. In most of the new 3D IC packages, TSV changes edge wiring with the help of body of packaging chip through vertical connections and the package formed like this does not have extra length and width. The devices which are stacked may be heterogeneous in nature, e.g., combining DRAM, III-V group materials, and CMOS logic into a single chip. The chip density is increasing day by day because the interconnects are becoming higher and higher. The greatest challenge in the packaging of flip chip is ultra-fine pitch wiring fabrication with substrate material for organic IC. The reliability enhancement is an issue which directly interlink with interconnects. TSV interposer is the best technique that can have high-density interconnect wires, minimizing CTE and improving electrical performance.

Xilinx has provided the solution for higher performance on 28 nm FPGA technology, in which dies placed directly on big TSV interposer. Same FPGA is interfaced with I/O and TSV interposer. The TSV level [14], schematic supporting to package configuration and assembly of SEM images, is shown in Fig. 25.12a–c. The simulation was carried out for strain, fatigue, and inelastic energy to evaluate the μ bump and C4 solder reliability.

Table 25.2 summarizes that the evaluated values for strain, fatigue, and inelastic energy are under consideration. Several designs on experiments (DOEs) [10] are completed for optimal design, material selection, and reliability. The stress of TSV process is detailed with the help of one local model. The overall stress in Cu, Si, and SiO_2 is found less than the fracture toughness of the materials. The results of corresponding materials are shown in Table 25.3. The samples are taken at temperature 50–130 °C to improve gap height and plasma cleaning. The value of stress in Si, SiO_2, and Cu materials below toughness signifies that there is no fracture risk for TSV interposer. No fracture was observed at 110 °C either in level 5 preconditioning or after 264 h of HAST [1] and it is suggested that TSV interposer is much reliable for high-density interconnects and helpful for high-volume production.

The development of the TSV interposer was done by leading manufactures such as ASE, TSMC, and Samsung Electronics. DRAM and I/O memory applications are developed by Samsung Electronics. An interposer is required to form TSV on logic device or CPU. As TSV is formed on active device, a lot of reliability issues are arising. First issue is of TSV stress, between Cu and Si because CTE mismatch for Cu ($\sim$17 ppm) and Si ($\sim$3 ppm). Another issue is that Cu contamination can damage the CMOS layer when Cu diffusion is carried. TSV stress may also cause low value of Inter metal Dielectric (K- IMD) and fracture can happened. ASE and TMSC developed 2.5D and 3D IC technologies, in which passive interposer was used. TSMC developed 2.5D IC interposer for Xilinx FPGA with good reliability. Synopsys extraction tools are upgraded to provide front and back sides of the wafers to account for μbumps, C4 bumps, TSVs, and RDL. Synopsys HSPICE® and Fast SPICE tools are used to understand the 3D structures. TSMC and Altera Company developed world's first 3D IC test vehicle using Chip on Wafer-on-Substrate (CoWoS) process in 2012. TSVs are reducing extra cost of wafer thinning and advanced logic, only needed on the silicon interposer. The overall assembly discussed here is used in a SEM cross section of Xilinx [15] ISE FPGA Virtex 7, which is made of multiple die (slices) of FPGA with 6.8 B

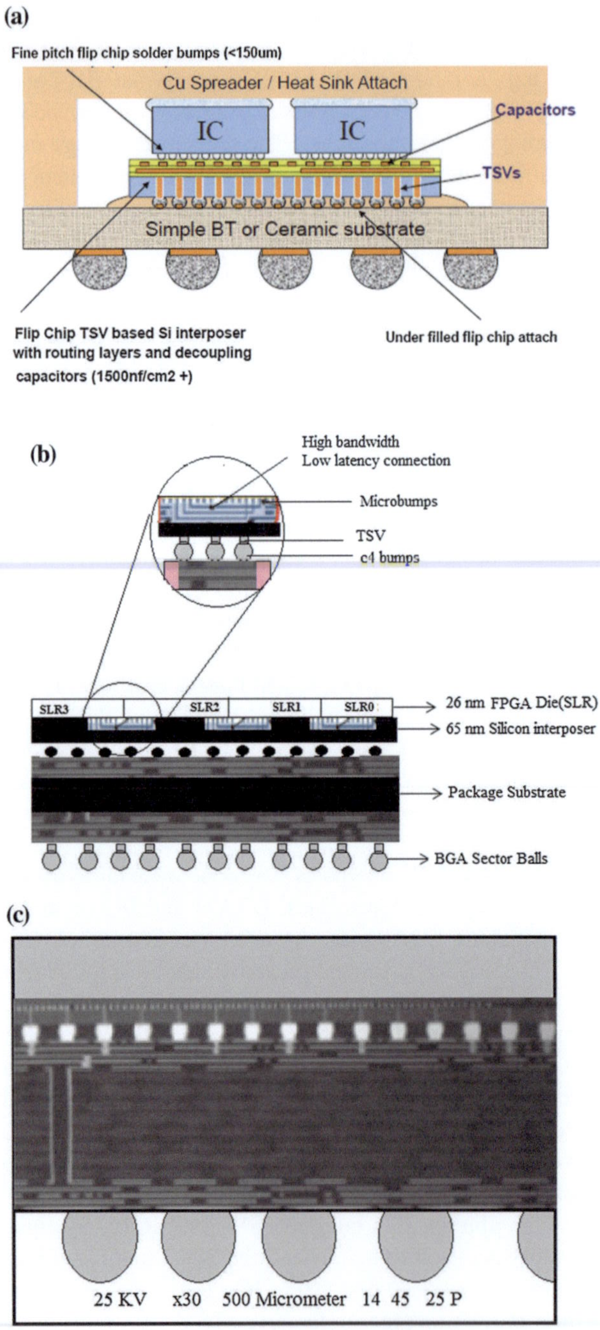

Fig. 25.12 **a** Level of TSV interposer [6], **b** the schematic package configuration [1, 4, 10], **c** overall assembly of TSV interposer [1, 10]

Table 25.2 Summary of solder fatigue taken from Xilinx interposer [1, 10]

	After *x* cycles at 130 °C	After *x* cycles at -50 °C
Inelastic energy for μ bump solder	0.14	0.38
Inelastic strains for μ bump solder	2.79	3.15
Inelastic strains for C4 solder	0.1	0.18

Table 25.3 Stress analysis of TSV in silicon interposer at (−50 to 130 °C) [1, 10]

	Maximum in-plane stress (MPa)	Maximum shear stress (MPa)	Maximum out of plane stress (MPa)
Si	129.5	4.89	83.80
SiO_2	121.4	18.12	79.09
Cu-TSV	145.5	20.65	110.3

transistors (TSMC 28 nm HPL) stacked onto a TSMC 65-nm passive silicon interposer having 4 layers of metal. Only the silicon interposer goes through TSV etching/drilling and wafer back-grinding/thinning. It is manufactured using a mature 65 nm process—features 100–1000×—and it is smaller than 65 nm at the package substrate ones. Silicon interposer has more interconnects, dramatically power consumption and latency. Silicon interposers are sampled now, and the current volume production is too high. With low-temperature bonding and TSV process, Xilinx platform increases reliability, optimized design, and material selection to reduce the induced stress and warpage.

For a packaging company, cost, supply chains, and heat management are the real challenges. The cost of TSV manufacturing process is still more for most of the production. In a 5 μ diameter, each wafer (i.e., each tier) bears an extra cost of $80–100 with 10:1 aspect ratio via middle TSV process technology, or 2–3% extra cost. The enabling point of 2.5D and 3D IC is truly dependent on design and application. Heat management and power management are critical challenges to resolve. Inventory management and delayed payments are major issues for supply chain management and business point of view. The real scenario will come into picture in the second half of the decade. Some companies believe that silicon interposers may represent an acceptable solution for large number of applications to boost performance and integration process to unprecedented levels. 3D-IC integration is certainly warming up, and the runway is still very long. So, 3D IC integration is not suiting for all applications especially at the same stage of maturity.

25.6 Conclusion

The paper listed the several steps on how semiconductor companies would be ready to analyze the next generation of IC packaging methods by reviewing the latest published works. The paper studies the fundamental principles of IC packaging, associated radiations, and reliability. It has been reported that the increase in number of stacked chips reduced the reliability. It has also been reported that solder ball reliability can be increased by adding under fill between interposer and test board. For a 3D IC, the key technologies are bonding technologies and TSV. The reliability of TSV process is evaluated before the real production of IC. The paper reviews the reliability of TSV with adhesive bonding and stacking WoW process. The reliability of TSV interposer is much reliable for 3D IC packaging. The detailed comparative study is carried for Xilinx 3D IC [16] packages with the technologies Virtex Scale +, Virtex Ultra Scale, Kintex Ultra Scale, and Virtex −7 family FPGA. The families of the technologies are compared with hardware parameters such as logic cells, Block RAM, DSP slices, PCI Express, Transreceiver, and I/O Pins. It is estimated that Virtex Ultra Scale architecture has maximum I/O supports and logic cells for IC integration and packing.

References

1. Banijamali B, Ramalingam S, Nagarajan K, Chaware R (2011) Advanced reliability study of TSV interposers and interconnects for the 28 nm technology FPGA. In: Proceedings of Electronic components and technology conference (ECTC)
2. Kitada H, Maeda N, Fujimoto K, Mizushima Y, Nakata Y, Nakamura T et al (2010) Stress and diffusion resistance of low temperature CVD dielectrics for multi- TSVs on bumpless wafer-on-wafer (WOW) technology. In: Advanced metallization conference
3. Fujimoto K et al (2010) Development of multi-stack process on wafer-on-wafer (WoW). In: The IEEE CPMT symposium Japan
4. Nimura M, Mizuno J, Shigetou A, Sakuma K, Ogino H, Enomoto T, Shoji S (2013) Study on hybrid Au–underfill resin bonding method with lock-and-key structure for 3-D integration. IEEE Trans Compon Packag Manuf Technol 3(4):558–565
5. Beyne E (2007) IMEC Kapeldreef 75,3001 Leuven "Tutorial T7A: Advanced IC packaging". In: 20th International Conference on VLSI Design (VLSID'07), 2007, p 1
6. Fukushima T, Iwata E, Ohara Y, Murugesan M, Bea J-C, Lee K-W, Tanaka T, Koyanagi M (2012) IEEE Trans Electron Devices 59:2956 (PDF by Alliva TSV and interposer technologies at Alliva, 2010, p 11)
7. Huang NKH, Jiang LJ, Yu H, Li J, Xu S, Ren H (2012) Fundamental components of the IC packaging electromagnetic interference (EMI) analysis. IEEE 141–144
8. Ito Y, Fukushima T, Lee K-W, Choki K, Tanaka T, Koyanagi M (2012) Flux-assisted self-assembly with microbump bonding for 3D heterogeneous integration. In: 2013 Electronic Components & Technology Conference, IEEE, pp 891–897
9. AbouGabol M (2011) Importance of integrated circuit (IC) packaging in semiconductor Companies design flow. In: IEEE GCC conference and Exhibition (GCC), Feb 2011 Dubai, United Arab Emirates, pp 355–356
10. Ko CT, Chen KN (2012) Reliability of key technologies in 3D integration. In: Microelectronics reliability, Elsevier, pp 7–12
11. Chen W-H, Yu C-F, Cheng HC, Tsai Y, Lu S-T (2012) IMC growth reaction and its effects on solder joint thermal cycling reliability of 3D chip stacking packaging. In: Microelectronics reliability. Elsevier, pp 30–40
12. Tu KN, Hsiao HY, Chen C (2012) Transition from flip chip solder joint to 3D IC micro bump: its effect on microstructure anisotropy. Elsevier, pp 2–6

13. Sakuma K, Smith K, Tunga K, Perfecto E, Wassick T, Pompeo F, Nah J-W (2012) Differential heating/cooling chip joining method to prevent chip package interaction issue in large die with ultra low-k Technology. 978-1-4673-1965-2/12/$31.00 ©2012 IEEE, pp 430–436
14. Ohara Y, Noriki A, Sakuma AK, Lee K-W, Murugesan M, Bea J-C, Yamada F, Fukushima T, Tanaka T, Koyanagi M (2009) Proceedings of 3D System Integration, p 389
15. www.xilinx.com/products/silicon-devices/fpga/virtex-ultrascale-plus.html
16. Lee Y-J, Su Y-F, Hung T-Y, Chiang K-N {2012) Reliability analysis of 3D IC integration packaging under drop test condition. IEEE Catalog Number: CFP1259B-ART, pp 299–302

Chapter 26
Modeling and Simulation of the Dynamic Response of a Generic Mechanical Linkage for Control Application Under the Consideration of Nonlinearities Imposed by Friction

Jitendra Yadav and Geeta Agnihotri

26.1 Introduction

In many industrial applications, involving a control mechanism using open kinematic chains along with spring or similar flexible elements, the positioning and dynamics of the target are prone to errors due to the presence of friction. Some of such applications are flight control mechanism, mechanical systems using single-acting hydraulic and pneumatic actuators, hydraulic copy turning, and robotics. In control applications, it is essential to model the dynamic response of the mechanical system to a sufficient accuracy, in order to obtain desired precision level. The mathematical model developed for obtaining the dynamic response would include the friction behavior of components. Further, the force of friction would depend upon the state variables such as velocity and displacement as has been shown in earlier research work. The objectives of the present work are to understand the system dynamics identifying the relevant parameters, to understand the inter-relationship among them, to determine their relative significance, and to estimate the extent of errors in system performance due to friction.

Researchers have attempted in past to model friction by including different effects and experimental observations, such as Coulomb friction, viscous friction, hysteresis, elastoplastic deformation at asperities, stiction, and Stribeck effect. The classical models fail to model behavior such as hysteresis and break-away.

J. Yadav (✉)
Mechanical Engineering Department, University of Petroleum and Energy Studies, Dehradun, Uttarakhand, India
e-mail: jyadav@ddn.upes.ac.in

G. Agnihotri
Mechanical Engineering Department, Maulana Azad National Institute of Technology, Bhopal, Madhya Pradesh, India
e-mail: dr.gagnihotri@gmail.com

V. Nath (ed.), *Proceedings of the International Conference on Nano-electronics, Circuits & Communication Systems*, Lecture Notes in Electrical Engineering 403,
DOI 10.1007/978-981-10-2999-8_26

A method for using the classical Coulomb model is described in detail [1, 2]; this is due to the fact that the Coulomb model is discontinues at zero velocity. Due to this disadvantage, it is not widely used for simulation purposes. de Canudas et al. [3] proposed a dynamic model for friction that includes most of the friction phenomena, i.e., stiction, Stribeck effect, hysteresis, and varying break-away force. Nguyen et al. [4] simulated a dynamic system using LuGre model for friction. Researchers proposed many friction models [5–8]. Tripathi et al. [9] performed simulation of dynamic response of a mechanical system using the Coulomb and LuGre model for friction. Gafvert [10] compared friction models on rate dependency, model order, and damping.

26.2 Modeling of Friction

LuGre model
The complete LuGre model of dynamic friction is described as follows:

$$F_f(t) = \sigma_0 Z + \sigma_1 \dot{Z} + \sigma_2 V \tag{26.1}$$

Here,

$$\dot{Z} = V - \frac{|V|}{g(V)} Z \tag{26.2}$$

and

$$g(V) = \frac{1}{\sigma_0}\left[F_C + (F_S - F_C)e^{-\left(\frac{V}{V_s}\right)^2}\right] \tag{26.3}$$

Therefore,

$$F_f = \sigma_0 Z + \sigma_1 V - \frac{\sigma_0 |V| Z}{\left[F_C + (F_S - F_C)e^{-\left(\frac{V}{V_s}\right)^2}\right]} + \sigma_2 V \tag{26.4}$$

where,

Z The lateral deflection of the bristle, expressed in microns
σ_0 Average bristle stiffness
σ_1 Micro-viscous friction coefficient
σ_2 Constant for viscous friction
F_C Static friction force, corresponding to Coulomb friction
F_S Dynamic friction force
V_S Stribeck velocity

Dahl Model

It is based on the stress–strain curve of classical mechanics which is governed by differential equation:

$$\frac{\mathrm{d}Z}{\mathrm{d}t} = v - |v|\frac{\sigma_0 Z}{F_\mathrm{C}} \tag{26.5}$$

$$F_\mathrm{f} = \sigma_0 Z \tag{26.6}$$

where Z is a state variable.

Dahl model is generalization of ordinary Coulomb model by introducing the system dynamics, i.e., presliding displacement, but it neither captures Stribeck effect nor stiction.

Coulomb Model

The classical Coulomb friction model is depicted in Fig. 26.1. The model shows discontinuity at zero velocity, which possess problem in the analytical solution of the dynamic system. For numerical computation using computer, this problem has been dealt with by obtaining piecewise solution for different sections and then integrating them. However, this is quite tedious and the method has to be devised separately for every individual system. The governing equation for the model is shown by Eq. 26.9.

$$F_\mathrm{f} = F_\mathrm{C}\mathrm{sgn}\,(v) \tag{26.7}$$

The problem of discontinuity in Coulomb model can be circumvented by the use of sigmoid function, as shown in Fig. 26.2. The sigmoid function is a monotonically increasing function with, asymptotically bound upper and lower limits of 0 and 1, respectively. The shape of the curve is controlled by the index n.

Sigmoid function is defined by Eq. 26.8 as a function of velocity.

$$F = \frac{1}{1 + \mathrm{e}^{-n.\mathrm{velocity}}} \tag{26.8}$$

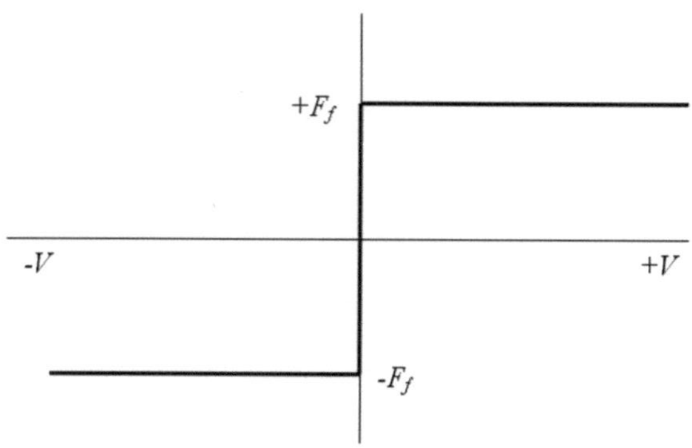

Fig. 26.1 Coulomb friction model

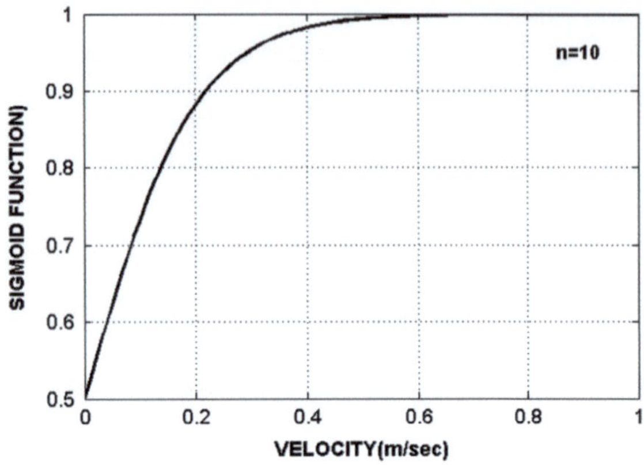

Fig. 26.2 Sigmoid function versus velocity plot

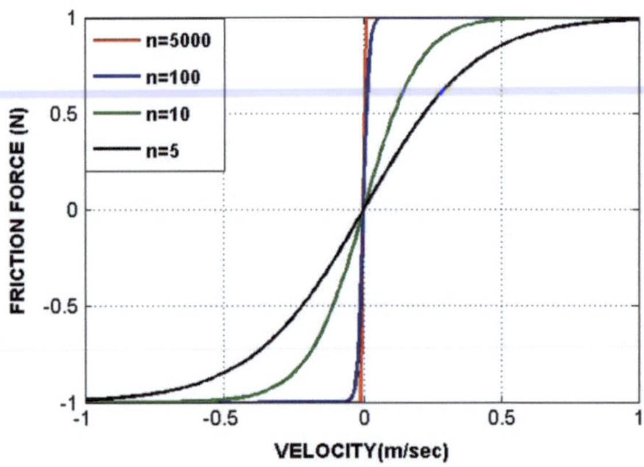

Fig. 26.3 Friction force versus velocity plot with the help of sigmoid function

The sigmoid function-based friction model can be expressed by Eq. 26.9.

$$F_{\mathrm{f}} = \mu mg\left[2\left(\frac{1}{1+\mathrm{e}^{-n.\mathrm{velocity}}} - 0.5\right)\right] \tag{26.9}$$

The friction force characteristic as a function of velocity by Eq. 26.11 for different values of index n is shown in Fig. 26.3.

The curves in Fig. 26.3 are confined to the velocity range −1.0 to +1 m/s. The velocity range is arbitrarily selected and can be extended to any value, as required by a particular dynamic system without any need for change in the computer program. It can be observed that for the value of index n = 5000 the curve can be compared quite well with the classical Coulomb model.

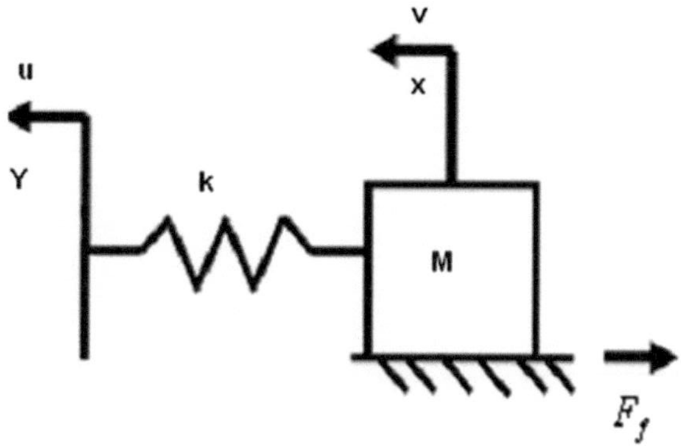

Fig. 26.4 Spring–mass system

Validation of Sigmoid Function-Based Friction Model

The proposed sigmoid function-based Coulomb model is validated by comparing the results with the results of Dahl model as both the models are the static models. For that a simple spring–mass system as shown in Fig. 26.4 is selected for the simulation.

For the system, the mass is driven with constant velocity on a rough surface in contact. The mathematical modeling of the identified problem can be a single degree of freedom, second-order spring–mass system dragged with a constant velocity (u), and the governing mathematical equation for the system is given by Eq. 26.10.

$$m\ddot{x} + k(y - x) - F_{\mathrm{f}} = 0 \qquad (26.10)$$

In the equations, $\dot{y} = u$ is the driving velocity and $\dot{x}$ is the velocity of mass.
where

- m is the mass of the moving body,
- x is the position of the body,
- y is the independent position of the wall,
- k is the spring constant, and
- F_{f} is the friction force.

The response of the system under the influence of sigmoid function-based model for the selected parameter as in Table 26.1, in terms of displacement, velocity, and friction force verses time plot, is depicted in Fig. 26.5.

The response of the system for Dahl model for the same parameters is shown in Fig. 26.6.

It is concluded from the comparison of responses shown in Figs. 26.5 and 26.6 that in case of both the models, there are friction-induced vibrations in the form of sinusoidal wave of constant frequency and amplitude and that are identical to both the models.

Though there is difference in friction force pattern as in the case of Coulomb model friction force in continuous and constant with time, it is intermittent with time in case of Dahl model that may be because of presliding displacement in case

Table 26.1 Values used in simulation for spring–mass system

Parameter	Value
Mass (m)	1 kg
Stiffness (K)	5 N/m
Micro-stiffness of bristle (σ_0)	100,000 N/m
Coulomb friction coefficient (μ)	0.1
Coulomb kinetic friction (F_C)	$\mu * \mathrm{m\,g\,N}$
Driving velocity (u)	0.1 m/s
Sigmoid index (n)	5000

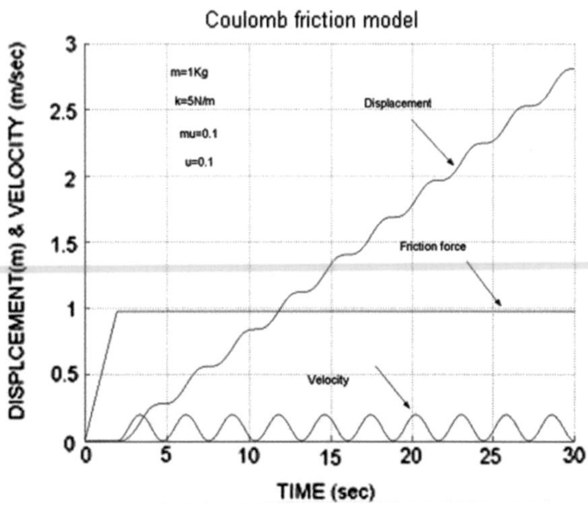

Fig. 26.5 Results of simulation using sigmoid function-based friction model

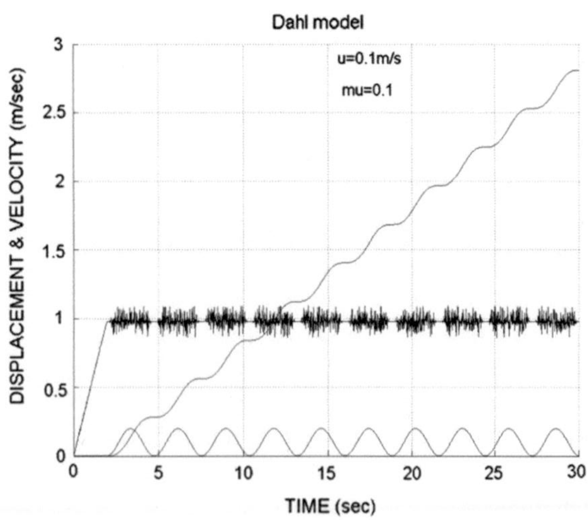

Fig. 26.6 Displacement, velocity, and friction force versus time plot for Dahl model

of Dahl model. So the sigmoid function-based model is validated by the result of simulation of the Dahl model.

26.3 Generalized Mechanical System for the Generic Linkage

The mechanical system used for carrying out simulation consists of a lever with moment of inertia as *I*, mass *m*, and a linear spring with stiffness *k*. The mass is subjected to force of friction. The spring is attached at one end to the mass, and at other with to a fixed support.

The lever is representative of a pedal, on which a time-varying force *F*(*t*) is applied. The other end of the lever is connected to the cable. The schematic diagram of the system is shown in Fig. 26.7. The simulation is performed considering that the lever is initially pressed down completely. Under this situation, the spring is under tension, and the cable mass is displaced toward the left. Upon release of the pedal, the mass shall move toward right, under the spring force. The tension in the cable is *T*(*t*). The governing equation is represented by a mathematical model depicted in Eq. 26.11.

$$\left[\frac{I}{\sqrt{l_2^2 - x^2}} + m\sqrt{l_2^2 - x^2}\right]\ddot{x} = \frac{-F(t)l_1\sqrt{l_2^2 - x^2}}{l_2} - F_f(t)\sqrt{l_2^2 - x^2} + kx\sqrt{l_2^2 - x^2} - \left[\frac{Ix(\dot{x})^2}{\left(l_2^2 - x^2\right)^{\frac{3}{2}}}\right] \quad (26.11)$$

Defining the Forcing Friction
The forcing function *F*(*t*) has been defined as a variable force, which is maximum in the beginning, i.e., when the pedal is completely pressed down. Then this force is

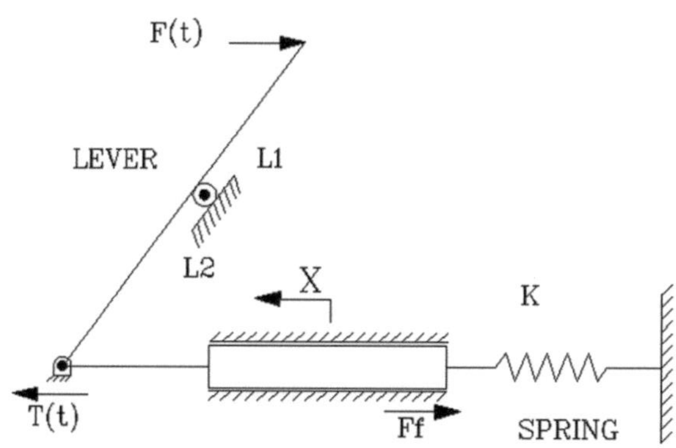

Fig. 26.7 Schematic diagram of generalized mechanical system

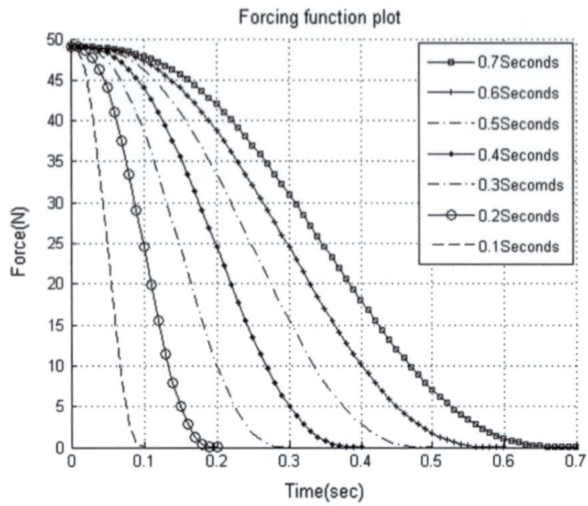

Fig. 26.8 Trajectory of applied force $F(t)$

gradually released, as is the case when the operator starts to release the pedal. Figure 26.8 shows the nature of forcing function, used in the simulation.

26.4 Results and Discussion

The dynamic model of the mechanical system shown in Fig. 26.7, the model of friction force, i.e., LuGre model and sigmoid function-based Coulomb friction model, and the model of forcing function described in the previous sections were

Table 26.2 Values used in simulation

Parameter	Value
Mass (m)	1 kg
Moment of inertia of lever	0.01234 kg m^2
Length l_1	0.25 m
Length l_2	0.06 m
Stiffness (K)	6600 N/m
σ_0	100,000 N/m
σ_1	$2\zeta\sqrt{(\sigma_0 \times m)}$ N s/m
σ_2	0.04 N s/m
F_C	Mass $*\ \mu$ g N
F_s	$1.5 * F_C$ N
V_s	0.01 m/s
ζ	0.2–0.7
ω_n	5 rad/s
ω_d	4.33 rad/s
Coulomb friction coefficient	0.1
Sigmoid index (n)	5000

used for simulation of dynamic response of the mechanical system. The values for various parameters used for this simulation are given in Table 26.2.

Comparison of the Models

The simulation was carried out using two models of friction, the LuGre model and sigmoid function-based Coulomb friction model. Figure 26.9 shows the comparison between LuGre and Coulomb models on the basis of total settling time taken by the target mass to settle after the force is removed. It can be clearly seen from the figure that the Coulomb model takes longer settling time than the LuGre model. The time taken for lower values of release time is high for both the models as the

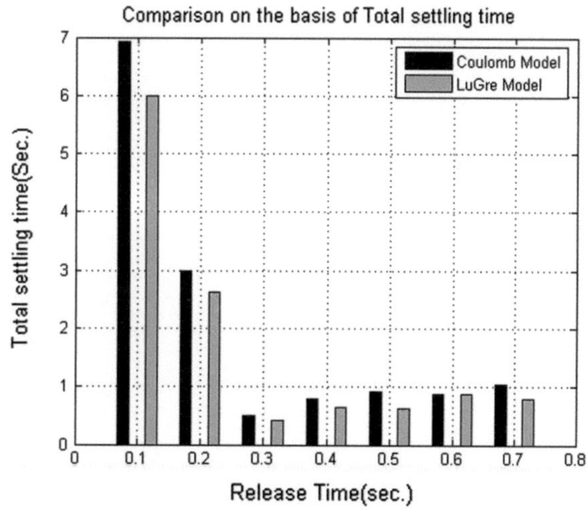

Fig. 26.9 Comparison on the basis of total settling time

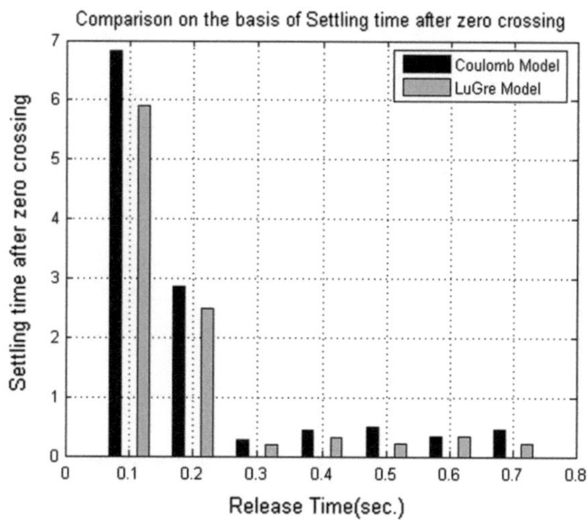

Fig. 26.10 Comparison on the basis of settling time after zero crossing

velocity of the system is high. The same phenomenon can be observed more clearly in the post-zero-crossing part of the dynamic response, as shown in Fig. 26.10. The time taken by the LuGre model to settle after zero crossing is less than Coulomb model for increasing values of release time. In the pre-zero-crossing part, the dynamics of the system is controlled mainly by the release force trajectory. Therefore, both the models show almost the same behavior, as evident from Fig. 26.11. The value for settling time before zero is increasing by a same value for both the models used.

In the present study, the computation time is the time required to obtain the dynamic response of the system numerically, for certain time duration. The time duration for all the results was taken as 8 s in this study. The computation time is

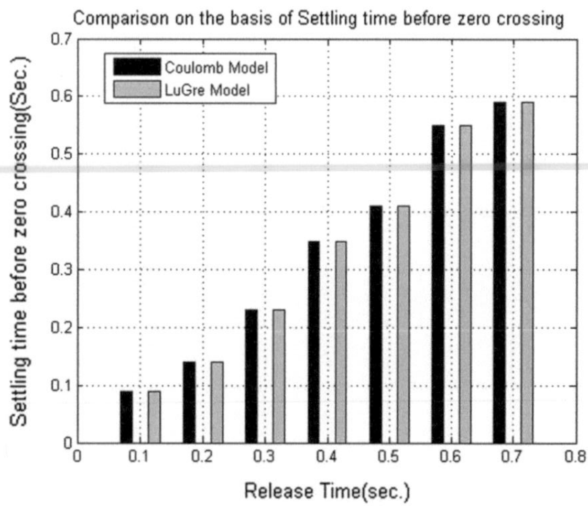

Fig. 26.11 Comparison on the basis of settling time before zero crossing

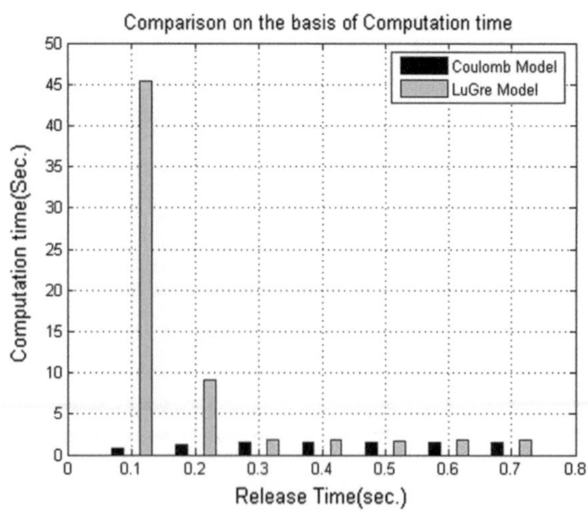

Fig. 26.12 Comparison on the basis of computation time

smaller in the case of Coulomb model for all the values of release time. For LuGre model, the value is very high for release times of 0.1 and 0.2 s. The values after 0.2 s are comparable to those of Coulomb model as shown in Fig. 26.12. This is due to the fact that the dynamics of bristles is much faster than that of the mechanical system. At higher velocities, the frequency of contact between the bristles increases, which causes the bristles to move faster. Numerical stiffness is caused due to difference between the speeds of friction dynamics and the dynamics of mass–spring system [10]. Due to this difference, smaller computational time steps are required, which results into longer computation times.

The time period for oscillations is higher for LuGre model for all values of release time as shown in Fig. 26.13 and the behavior is just opposite for frequency.

The value of maximum amplitude is almost same, as shown in Fig. 26.14.

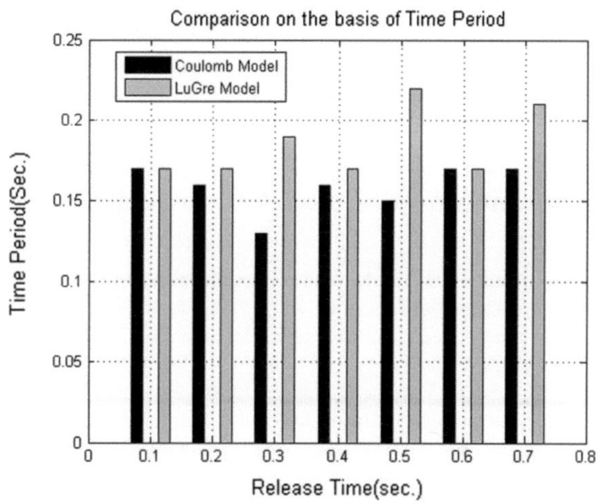

Fig. 26.13 Comparison on the basis of time period

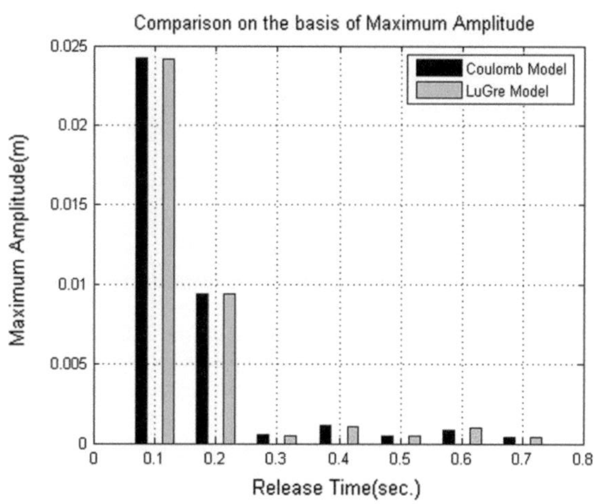

Fig. 26.14 Comparison on the basis of maximum amplitude

26.5 Conclusion

A typical mechanical actuation system for control applications was analyzed, with an objective to understand its dynamic behavior, in light of friction present in the system. The performance parameters were determined for different force trajectories, under which the system which moved from a starting condition. The friction force was modeled using LuGre model of dynamic friction. The dynamic response was compared with that obtained using the classical Coulomb model. Both the models predict dynamic response typical to an oscillating system. In a practical situation, where the motion of target mass is being controlled, it would be desirable that the target mass follows a smoother trajectory. The system parameters may be modified accordingly, and the model presented in this paper may be used for the design of such a system. The method of analysis and results presented in this paper may be useful in designing similar mechanical control systems in mechatronic systems such as hydraulic copy-turning machines, flight control mechanisms, single-acting hydraulic and pneumatic actuators, and robotics. This work can be extended for higher order mechanical systems.

References

1. Tripathi K, Mishra H, Yadav J, Agrawal MD (2007) Modeling of dry friction characteristics for computational simulation of dynamic system. In: Proceedings of international conference on recent trends in mechanical engineering, ICRTME, 4–6 Oct 2007, Ujjain Engineering College, Ujjain
2. Kikuuwe R, Takesue N, Sano A (2005) Hiromi Mochiyama, and Hideo Fujimoto, "Fixed-step friction simulation: from classical coulomb model to modern continuous models", IEEE/RSJ international conference on intelligent robots and systems, pp 3910–3917
3. de Canudas W, Olsson CH, Åström KJ, Lischinsky P (1995) A new model for control systems with friction. IEEE Trans Autom Control 40:419–425
4. Nguyen BD, Aldo AF, Oliver AB (2007) Efficient simulation of a dynamic system with LuGre friction. J Comput Nonlinear Dyn 2:281–289
5. Pierre ED, Eric PD (1993) Friction modeling and control in boundary lubrication. In: Proceedings of the 1993 American control conference, San Francisco, CA, June 1993, pp 1910–1914
6. Dupont P, Hayward V, Armstrong B, Altpeter F (2002) Single state elasto-plastic models. IEEE Trans Autom Control
7. Andersson S, Soderberg A, Bjorklund S (2007) Friction models for sliding dry, boundary and mixed lubricated contacts. Tribol Int 40:580–587
8. Swevers J, Al-Bender F, Ganseman C, Prajogo T (2000) An integrated friction model structure with improved presliding behavior for accurate friction compensation. IEEE Trans Autom Control 45:675–686
9. Tripathi K, Mishra H, Yadav J, Agrawal MD (2008) Simulation of dynamic response of a second order mechanical system with boundary friction. In: Proceedings of 15th ISME international conference on new horizons of mechanical engineering, Rajiv Gandhi Technological University, Bhopal, 18–20 March 2008, pp 519–525
10. Gafvert M (1997) Comparisons of two dynamic friction model. In: The proceedings of the 1997 IEEE international conference on control applications Hartford, CT. Oct 5–7, pp 387–391

Chapter 27
Design of a Third Order Self Biased Adaptive Bandwidth Phase Lock Loop

J. Dhurga Devi

27.1 Introduction

The clock generator plays a crucial role in deciding the performance of a microprocessor. For more than a decade, the self-biased adaptive PLL [1] architecture has persisted as a defacto choice of clock generators in most of the microprocessors. This is due to its inherent features such as excellent supply immunity, wide capture range, process-, voltage-, and temperature-insensitive loop parameters. Unlike the older scenario with a single-core processor, the current state-of-the-art microprocessors with multiple cores impose stringent requirements on this clock generator. In multicore systems, there is a strong requirement to operate at optimum clock frequencies while always satisfying the power envelope constraint. To maintain the power envelope, currently the clock generators are switched across frequencies and supply voltages based on a lookup-table approach depending on the number of architectural events active [2, 3]. This requirement of fast switching across clock frequencies demands fast settling of the switched clock, wide capture range, and operation under low supply voltages. Also, since these PLLs operate with highest permissible operating frequency and lower supply voltages that satisfy the power envelope constraint, the jitter generated from these PLLs has to be as minimum as possible. Moreover, the PLLs in multicore systems directly feed a number of digital circuits and delay-lock loops (DLLs). In order that the DLL operates within its loop bandwidth and the digital circuits meet their timing constraints, the feeding clock PLLs are not expected to experience severe frequency overshoot/undershoot [2–4]. This calls for minimizing the capture transients from the PLL clock generators.

The conventional self-biased adaptive bandwidth PLL, even though is well known for its wide capture range, process parameter insensitivity, and high supply

J. Dhurga Devi (✉)
Department of ECE, College of Engineering, Anna University, Guindy, Chennai 600 025, India
e-mail: j_dhurgadevi@annauniv.edu

V. Nath (ed.), *Proceedings of the International Conference on Nano-electronics, Circuits & Communication Systems*, Lecture Notes in Electrical Engineering 403,
DOI 10.1007/978-981-10-2999-8_27

noise immunity, being a second-order loop, it has its own inherent capture transients. The second-order PLL's damping factor can be chosen to minimize capture transients (overshoot/undershoot), but this trades off with the settling time of the loop. It is well known that a third-order PLL provides an additional degree of freedom in its loop parameters, and this enables independent control of overshoot/undershoot transients and settling time. In a third-order PLL, the phase margin of the system can be chosen [5] to independently minimize acquisition time and capture transients. This characteristic of the third-order PLL is ideally suited for the multicore environment. Moreover, with the third-order loop incorporated with a type-III system, it can suppress noise within its loop bandwidth to a larger extent than the second-order type-II system. Even though the realization of type-III PLL was considered impractical and unstable, the present-day literature had shown practical realization [6, 7] of these PLLs and the analysis carried out in [8] shows that the type-III PLLs have capability to attain higher phase margin than its type-II counterpart. This feature enables type-III PLL to present less peaking in its jitter transfer function and hence operates with minimum jitter. Third-order PLLs designed specifically for jitter reduction have been reported in [9]. Also, the third-order PLL with fast settling time and reduced overshoot/undershoot transient characteristics simulated in MATLAB is reported in [10]. Even though the analysis on type-III third-order PLL is found to be reported periodically in literature, circuit design aspects of these PLLs are very rarely reported. With these motivations, the present work proposes to modify the well-known self-biased adaptive bandwidth PLL into a third-order type-III PLL that incorporates self-biased adaptive bandwidth architecture. The circuit in this proposed work is designed to meet all the salient features of the well-proven and widely applied traditional self-biased PLL proposed in [1].

As part of the present work, detailed simulations have been carried out on the traditional second-order self-biased adaptive bandwidth PLL and the proposed third-order loop PLL to compare their performance parameters in terms of jitter, capture transients, and acquisition time. Section 27.2 explains the system functionality of the proposed third-order PLL system. The simulation results are discussed in Sect. 27.3. Finally, conclusion is presented in Sect. 27.4.

27.2 Design of Third-Order Self-Biased PLL

A generic type-III third-order PLL architecture can be derived from a general second-order PLL by introducing an additional transconductance Gm stage and a first-order loop filter section as described in [5] and as implemented in [6] for realizing FMCW transmitter. The functional block diagram of this type-III third-order PLL is shown in Fig. 27.1. The VCO chosen is a ring oscillator with symmetric load delay elements along with appropriate bias generators. These bias generators provide the required biases for the delay elements, the charge pump (CP), and the transconductance Gm stage. The resistors R shown in Fig. 27.1

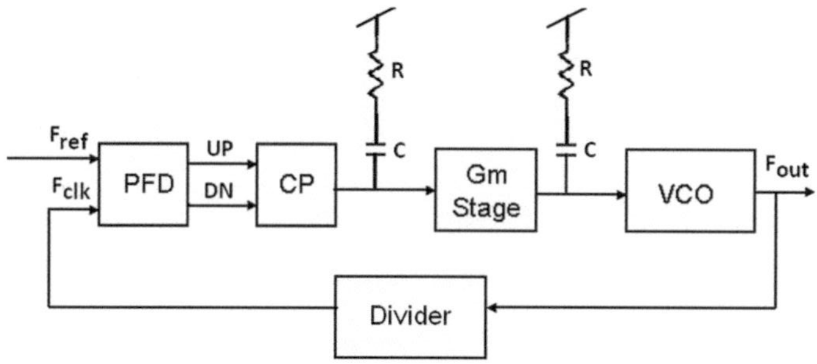

Fig. 27.1 Block diagram of type-III self-biased PLL

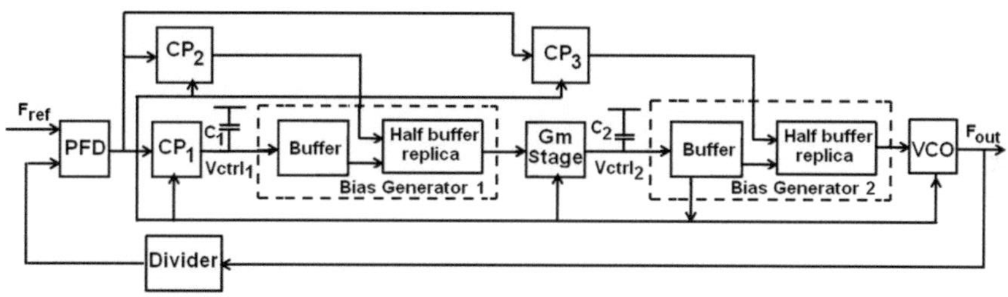

Fig. 27.2 Block diagram of the proposed third-order PLL

associated with the first- and second-loop filters are realized using separate charge pumps, and these are also biased from the bias generators. The modified third-order self-biased adaptive bandwidth PLL system functionality and the selection of the loop parameters of the modified system are explained in the following sections.

A. *Impact of system parameters on jitter*

The functional blocks of the third-order self-biased adaptive bandwidth PLL are shown in Fig. 27.2. The system comprises of a phase frequency detector (PFD); three charge pumps CP_1, CP_2, and CP_3; loop filter capacitances C_1, C_2; a Gm stage; two bias generators; a VCO; and a divider.

The PFD employed is a tristate PFD that compares the reference signal F_{ref} and the feedback clock signal from the divider. The pulses generated from the PFD drive the charge to drive C_1, and CP_2 and CP_3 are used for realizing the resistors for the loop filter sections. The first bias generator buffers the integrated voltage from the capacitor C_1 and combines the current from CP_2 to derive an output voltage $Vctrl_1$. This output in turn drives the Gm stage and which in turn proportionately drives the capacitor C_2. Next, the second bias generator buffers the integrated voltage from C_2 and combines the current from CP_3 to generate $Vctrl_2$ which finally controls the VCO to generate the required VCO frequency.

The bias currents for CP_1, CP_2, CP_3, and Gm stage are derived from the second bias generator which also delivers bias current to the delay elements in the VCO.

The charge pump and Gm-stage bias currents are adjusted as the VCO operating frequency is varied. Since the bias current in CP_2 and CP_3 are also adjusted, the resistors vary accordingly. As the charge pump currents and the associated resistances change with operating frequency, the system loop bandwidth tracks the reference frequency as in [1]. The phase margin also gets adjusted to remain constant over the entire operating frequency range.

B. *Description of third-order PLL system parameters*

The transfer function $H_1(s)$ of the generic type-III third-order system shown in Fig. 27.1 is given in Eq. (27.1). This transfer function has an additional integrator and a zero which is obtained by introducing an additional loop filter in a type-II second-order PLL as given in Eq. (27.2).

$$H_1(s) = \frac{I_{cp} \cdot \frac{K_{VCO}}{S} \frac{(sRC+1)^2}{(sC)^2}}{1 + \frac{I_{cp} \cdot K_{VCO}}{N \cdot s} \frac{(sRC+1)^2}{(sC)^2}} \tag{27.1}$$

$$H_2(s) = \frac{I_{cp} \cdot \frac{K_{VCO}}{S} \frac{sRC+1}{sC}}{1 + \frac{I_{cp} \cdot K_{VCO}}{N \cdot s} \frac{sRC+1}{sC}} \tag{27.2}$$

The loop parameters of the generic type-III third-order PLL, namely its 3-dB bandwidth ω_{3dB} and the phase margin, are derived similarly, by following the procedure given in [5], and are reproduced in the below equation.

The generic third order PLL when modified to incorporate self biased adaptive bandwidth feature, the charge pump currents are derived in a certain ratio with the current defined by the second bias generator shown in Fig. 27.2. These ratios are described below.

Assuming I_{bias}, I_{cp1}, I_{cp2}, I_{cp3}, and I_{gm}, respectively, represent the bias currents of the second bias generator, CP_1-, CP_2-, CP_3-, and the Gm stage, their ratios are fixed as given in Eq. (27.3).

$$I_{cp1} = x \cdot I_{bias}; I_{gm} = z \cdot I_{bias}; I_{cp2} = y \cdot I_{cp1}; I_{cp3} = y \cdot I_{gm} \tag{27.3}$$

where x, y, and z are constants, with y deciding the resistor R value of the loop filter, and z determining the value of transconductance in the Gm stage. The expression for the resistors that are derived from CP_2 and CP_3 is given in Eq. (27.4).

$$R = \frac{y}{\sqrt{8KI_{bias}}} \tag{27.4}$$

The ratio of the charge pump currents and the expressions for the resistors from Eqs. (27.1) and (27.2) are applied in the expressions for ω_{3dB} and the phase margin that are derived in [5]. These expressions for ω_{3dB} (crossover frequency) and its phase margin PM derived in [5] are reproduced below in Eqs. (27.5) and (27.6).

$$\omega_{3\mathrm{dB}} = \frac{RI_{\mathrm{cp1}}K_{\mathrm{VCO}}}{2\pi}(\mathrm{Gm}R) \tag{27.5}$$

Here, K_{VCO} is the gain of the VCO in the third-order PLL.

$$PM = -90^\circ + 2\tan^{-1}(RC_1\omega_{3\mathrm{dB}}) \tag{27.6}$$

The ratio of $\omega_{3\mathrm{dB}}$ to the reference frequency ω_{ref} that are expressed using the factors x and z is derived similar to [1] as given below in Eq. (27.7).

$$\frac{\omega_{3\mathrm{dB}}}{\omega_{\mathrm{ref}}} = \frac{y^2 x\sqrt{z}}{16\pi^2} \tag{27.7}$$

From the above, it can be observed that the ratio remains constant and depends only on the ratio of the currents.

Similarly, the phase margin is derived using the factors x, y, and z similar to the procedure employed in [1] and is given below in Eq. (27.6).

$$\mathrm{PM} = -90^\circ + 2\tan^{-1}\left(\frac{xy\sqrt{z}C_1}{16\pi NC_B}\right) \tag{27.8}$$

where C_1 represents the LPF capacitance and C_2 (Fig. 27.2) is chosen to be of the same value as C_1. The effective capacitance C_{B} is obtained from the oscillator and N is the prescalar division factor. From Eq. (27.8), it can be observed that the phase margin is dependent on the ratio of capacitances and therefore is a constant for the entire capture range of the third-order PLL.

From Eqs. (27.7) and (27.8), it can be noted that the $\omega_{3\mathrm{dB}}$ to ω_{ref} and phase margin can be independently chosen since $\omega_{3\mathrm{dB}}$ to ω_{ref} factor is independent of C_1, and C_1 can then be chosen for a required phase margin without influencing $\omega_{3\mathrm{dB}}$ to ω_{ref}. This permits the independent control of settling time and capture transients in the third-order PLL.

The performance improvement that can be obtained from this type-III third-order PLL in comparison with type-II second-order PLL is evident from the expressions for $\omega_{3\mathrm{dB}}$ and phase margin given in Eqs. (27.5) and (27.6). These expressions show that the type-III third-order PLL has the capability to operate with a wider bandwidth and large phase margin and hence has the capability to obtain improvement in settling time and jitter performance when compared with the type-II second-order PLL.

27.3 Simulation Results

The circuit was deigned in 0.18-μm CMOS process UMC technology library. The simulations were carried out using Cadence Spectre tool. The system stability conditions and step response were determined using MATLAB.

The VCO circuit is designed for a frequency range of 1–3 GHz. The prescalar divide ratio N is chosen as 16. The values of loop parameter constants *x, y*, and *z*are chosen depending on the required PM and the ω_{3dB} to ω_{ref} ratio. The value of C_1 is chosen as 100 pF and the effective capacitance C_B obtained from VCO equals 0.248pF. With C_1 chosen to be a larger and a circuit realizable value that maximizes phase margin and the factors *x, y*, and *z* values were chosen such that it can give reduced settling time and minimum overshoot/undershoot transients.

In order to perform a comparative study, the traditional second-order and the proposed third-order self-biased adaptive bandwidth PLLs have been designed with similar functional block specifications. The values chosen for the factors *x* and *y* defined in Eq. (27.1) remain the same for both the cases. The values for *x* and that of *y* are chosen as 0.2 and 5, respectively.

With the above constants, the loop bandwidth ω_n to ω_{ref} ratio in the second-order system turns out to be 1/26, and the damping factor was found to be 3. These specifications maintain that the second-order system is to be wide bandwidth and overdamped and hence enabling it to settle with minimum overshoot/undershoot transients.

The step response of the third-order self-biased adaptive bandwidth PLL for the chosen *x* and *y* and for the three sets of *z* values was computed using MATLAB. The resulting plots are given in Fig. 27.3. The third-order PLL step response plot that corresponds to a PM of 40°, 58°, and 80° was obtained by choosing *z* as 0.25, 0.35, and 0.6, respectively. Finally, the value of *z* is chosen as 0.6 since it gives a

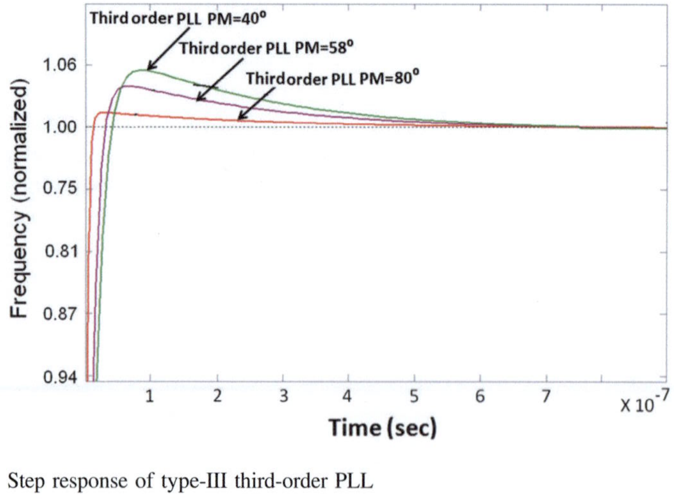

Fig. 27.3 Step response of type-III third-order PLL

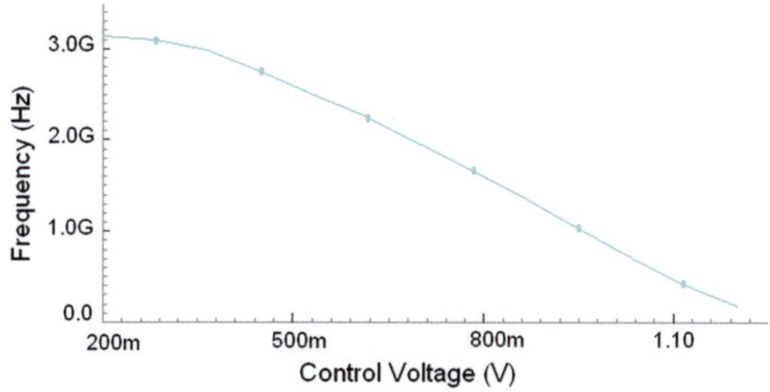

Fig. 27.4 VCO gain of the third-order PLL

realizable circuit implementation of the Gm stage to obtain a transconductance value of the stage as 630 μS. With this chosen z value, the phase margin of the third-order PLL was calculated using Eq. (27.6) and was found to be 80°. Further for this case, it may be also be clearly seen from the plot that the settling time, undershoot/overshoot happened to be the least in comparison with the step responses corresponding to the phase margin of 40° and 58°.

Having derived the system-level specifications using MATLAB, the circuit simulations were next carried out for both the second- and the third-order systems. The proportional charge pumps CP_1 and CP_2 were replaced by appropriate value of resistance for the sake of simplicity. This is done to verify the circuit realization of mathematical modeling of third-order PLL.

The VCO operating frequency range is shown in Fig. 27.4, and the plot shows a linear operating range from 500 MHz to 3 GHz with a gain of 3.3 GHz/V. The capture range of the resulting third-order PLL was observed to be from 1 GHz to 2.72 GHz.

A comparison of the control voltage transients for the second- and third-order systems at 2.4 GHz output frequency (with a step frequency of 700 MHz) is shown in Fig. 27.5. It can be observed that the third-order PLL while acquiring the final frequency shows negligible capture transients with fast settling characteristic when compared with the second-order PLL. The second-order PLL acquires within 2% of its final frequency after 190 ns (29 reference clock cycles) whereas the third-order PLL acquires similarly after 111 ns (17 reference clock cycles), showing 40% improvement in its settling time.

The jitter performance of both the PLLs at 2.4 GHz is shown in Fig. 27.6 by means of an eye diagram. The peak to peak jitter measured from the eye diagram is 16.6 ps (0.04UI) for the second-order PLL and is found to be 8.7 ps (0.02UI) for the third-order PLL. Thus, the third-order PLL shows 48% improvement in its jitter performance.

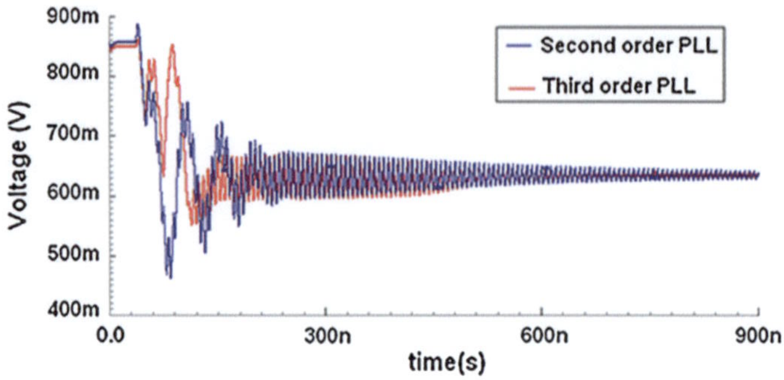

Fig. 27.5 Comparison of control voltage transients of the second- and third-order PLL at 2.4 GHz

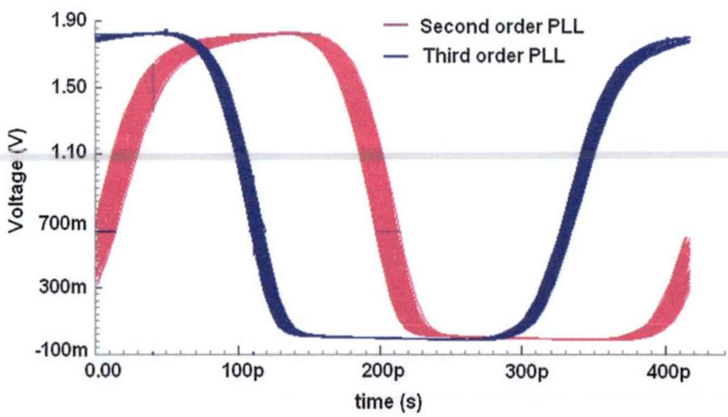

Fig. 27.6 Jitter comparison at 2.4 GHz

The control voltage transients at the extreme operating frequencies of 2.72 and 1 GHz for both the second- and third-order systems are shown in Fig. 27.7. From the plots, it can be noted that the settling time at both the extreme frequencies are of the same-order and the same observation holds good for overshoot/undershoot capture transients. It may be noted that this feature is primarily possible due to the adaptive bandwidth nature of the system under consideration.

The stability of the system is verified over process corners. The capture transients simulated for various process corners at the highest extreme operating frequency of 2.7 GHz are shown in Fig. 27.8. The plot shows the capture transients of the system for the typical, as well as for the worst-case process corners of slow slow and fast fast. Though not shown here, similar behavior is observed for the other corners. For all the process corners, the capture transients settle within 1 μs and is confirmed that the design of third-order PLL is stable at all the corners.

The performance comparison of the proposed third-order PLL with the second-order PLL designed in the present work and with a recently reported adaptive self-biased second-order PLL designed with similar specification and TT

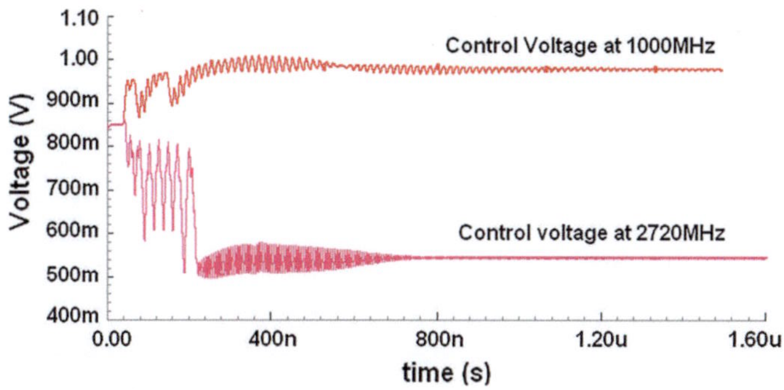

Fig. 27.7 Control voltage transients of the type-III third-order PLL at extreme operating frequencies

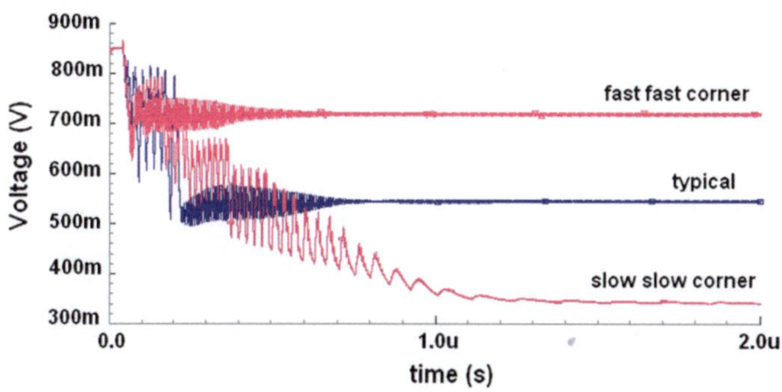

Fig. 27.8 Control voltage transients of the proposed type-III third-order PLL at process corners at 2.7 GHz

corner simulation is shown in Table 27.1. The third-order PLL shows better jitter performance of about 67% when compared with the second-order PLL. The type-III third-order PLL shows similar jitter performance but shows significant improvement in its settling time. For the reported settling time in [11] as shown in the tabulation, the type-III PLL in the proposed work can be designed with better jitter performance by relaxing its settling time specification.

Moreover, the third-order system also shows negligible overshoot/undershoot capture transients without any degradation in its settling time when compared with the second-order counterpart and the recently reported second-order self-biased adaptive bandwidth PLL. The performance improvement thus obtained in the third-order loop is with less additional power consumption than that of the traditional second-order PLL which is about 11% at 2.7 GHz but shows degradation in the power consumption measure reported in [11] which in this case is reported for the lowest extreme operating frequency of 1 GHz.

Table 27.1 Performance comparison of second-order PLL and third-order PLL

Performance parameters	Second-order PLL	Proposed type-III third-order PLL	Self-biased adaptive bandwidth PLL [11]
VCO operating range (MHz)	500–3000	500–3000	800–3200
Capture range	500–2700	1000–2700	–
Jitter (ps)	16.2 (0.04UI) @2.7 GHz	5.4 (0.01UI) @2.7 GHz	<1%
Settling time (μs)	0.190 @2.7 GHz (1 GHz step frequency)	0.190 @2.7 GHz (1 GHz step frequency)	15 @ 2.0125 GHz
Peak undershoot magnitude @ 2.7 GHz	557 MHz (170 mV of undershoot in its control voltage)	2.3 kHz (690 nV of undershoot in its control voltage)	–
Peak power (mW)	36 @2.7 GHz,	40 @2.7 GHz,	2.16 @ 1 GHz
Technology (nm)	180	180	65 and 45

27.4 Conclusions

The present work shows a feasible circuit realization of the third-order self-biased adaptive bandwidth PLL. It is shown that the system is stable for a wide capture range. It was also demonstrated that the third-order PLL performs improvement in (a) jitter, (b) settling time, and (c) undershoot/overshoot transients. Even though the circuit was not designed optimally for power, the third-order system exhibits these performance improvements over the second-order system only on a marginal increase of 10% in its power consumption.

References

1. Maneatis John G (1996) Low-jitter process-independent DLL and PLL based on self-biased techniques. IEEE J Solid State Circ 31(11):1723–1732
2. Allen A, Desai J, Verdico F, Anderson F, Mulvihill D, Krueger D (2009) Dynamic frequency-switching clock system on a quad-core titanium processor. In: IEEE international solid state circuits conference, pp 62–63, Feb 2009
3. Stackhouse B, Bhimji S, Bostak C, Bradley D, Cherkauer B, Desai J, Francom E, Gowan M, Gronowski P, Krueger D, Morganti C, Troyer S (2009) A 65 nm 2-billion transistor quad-core itanium processor. IEEE J Solid State Circ 44(1):18–31
4. Kurd N, Mosalikanti P, Neidengard M, Douglas J, Kumar R (2009) Next generation Intel® Core™ micro-architecture (Nehalem) clocking. IEEE J Solid State Circ 44(4):1121–1129
5. Duarte D, Adrang H, Naeimi HM (2009) A Type III fast locking time PLL with transconductor-C structure. In: IEEE international Midwest symposium on circuits and systems, pp 58–61, September 2009
6. Velner J, Klumperink EAM, Nauta B, van Vliet FE, Multi-band linear chirp generation based on a type-III PLL. In: IEEE MTT-S international microwave symposium digest (IMS), pp 1–4, June 2013

7. Sai A, Kobayashi Y, Saigusa S, Watanabe O, Itakura T (2012) A digitally stabilized Type-III PLL using ring VCO with 1.01psrms integrated jitter in 65 nm CMOS. In: IEEE international solid state circuits conference, pp 248–249, Feb 2012
8. Ugarte M, Carlosena A (2014) Performance trade-offs between Type II and Type III PLLs. In: 11th international multi-conference on systems, signals & devices (SSD), pp 1–5, Feb 2014
9. Mansuri M, Ken C-K (2002) Jitter optimization based on phase-locked loop design parameters. IEEE J Solid State Circ 37(11):1375–1382
10. Stephens DR (2002) Phase locked loops for wireless communications digital, analog and optical implementations, 2nd edn. Kluwer Academic Publishers, Dordrecht
11. Hengzhou Y, Yang G, Zhuo M (2014) A 40/65 nm process adaptive low jitter phase-locked loop. In: 14th international symposium on integrated circuits, pp 500–503, December 2014

Chapter 28
Operational Transconductance Amplifier-based Sinusoidal Oscillator with Grounded Capacitors

Srinivasulu Avireni, G. Sowjanya, S.H. Gautham, Pitchaiah Telagathoti and V.S.V. Krishna V

28.1 Introduction

Sinusoidal oscillators with grounded capacitors play a vital role in the implementation of communication systems, instrumentation applications, measurement devices, and signal processing. The oscillators are mainly designed keeping in view the importance of sinusoidal signal and its requirement in various operations. Attempts have been made to fabricate a sine wave oscillator to work on current as well as voltage mode and to serve its purpose for which it is synthesized. The sinusoidal oscillator realized with OTAs, transconductance-C, or g_m-C technique provides highly linear electronic tunability and have more reliable high-frequency performance than an operational amplifier-based oscillators. For the purpose of integrated circuit implementation both in bipolar and CMOS technologies, the g_m-C sinusoidal oscillator circuits are found to be more suited as well as cost-effective. The operational transconductance amplifiers and capacitors have found birth in the imposition of circuits. These circuits (OTA-Cs) have proved to be having many advantages in the synthesis of high-frequency monolithic analog operators [1]. In the recent past, the electronic industry has achieved a wide range of signal processing circuits right from a vacuum type to a semiconductor. This could be in the form of integrated circuit or so called IC [2]. These techniques help improve the existing circuits for better performance, more resolution, and more accurate so as to gain high-frequency response by designing a circuit with simplest structure.

S. Avireni (✉) · G. Sowjanya · P. Telagathoti · V.S.V. Krishna V
Department of Electronics & Communication Engineering,
Vignan's University, Vadlamudi, Guntur 522213, India
e-mail: avireni_s@yahoo.com

G. Sowjanya
e-mail: souji.gaddam9@gmail.com

S.H. Gautham
Analog Mixed Signal Team, Freescale Semiconductor Pvt., Ltd., Noida, India

V. Nath (ed.), *Proceedings of the International Conference on Nano-electronics, Circuits & Communication Systems*, Lecture Notes in Electrical Engineering 403,
DOI 10.1007/978-981-10-2999-8_28

The circuit can then be simply and cost-effectively developed in an integrated circuit. The wireless devices that run on compact batteries needed reduced voltage consumption in the circuit as it was the case in the last decade. In order to overtake such situations, the design of current-mode circuits has become a necessity as they have potential advantages such as inherently wide bandwidth, higher slew rate, greater linearity, wider dynamic range, and simple circuitry [3–6]. From the survey conducted literally, the implementation of oscillator employed OTAs has been introduced in the electronic industry [7–21]. Grounded capacitor circuits are rated as the most favored ones in both monolithic IC technology and their thin film fabrication technology as well.

28.1.1 Basic Concept of OTA

An ideal OTA has infinite input and output impedances. The output current of an OTA is given by

$$I_O = g_m(V_+ - V_-) \tag{28.1}$$

The transconductance of the OTA is g_m. The external input bias current (I_B) is used in this circuit. This bias current is used to tune the g_m. For a BJT OTA, the transconductance can be formulated as

$$g_m = I_B/2V_T \tag{28.2}$$

where V_T is the thermal voltage. The symbol and equivalent circuit of the OTA are illustrated in Fig. 28.1a, b, respectively [2].

OTA is accomplished with both inverting and non-inverting outputs, where V_+ and V_- act as two power supply lines besides I_{out} being a single output. The bias current (I_{bias}) is the additional input. OTA is an amplifier whose differential input voltage produces an output current. So it is treated as voltage-controlled current source. This is usually an additional input for a current to control the amplifiers' transconductance. The OTA is similar to standard operational amplifier in that it has a high impedance differential input stage and also that it could be used with negative feedback. The bias current I_B could be made use of to bring back

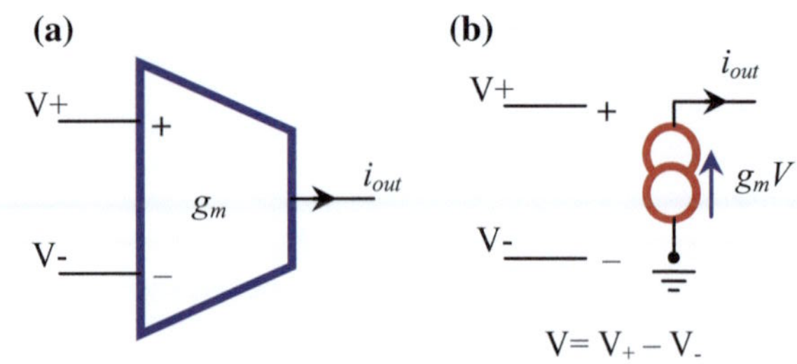

Fig. 28.1 **a** OTA symbol and **b** its equivalent circuit

transconductance of the OTA under control. OTA-grounded capacitor circuits have an added advantage possessing with high-frequency performance as well as minimum chip area utilization. This paper introduces a new electronically tunable current-mode sinusoidal oscillator employing three single-ended OTAs and two grounded capacitors. In general, the single-ended OTA configuration is simpler than its multiple output OTA counterpart warranting requirement of grounded capacitors as preferable for the integrated circuit fabrication process. Therefore, the proposed circuit is considered to be a favored one as it provides simple circuitry using single-ended OTA and also most sought after in integrated circuit (IC) implementation with grounded capacitors.

28.1.2 Proposed Circuit Diagram

The sinusoidal oscillator circuit proposed makes use of three OTAs and two grounded capacitors as shown in Fig. 28.2. In the above said oscillator, the transconductance of three OTAs has different functionalities charged by a single external bias control current I_B by implementing device level circuit arrangement (Note: reflected in Fig. 28.2). By fluctuation of bias current of OTAs or grounded capacitors, oscillator frequency could be differed. The condition of oscillation and frequency of oscillation of the circuit can be set orthogonally.

28.2 Description and Analysis of Circuit

In this paper, we have focused mainly on an oscillator which can be ideally described by a second-order characteristic equation:

$$s^2 - bs + \omega_0^2 = 0 \tag{28.3}$$

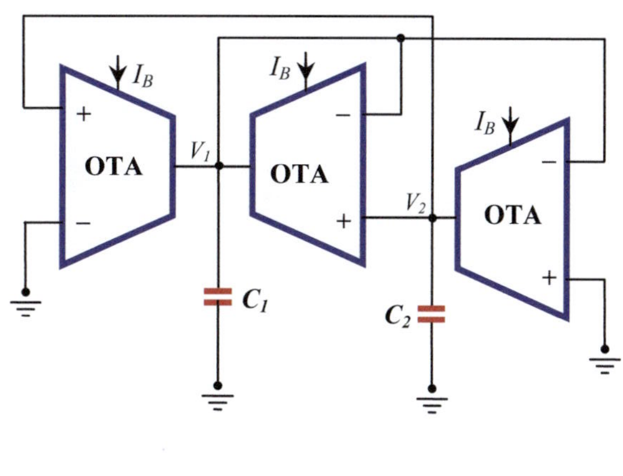

Fig. 28.2 Proposed sinusoidal oscillator circuit using three OTAs with two grounded capacitors

Parameters b and ω_0^2 are given as functions of the OTA transconductance gain and capacitor values. The relevant expressions for b and ω_0^2 are stated below. The performance of transconductance amplifier capacitor oscillators (TACO's) oscillation condition b and the frequency of oscillation ω_0^2 are independent of the outcome of transconductance gains. It indicates that any intent to change ω_0^2 by any transconductance gain will bring out a change in b [1].

From the equivalent topology of Fig. 28.3 and for ideal behavior of the OTA given by (28.1) and (28.2), it is easy to derive the characteristic equation in the form

$$s^2 - s[g_{m2}(C_1 + C_2)/C_1C_2] + [g_{m3}(g_{m1} + g_{m2}) - g_{m1}g_{m2}]/C_1C_2 = 0 \quad (28.4)$$

From (28.4), the condition of oscillation and frequency of oscillation are summarized below.

If the transconductance of the OTAs is varied simultaneously through a single external current, then oscillation conditions are as follows:

$$g_{m1} = g_{m2} = g_{m3}, \quad \text{from} \quad I_{B1} = I_{B2} = I_{B3} \quad (28.5)$$

and

$$\omega^2 = [g_{m3}(g_{m1} + g_{m2}) - g_{m1}g_{m2}]/C_1C_2 \quad (28.6)$$

It is found from (28.3) to (28.4) that the frequency of oscillation (FO) is written as (28.6) reduces to

$$\omega = (g_m/\sqrt{C_1C_2}) \quad (28.7)$$

The transconductance g_m of an OTA can be brought under control by applying an external bias current I_B, i.e.,

$$g_m = I_B/2V_T \quad (28.8)$$

From the Eq. (28.7) cited, we may conceive and reflect the oscillation frequency as

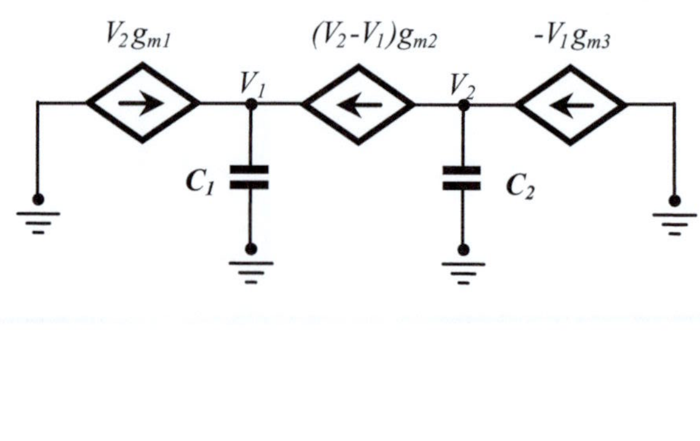

Fig. 28.3 Equivalent topology of proposed sinusoidal oscillator of Fig. 28.2

$$f = \frac{g_m}{2\pi\sqrt{C_1 C_2}} \quad \text{or} \quad f = \frac{I_B}{4\pi V_T\sqrt{C_1 C_2}} \tag{28.9}$$

From (28.9), it can be seen that the frequency of oscillation current as shown in Fig. (28.2) can be linearly controlled by imposing single external current. Thus, the oscillation results can be satisfactorily re-trilled by proper adjustment of capacitors C_1 and C_2. The use of grounded capacitors is particularly attractive for integrated circuit implementation. The frequency of oscillation can also be rearranged by adjusting grounded capacitors. As seen from the above equation, it can thus be concluded the Condition of Oscillations (CO) could be adjusted independently of FO by the employment of different capacitors. The proposed circuit enables non-interactive dual current control for both the condition of oscillation and frequency of oscillation.

28.3 Simulation Results

The circuit shown in Fig. 28.2 was simulated with the *Cadence* and the model parameters of a 180 nm technology. For component values C_1 = 1 pF, C_2 = 5 pF, and I_B = 110 mA, the generated sine waveform is of T = 3.06 ns, as shown in Fig. 28.4 and the calculated value from Eq. (28.9) is 3.0 ns. For all measurements on tunability, the supply rail of ±5 V was used.

The circuit was tested for the stability of its temperature ranging from 0 to +125 °C and the simulation profile produced to be less than 0.02% variation of sine waveform. Figure 28.5 shows the temperature sensitivity of the proposed configuration.

The tunability of the frequency range is shown in Fig. 28.6. From the results obtained, it has been observed that the sinusoidal oscillator using OTA has improved performance and generated better results. Figure 28.6 shows the frequency of oscillation for different capacitor values and bias currents (C_1 = 1 pF, C_2 = 5 pF, and I_B = 110 mA and C_1 = 2 pF, C_2 = 3 pF, and I_B = 110 mA). A good agreement of the theoretical analysis with the simulated results is evident.

28.4 Sensitivity

Classical sensitivity (S_x^y) for parameter y with respect to parameter x has been defined [22] as

$$S_x^y = \frac{\frac{\delta y}{y}}{\frac{\delta x}{x}} = \frac{x}{y} \cdot \frac{\delta y}{\delta x} \tag{28.10}$$

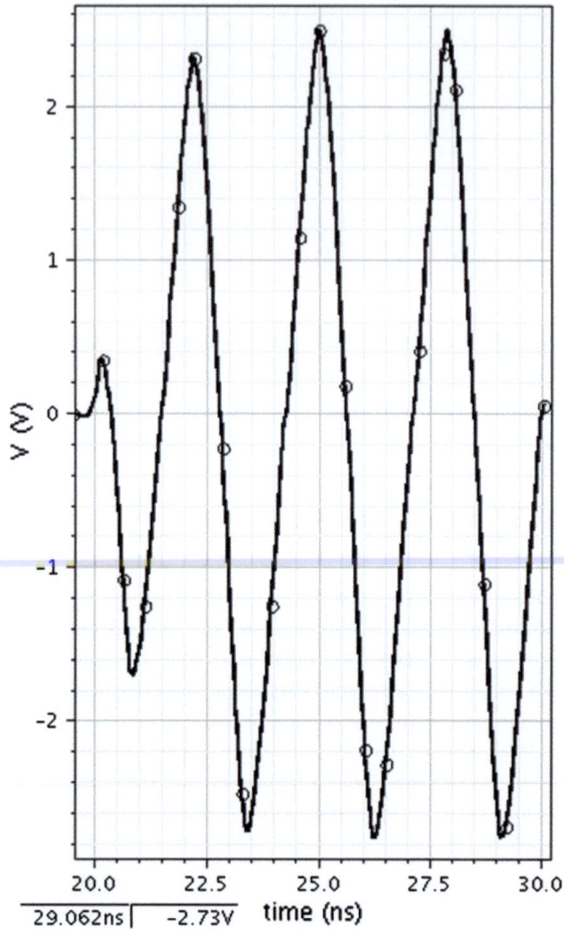

Fig. 28.4 Proposed OTA oscillator output waveform

Applying this for (28.9), it is easy to find $S^T_{g_m}$, $S^T_{c_1}$, and $S^T_{C_2}$ which are given by

$$S^T_{g_m} = -1 \tag{28.11}$$

$$S^T_{C_1} = S^T_{C_2} = 0.5 \tag{28.12}$$

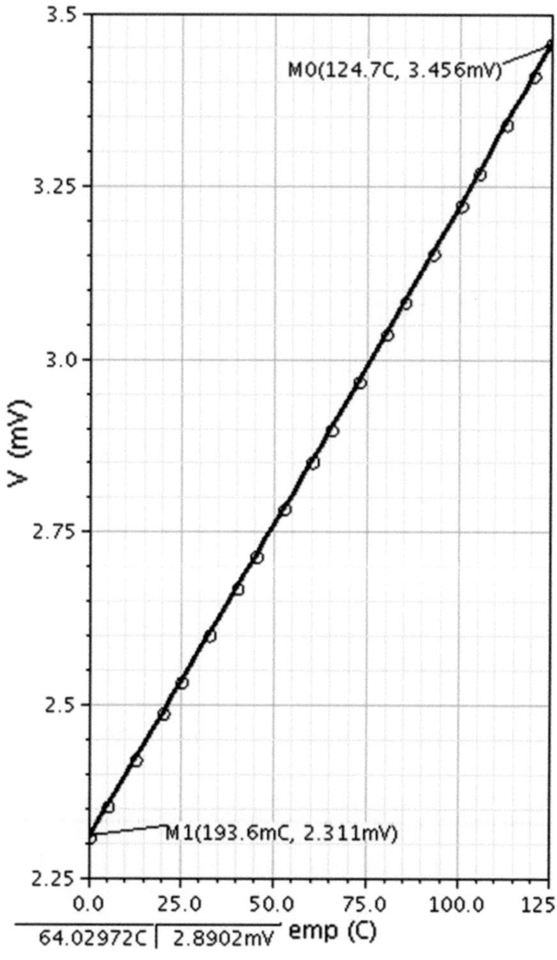

Fig. 28.5 Simulated temperature stability characteristics of Fig. 28.2

Obviously, in the summation to find the net sensitivity $\left(S = \sum_{x=g_{\rm m},C_1,C_2} S_x^T\right)$, it can be easily verified that the sensitivities for $g_{\rm m}$, C_1 and C_2 completely cancel out. As such, the proposed configuration can be stable against component sensitivity as per the classical definition [22–30].

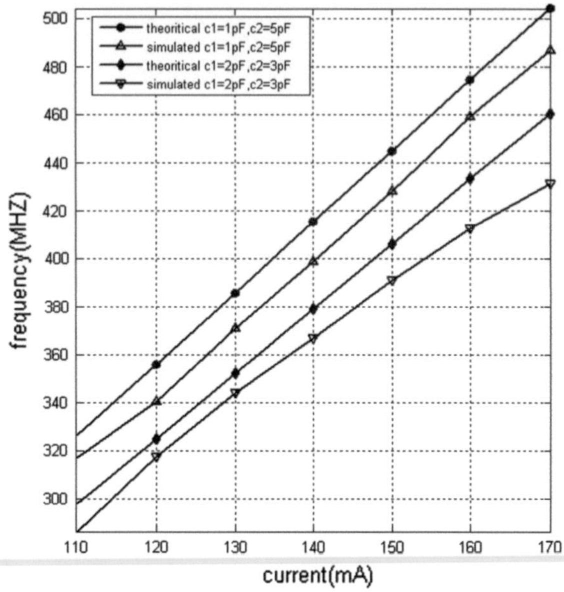

Fig. 28.6 Oscillation frequencies versus bias currents

28.5 Conclusion

In this paper, a sinusoidal oscillator circuit has been proposed using 180 nm technology with ±5 V supply voltage. The proposed oscillator is provided with three OTAs and grounded capacitors. The frequency of waveform generator can be adjusted by either tuning grounded capacitors C_1 and C_2 or bias current I_B. Tunability of grounded capacitors is an attractive feature of the design as it provides an easy option for digital control by the programmable switched capacitor array. The component's sensitivity is very good; making it there by a candidate can easily design for integration. This circuit may gain greater importance as it is bestowed with wider applications in many electronics and communication systems as well as signal processing, instrumentation, and measurement systems.

References

1. Linares-Barranco B, Rodriguez-Vazquez A, Sancnez E, Jose LH (1991) CMOS OTA-C high-frequency sinusoidal oscillators. IEEE J Solid-State Circuits 26(2):160–165
2. Summart S, Thongsopa C, Jaikla W (2012) OTA based current-mode sinusoidal quadrature oscillator with non-interactive control. Przeglad Elektrotechniczny (Electr. Review), pp 14–17. ISSN 0033-2097, R. 88 NR 7a/2012
3. Toumazou C, Lidgay FJ (1986) Universal active filter using current conveyors. Electron Lett 22:662–664

4. Abuelma'atti MT, Al-zaher HA (1999) Current-mode sinusoidal oscillators using single FTFN. IEEE Trans Circuits Systems-II: Analog Digital Signal Proc 46:69–74
5. Jaikla W, Lahiori A (2012) Resistor-less current-mode four-phase quadrature oscillator using CCCDTA and grounded capacitors. Int J Elec Commun 66:214–218
6. Jaikla W, Prommee P (2011) Electronically tunable current-mode multiphase sinusoidal oscillator employing CCCDTA-based allpass filters with only grounded passive elements. Radioeng J 30:594–599
7. Abuelma'atti MT (1990) Anew minimum component active-c OTA-based linear voltage (current)-controlled sinusoidal oscillator. IEEE Trans Instr Meas 39:795–797
8. Linares-Barranco B, Serrano-Gotarrendona T, Ramos-Martos J, Ceballos-Caceres J, Linares-Barranco A (2004) A precise 90 quadrature OTA-C oscillator tunable in the 50–130 MHz range. IEEE Trans Circuits Syst 51:649–663
9. Odame KM, Hasler P (2007) An efficient oscillator design based on OTA nonlinearity. In: Proceedings of IEEE International Symposium on Circuits and Systems-07 (IEEE ISCAS-07), pp 921–924
10. Pranayanuntana P (2009) An electronically adjustable amplitude of OTA-C based sinusoidal nonlinear oscillator. In: Proceedings of IEEE International Conference on Digital Image Processing, pp 335–340
11. Das BP, Watson N, Liu YH (2010) Bipolar OTA based voltage controlled sinusoidal oscillator. In: Proceedings of international conference on circuits, systems and signals, pp 101–105
12. Abuelma'atti MT, Almaskati RH (1998) Digitally programmable active-c OTA-based oscillator. IEEE Trans Instr Meas 37:320–322
13. Prommee P, Angkeaw K, Chanwutitum J, Dejhan K (2007) Dual input all-pass networks using MO-OTA and its application. In: Proceedings of ECTI International Conference, pp 129–132
14. Nacaroblu A, Ercelebi E (1998) Effects of parasitic elements on oscillation frequency. Turk J Elec Eng 7:39–44
15. Sotner R, Jerabek J, Petrzela L, Dostal T, Vrba K (2009) Electronically tunable simple oscillator based on single-output and multiple-output transconductor. IEICE Elec Exp 6:1476–1482
16. Bhasker DR, Abdulla KK, Senani R (2011) Electronically-controlled current-mode second order sinusoidal oscillators using MO-OTAs and grounded capacitors. Circuits Syst 2:65–73
17. Singh V (2006) Equivalent forms of duel-OTA RC oscillators. IEEE Proc Circuits Dev Syst 153:95–99
18. Linares-Barranco B, Rodriguez-Vazquez A, Huertas JL, Sanchez-Sinencio E, Hoyle JJ (1988) Generation and design of sinusoidal oscillators using OTAs. In: IEEE International Symposium on Circuits and Systems, pp 2863–2866, 1988
19. Bhasker DR, Tripati MP, Senani Raj (1993) A class of three-OTA-two capacitor oscillators with non-interacting controls. Int J Elec 74(3):459–463. doi:10.1080/00207219308925849
20. Thosdeekoraphat T, Summart S, Thongsopa C (2013) Current-mode sinusoidal oscillator using single dual-output current controlled current conveyor transconductance amplifier. Aust J Basic Appl Sci 7(8):230–236
21. Kumngern M, Junnapiya S (2012) Tunable quadrature Sinusoidal oscillator using single-ended OTA's. In: Proceedings of the 4th IEEE International Conference on Computational Intelligence, Communication Systems and Networks (CICSyN- 2012), 24–26 July 2012, Phuket, pp 74–77
22. Sedra AS, Smith KC (1998) Microelectronic circuits. Oxford University Press, 4th ed, pp 1002–1005
23. Srinivasulu Avireni (2011) A novel current conveyor based Schmitt trigger and its application as a relaxation oscillator. Int J Circuit Theory Appl 39(6):679–686. doi:10.1002/cta.669
24. Srinivasulu A (2012) Current conveyor based relaxation oscillator with tunable grounded resistor/capacitor. Int J Des Anal Tools Circuits Syst (Hong-Kong) 3(2):1–7, Nov 2012. ISSN: 2071-2987

25. Khan AA, Bimal S, Dey KK, Roy SS (2005) Novel RC sinusoidal oscillator using second-generation current conveyor. IEEE Trans Instrum Meas 54(6):2402–2406
26. Srinivasulu A (2009) Current conveyor-based square-wave generator with tunable grounded resistor/capacitor. In: Proceedings of IEEE Applied Electronics 2009 International Conference, Pilsen, Czech Republic, 9–10 Sept 2009, pp 233–236. ISSN: 1803-7232. INSPEC Accession No: 10975061
27. Pal D, Srinivasulu A, Pal BB, Demosthenous A, Das BN (2009) Current conveyor based square/triangular-waveform generators with improved linearity. IEEE Trans Instrum Meas 58 (7):174–2180, Jul 2009. doi:10.1109/TIM.2008.2006729
28. Cicekoglu MO, Kuntman H (1998) On the design of CCII + based relaxation oscillator employing single grounded passive element for linear period control. Microelectron J 29:983–989
29. Pal D, Srinivasulu A, Goswami M (2009) Novel current-mode waveform generator with independent frequency and amplitude control. In: Proceedings of IEEE International Symposium on Circuits and Systems-09, Taipei, Taiwan, pp 2946–2949, 24–27 May 2009. doi:10.1109/ISCAS.2009.5118420
30. Abuelma'atti MT, Al-Absi MA (2005) A current conveyor-based relaxation oscillator as a versatile interface for capacitive and resistive sensors. Int J Electron 92(8):473–477, Aug. 2005

Chapter 29
Robotics Research Visibility in BRICS Countries: A Scientometrics Study

M. Vijayakumar and S.M. Annapurna

29.1 Introduction

Scientometrics is the art of determining and examining science and technology. Key study issues comprise the measurement of impact, reference sets of papers to examine the impact of periodicals and universities/institutes, perception of scientific citations, mapping scientific fields, and the invention of indicators for use in guiding principle and administration circumstances. In tradition, there is a major overlap between scientometrics and other scientific fields such as informatrics and bibliometrics.

The developing countries or recently industrialised countries such as Brazil, Russia, India, China, and South Africa formed an alliance called BRICS. The BRICS federation is famous for its accelerating economies and noteworthy power for local and world affairs.

29.2 Literature Review

The literature review reflects that there are many studies that have been conducted on bibliometrics, scientometrics, and informatics of various subjects, but the researches have not made any attempt on scientometrics study on robotic especially in BRICS countries.

M. Vijayakumar · S.M. Annapurna (✉)
Birla Institute of Technology, Mesra, Ranchi, Jharkhand, India
e-mail: annapurna.vg@gmail.com

V. Nath (ed.), *Proceedings of the International Conference on Nano-electronics, Circuits & Communication Systems*, Lecture Notes in Electrical Engineering 403,
DOI 10.1007/978-981-10-2999-8_29

29.3 Objectives

The main objectives of the study are to know the robotic research visibility in BRICS countries, especially to know:

- citation style of research documents,
- reference citation style of research documents, and
- H index of the research documents.

29.4 Methodology

To collect the primary data for our study, we used the bibliographic database like Web of Science (Data accessed on 1–31 March 2015). Then, the primary data have been collected from Web of Science for a period 1990–2014 and then tabulated and extracted from the charts using MS Excel.

29.5 Results and Discussion

29.5.1 Research Publications on Robotics

When we look into Table 29.1, it clearly indicates the research community of China snatched the dominant portion of the research publications by publishing the 59.7114% documents (538), followed by India 17.8690% (161 documents) and Brazil 15.5383% (140 documents). The research community of South Africa claims 01.7114% on overall research publications on robotics (Chart 29.1).

29.5.2 Citation Style

A "citation" is the technique by which we pass the message to the other researchers or readers that certain information/source in our work has taken from other source.

Table 29.1 Research document published

BRICS countries	Document published	%
Brazil	140	15.5383
Russia	047	05.2164
India	161	17.8690
China	538	59.7114
South Africa	015	01.6648

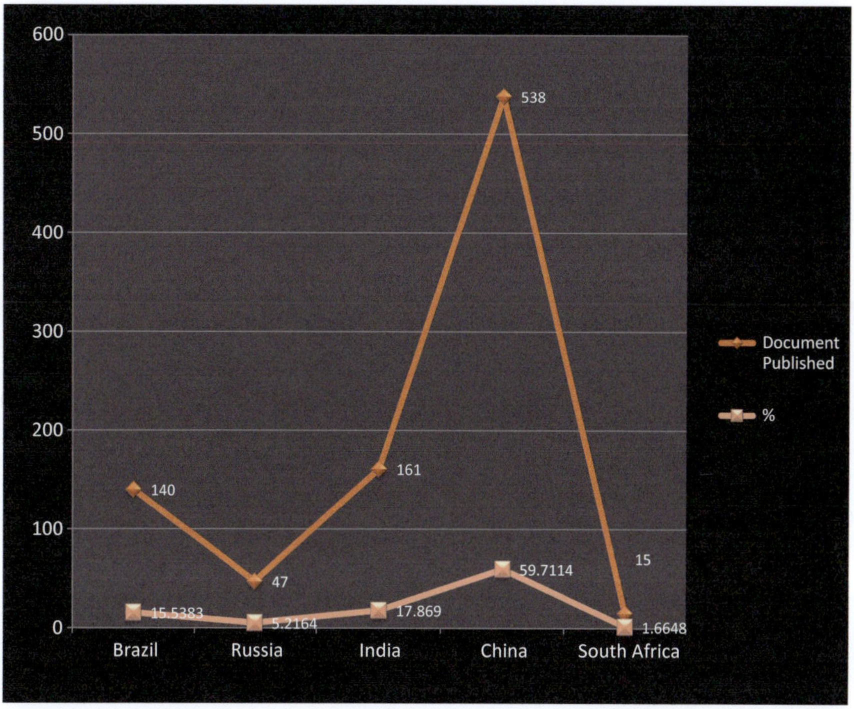

Chart 29.1 Research document published

The quality of research publication will be normally depended upon the number of citations gained for a particular publication. So we made an attempt to know the quality of the research papers of BRICS research community. Even though South African research community published less documents (15) but when we see average number of citation per document, South Africa tops first position by getting 9.9333 average of citation per research paper followed by Brazil with an average citation of 7.5643. Here, India claims last position with average citation of 4.0559.

One more interesting thing about the research community in general and specific to BRICS countries is they have the habit of citing the self-citation. It may be due to their area of research is same or to get more citation for their publications. However, the research community of China has more tendency (80.1829%) to cite their own publications followed by Indian research community (9.4512%). But the research community of South Africa has very less tendency to cite their own publications (0.3049%) (Chart 29.2 and Table 29.2).

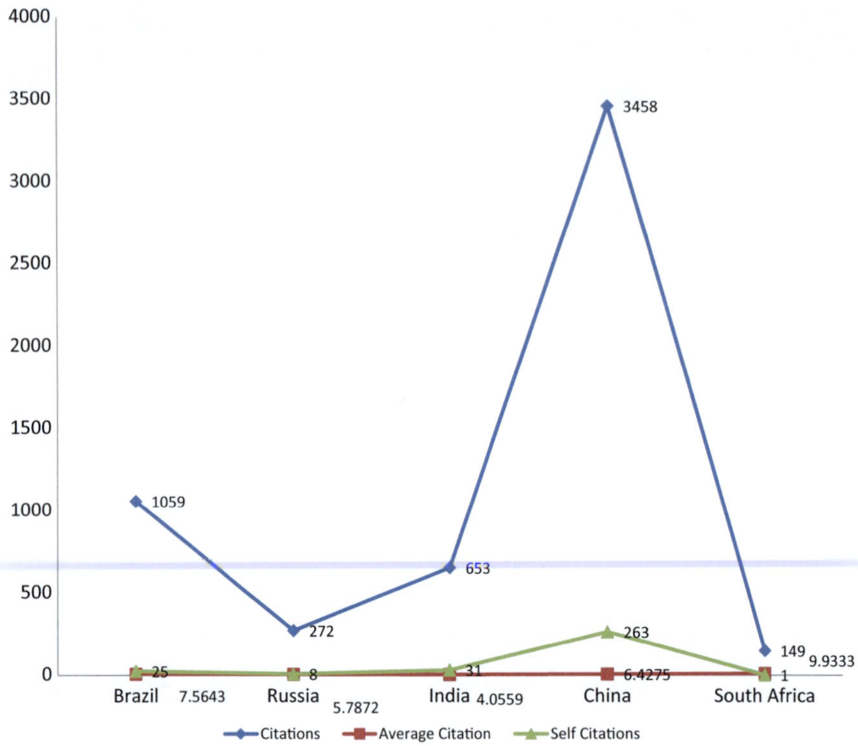

Chart 29.2 Citation style

Table 29.2 Citation style

BRICS countries	Document published	Citations	Average	%	Self-citations	%
Brazil	140	1059	7.5643	18.9412	025	7.6219
Russia	047	0272	5.7872	4.8650	008	2.4390
India	161	0653	4.0559	11.6795	031	9.4512
China	538	3458	6.4275	61.8494	263	80.1829
South Africa	015	0149	9.9333	2.6650	001	0.3049
Total	901	5591	6.2053			

29.5.3 *Reference Citation Style*

Similarly, in the way citation, we made an attempt know the attitude of BRICS research community in citing the articles. Table 29.3 provides the clear picture of

Table 29.3 Reference citation style

BRICS countries	Document published	Reference citation	Average	%	Self-reference citation	%
Brazil	140	0995	7.1071	20.3227	023	11.3861
Russia	047	0257	5.4681	5.2492	004	1.9802
India	161	0579	3.5963	11.8260	020	9.9010
China	538	2919	5.4256	59.6201	154	76.2376
South Africa	015	0146	9.7333	2.9820	001	0.4950

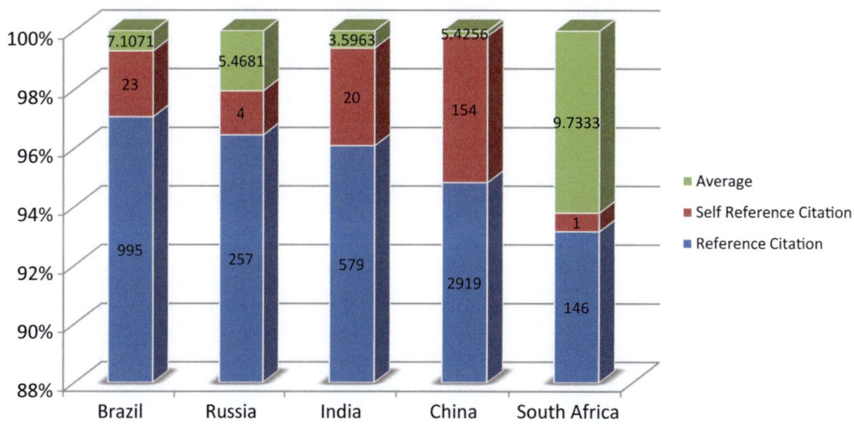

Chart 29.3 Reference citation style

the attitude of the BRICS research community. It shows that the maximum research community of China cited the 2919 articles for their research papers followed by Brazil research community (995) and Indian research community (579). Here, once again the research community of South Africa cited less articles (146). When we see the average reference citation, South African researchers claim top position with 9.7333 followed by Brazil researchers with the average 7.1071 and Russia (5.4681). Indian researchers claim last position with an average 3.5963.

Self-citing articles are a common attitude of all research community may it BRICS countries or others. In case of self-citing articles also, researchers of China claim top position with 76.2376% followed by Brazil researchers (11.3861%) and India (9.9010%) Here, once gain, the research community of South Africa claims least self-citing articles with 0.4950% (Chart 29.3).

Chart 29.4 *h* index

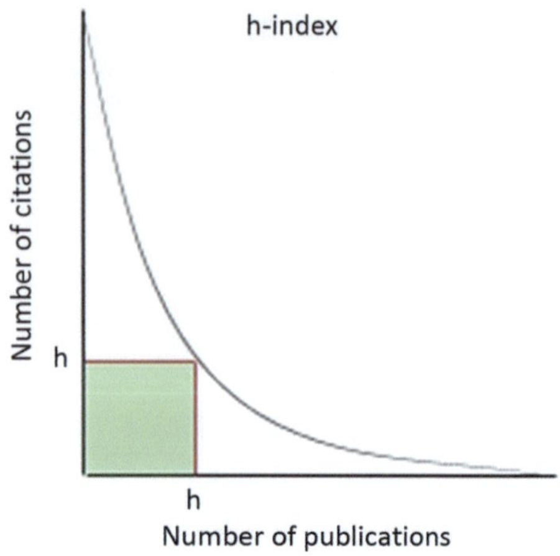

Table 29.4 H index of the research documents

BRICS countries	*H* index	%
Brazil	19	26.3889
Russia	09	12.5
India	13	18.0556
China	26	36.1111
South Africa	05	6.9444

29.5.4 h *Index*

The new tool called *h* index attempts to gauge the citation and productivity collision of the published work of a research or scientist community. The index can be calculated on the basis of research community's generally cited papers and total the number of citations that they have acquired with other published papers (Chart 29.4).

We made an attempt to know the *h* index of the research community of BRICS countries. Table 29.4 reflects that the research community of China have the highest *h* index of 26 which is more than one-third of *h* index of BRICS countries followed by Brazil with 19 *h* index which is more or less equivalent to one-fourth *h* index of BRICS countries. The South Africa claims minimum of 5 index (Chart 29.5).

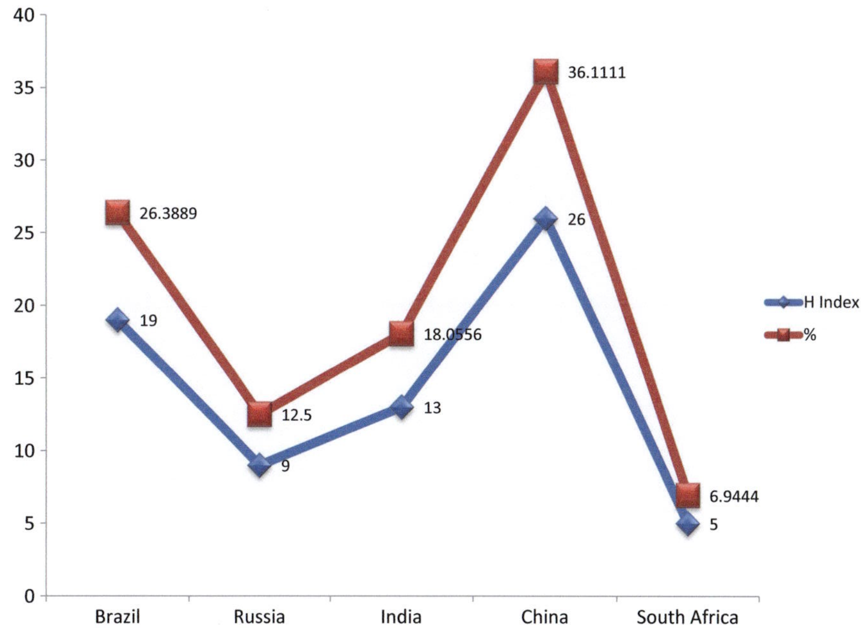

Chart 29.5 *h* index

29.6 Conclusions

The result of our study shows that the research community of China published the more research documents on robotics (538–59.7114%). Even in the case of citation and self, also the research documents published by China community get more citation (3458) and the same community have more self-cited (263–80.1829%), but when looking in terms of average of citation per document, South Africa possess more with 9.9333. In case of "*h* index", also the research community of China have the highest *h* index of 26 which is more than one-third of *h* index of BRICS countries.

Chapter 30
Analysis of a Self-compensating, Low-Noise, Low-Power PLL Circuit @ 45-nm Technology Node

Agnish Mal, Akshat Chitransh, Harsh Srivastava, Suraj Kumar Saw and Vijay Nath

30.1 Introduction

Phase Locked Loop is a feedback circuit that helps to build a synchronization between the input and output signals by generating a clock signal. Phase Locked Loop (PLL) circuits basically comprise a VCO, a Phase Detector, and a loop (low-pass) filter. It allows an oscillator to track or synchronize with another oscillator [1] and latches the input and output so that their frequency matches exactly (but there can be a phase mismatch) [2]. The VCO generates a signal whose frequency is controlled by the control voltage and the phase comparator (PC) compares the phase component of the input (reference) with that of the VCO, thereby generating an error voltage proportional to the phase difference. The cutoff frequency of the loop filter is calculated such that it blocks the unwanted high-frequency components in the error voltage signal to pass onto the next stage, causing a sharp transition between its pass-band and stop-band (Fig. 30.1).

A. Mal · A. Chitransh (✉) · H. Srivastava · S.K. Saw · V. Nath
VLSI Design Group, Department of Electronics and Communication Engineering, Birla Institute of Technology, Mesra, Ranchi 835215, Jharkhand, India
e-mail: akshatchitransh@gmail.com

A. Mal
e-mail: agnish.mal@gmail.com

H. Srivastava
e-mail: harsh.sri02@gmail.com

S.K. Saw
e-mail: srjsaw98@gmail.com

V. Nath
e-mail: vijaynath@bitmesra.ac.in

V. Nath (ed.), *Proceedings of the International Conference on Nano-electronics, Circuits & Communication Systems*, Lecture Notes in Electrical Engineering 403,
DOI 10.1007/978-981-10-2999-8_30

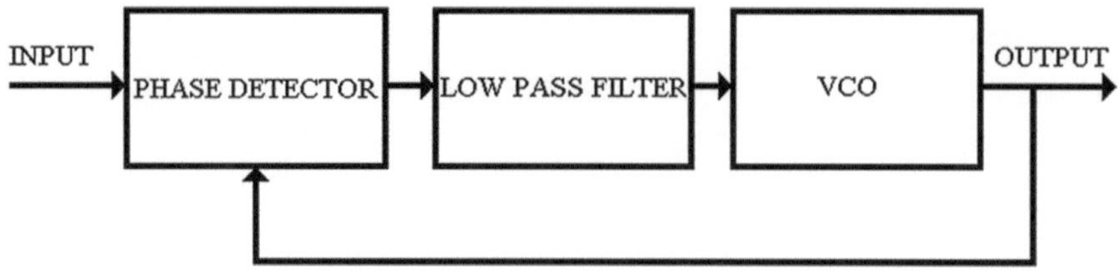

Fig. 30.1 Top-level block diagram of a Phase-Locked-Loop circuit

The Standard PLL Circuit that we have proposed comprises the following three blocks

(1) Phase Detector Comparator (PD).
(2) Loop Filter (LF).
(3) Voltage-Controlled Oscillator (VCO).

Monolithic CMOSPLLs are among the most important components of modern integrated systems (system-on-a-chip) [3]. One of the most demanded PLL functions is its on-chip clock synthesis. Other distinctive features of PLL circuits are that their output frequency is changeable or programmable in proportion to the input phase difference. This is the main reason why PLL finds applications in frequency synthesizer, clock generator, and frequency modulation and demodulation systems.

The main trade-offs of PLL design are given as:

(1) Stability of synthesized frequency (low jitter).
(2) Power consumption (power as low as possible).
(3) Area (lesser the area more is its efficiency).

Although all of the above three must be considered for an efficient PLL design, yet low jitter is the most important among all of them. Sometimes low jitter is obtained at the expense of power consumption, which is again undesirable for mobile applications. Another error that occurs in PLL circuit design is that the synthesized frequency comes to be time variant.

An important thing to note here is that as the frequency of the inputs of the PC increases, more and more error creeps in the measurement of phase difference [3]. Hence, to obtain stability in the operation of the PC at high frequencies, the detection sensitivity of the PC must be increased. A low detection sensitivity further leads to instability in the control voltage, thereby enhancing the noise component at the VCO output. The phase noise of the output signal of PLL again depreciates data error ratio [4].

In the following sections of the paper, we have analyzed the basic building blocks of the Phase-Locked-Loop circuit. These blocks have been tested for characteristic input(s) and are observed to produce the desired results. The primary merit of the discussed design is that it employs CMOS logic and hence is quite

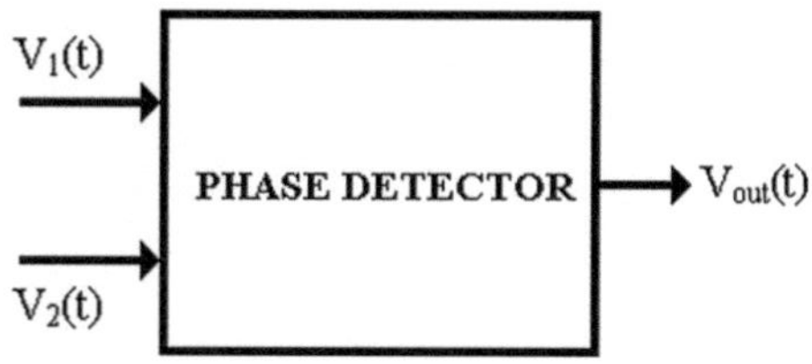

Fig. 30.2 Top-level block diagram of a Phase Detector circuit

more efficient compared to the other existing designs. The response for this circuit is quite commendable at higher frequencies also, thus depicting the versatility of the design (Fig. 30.2).

30.2 Phase Detector or Comparator

PC is a circuit arrangement, which is used to detect the difference in phase between two time-dependent signals ($V_1(t)$ and $V_2(t)$). The difference in phase is also obtained as a time-dependent signal 'V_{out}' [5]. It is an essential element of the Phase-Locked-Loop circuit.

Detecting the phase difference is very important in applications such as motor control, radar, and communication systems, servo-mechanism, and demodulators. The most common and easiest way to design a Phase Detector is to use a XOR-Gate [6].

If $\Phi_1(t)$ and $\Phi_2(t)$ are the respective phases of inputs to be compared, then the output voltage signal is given by

$$V_e(t) = K_D(\Phi_1(t) - \Phi_2(t)) \tag{30.1}$$

where K_D is the gain of Phase Detector (volt/radian).

As we know, if Y is the output of an XOR-Gate with inputs A and B, then the dependence of Y on A and B is given by the following Boolean expression

$$Y = AB' + A'B \tag{30.2}$$

Thus, by observing the values of Y, we find that Y is high when the inputs A and B are at different phases with respect to each other. Hence, a XOR-Gate can be used as a Phase Detector. Phase Detector can be made by mixer or analog multiplier, but this increases the complexity of the design, which is generally not prescribed for novice designers. Three cases have been shown for Phase Detector where the phase differences between the two input signals are 60°, 120°, and 180°, respectively. It is observed that the, if phase difference increases between the two inputs, the output (depicted by Y in the following plots) widens at those points, thus indicating the circuits sensitivity to phase mismatch (Fig. 30.3, 30.4, 30.5 and 30.6).

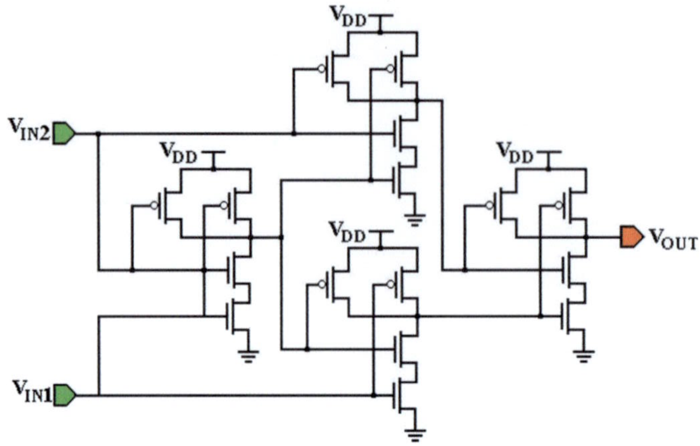

Fig. 30.3 Phase comparator circuit implemented using CMOS logic

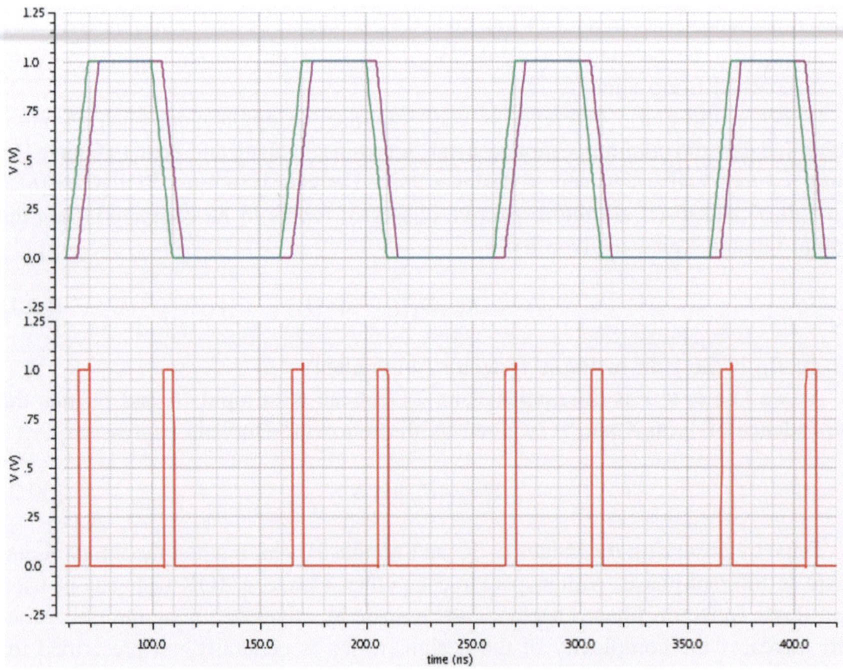

Fig. 30.4 Transient response at $\Delta\phi = \pi/3$

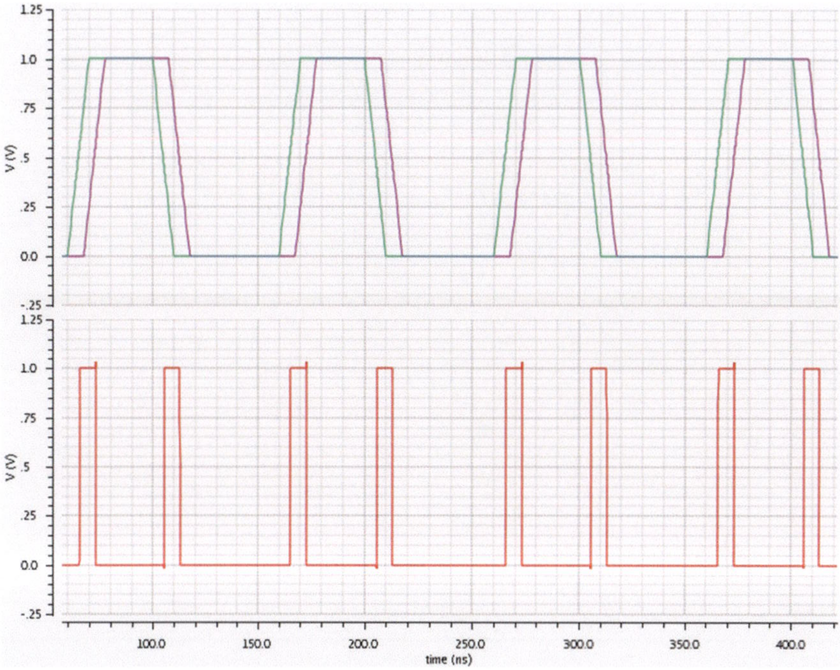

Fig. 30.5 Transient response at $\Delta\phi = 2\pi/3$

30.3 Voltage-Controlled Oscillator

A VCO as depicted by the name is an oscillator, which produces oscillations at a frequency proportionate to the input voltage (V_{in}).When the input voltage is zero, the oscillation frequency is set to a minimal value (ω_o), which can be adjusted to zero also.

The oscillation frequency for a VCO is given by the following expression:

$$\omega = \omega_o + K_o V_{in} \tag{30.3}$$

where K_o is the gain coefficient(rad/s/volt).

Thus, within the VCO tuning range, a finite V_{in} is required. Earlier VCOs were used for frequencies with lower values but by the use of swing control, frequencies of the order of MHz can be obtained [6].

Almost in analog and digital circuits, VCO is important basic building blocks [7]. The VCO played a very important role in wireless system.

The basic requirements to design a VCO are as follows:

1. Large electrical tuning range and phase stability.
2. Linearity of frequency versus control voltage.

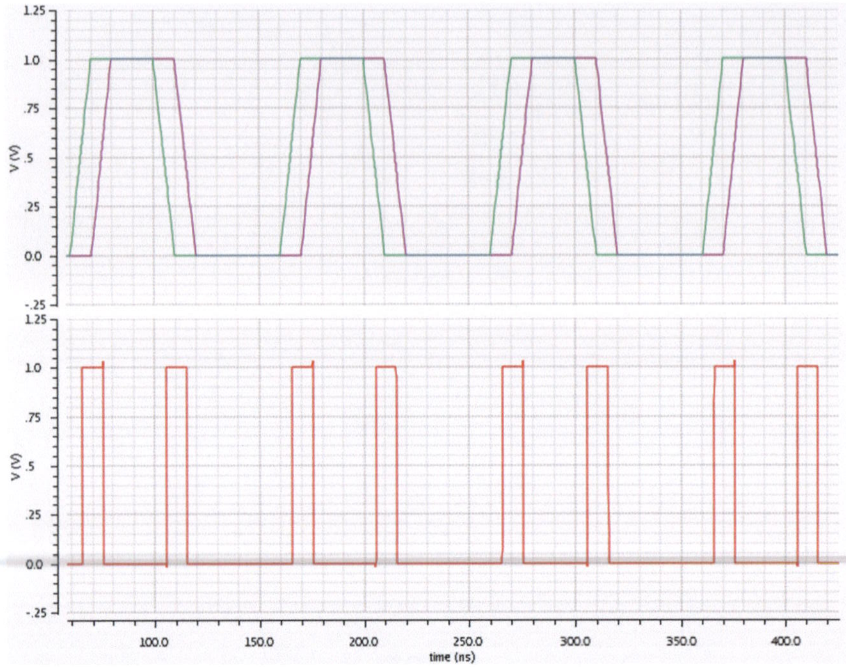

Fig. 30.6 Transient response at $\Delta\phi = \pi$

3. Large gain factor and low cost.
4. Capability of accepting wideband modulation.

The design that we have used here is of a 5-stage current starved VCO (CS-VCO). The 5-stage CS-VCO has been preferred over the 3-stage and 7-stage designs because of its optimum power consumption and minimal noise benefits.

30.4 Current Starved VCO

Figure 30.7 shows the schematic of five-stage CS-VCO. The name of this oscillator basically implies that the inverter used is hungry for current. The current starved VCO designed here uses a ring oscillator, based on the fact that it is working and the working of a ring oscillator resembles in many ways.

The ring oscillator is an enhanced form of oscillator's delay. It uses an odd number of inverters in cascade form, where output of last inverter is fed back through input. It behaves like a single inverting amplifier which gain is greater than 1. Its working is based upon the concept of gate delay. In this concept, when a signal passes through any CMOS network which comprises back-to-back connections of

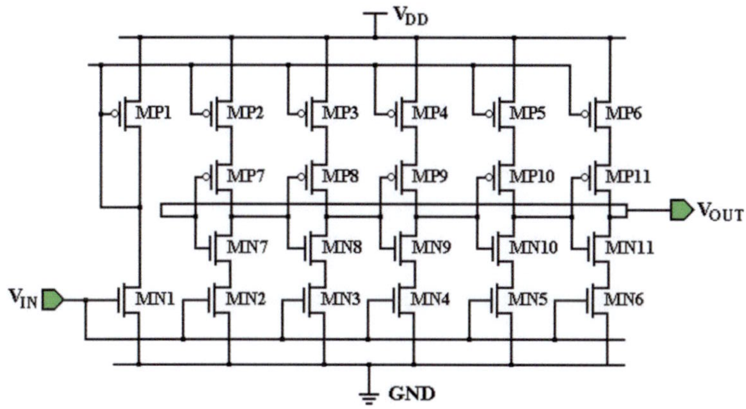

Fig. 30.7 Five-stage current starved voltage-controlled oscillator

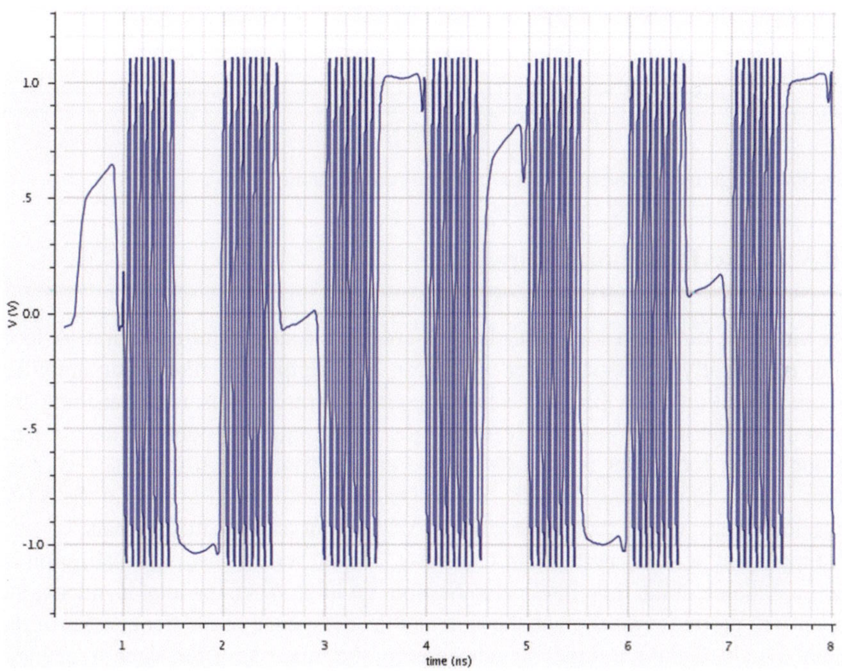

Fig. 30.8 Transient response for sinusoidal input of frequency 1M [CS-VCO]

logic gates, a certain time delay (gate delay) is introduced which causes the output to change after certain definite period of time when the input was changed. In this design, five such delay stages are cascaded (Fig. 30.8 and 30.9).

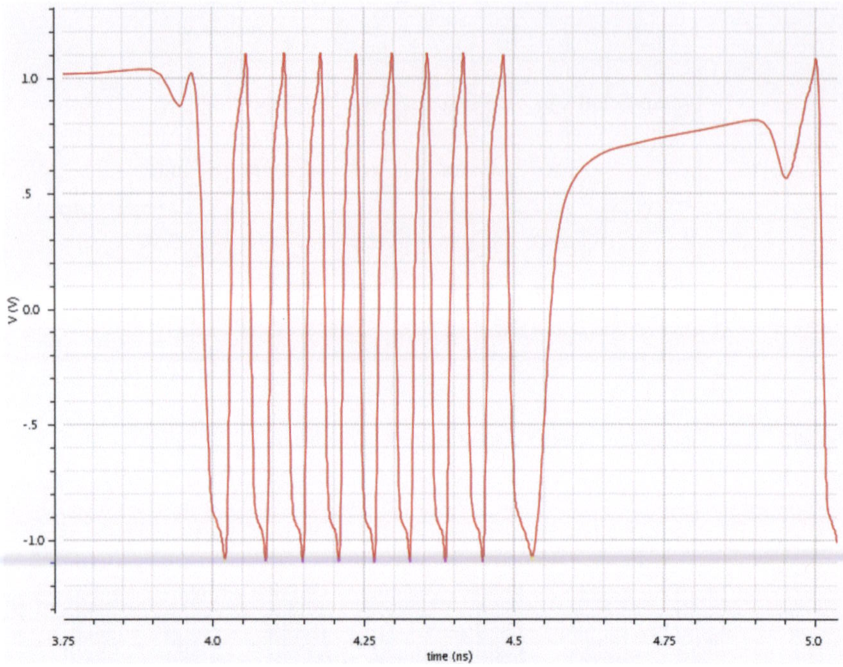

Fig. 30.9 Magnified response [CS-VCO] pointed-peak diagram

30.5 Low-Pass Loop Filter

For any PLL circuit, a loop filter is very crucial for successful operation. A loop filter is basically a low-pass filter which can be a simple RC filter or an amplifier whose operation is to block the high-frequency components from reaching the locked oscillator and selectively allow the low-frequency components to pass through. The loop filter is used to remove any unwanted high-frequency components, which might pass from the Phase Detector output and appear in the VCO tune line. They would then appear at the VCO input as unwanted signals.

Employing a mixer network in the Phase Detector circuit helps in realization of the importance of the loop filter network more vividly. When the loop is in lock, the mixer will produce two signals, the sum and the difference of the frequencies of the input signals. If the two input signals entering the mixer have the same frequency, then the frequency difference comes to be zero and as a result a DC voltage proportional to the phase difference is produced at the output. The frequency sum is also produced which is equal to twice the frequency of the reference signal. If this signal is not attenuated or eliminated, it reaches the VCO giving rise to unwanted (spurious) signals.

When other types of Phase Detectors are used, similar spurious signals can be produced and the loop filter becomes typically essential to eliminate them. If the loop filter has a very low cut-off frequency, then the changes in voltages will occur at a slow pace, and the VCO will not be able to change its frequency in a rapid manner. This is because the filter will only allow the low-frequency signals to pass through and correspondingly, the changes in voltage levels will be very slow. Conversely, a filter with a higher cut-off frequency will enable the changes to happen faster. However when using filters with high cut-off frequencies, care must be taken to ensure that unwanted frequencies are not passed along, thus creating a trade-off between the selectivity of the filter and operation speed.

The loop filter also governs the stability of the loop. If the filter is not designed properly, various oscillations can build up around the loop, and large magnitude signals will appear at the VCO. This will result in the VCO being forced to sweep over wide bands of frequencies. However, if the filter is implemented with proper design parameters of the constituting components, such cases can be eliminated to a large extent.

30.6 Simulation Results and Conclusion

All the simulations have been performed in Virtuoso Analog Design Environment of Cadence Design System @ 45-nm technology node. It has been observed that a five-stage current starved voltage-controlled oscillator provides the best performance when compared to its existing oscillator counterparts, viz., the source-coupled VCO [8], three-stage and five-stage CS-VCO in terms of noise performance, input tuning range, range of oscillation frequencies, area occupancy, and power dissipation factors. The simulation results attached demonstrate the merit of the design. The Phase Detector circuit analyzed above also justifies the fact that although there is a significant reduction in transistor count as well as system-on-a-chip area on utilization of depletion logic or pseudo-logic instead of CMOS logic for design purpose, there is a substantial difference in the overall power consumption which is efficiently reduced on incorporation of CMOS technology. The Phase Detector circuit responds perfectly when there is a phase difference between the two input signals. (Simulation results have been shown for phase difference = 60°, 120°, and 180°, respectively.) The loop filter circuit connected in series with the VCO has been designed using a simple RC low-pass filter network with appropriate specifications of the components, which is basically concerned here with the selective transmission of low-frequency components over the unwanted high-frequency components. Finally, connecting all the three blocks in sequence, the desired PLL output is obtained.

Acknowledgements We are thankful to DST, New Delhi, and DRDL, Hyderabad, for funding the IC design tools like Cadence Virtuoso to Dr. Vijay Nath. We are also thankful to Dr. M.K Mishra, Vice Chancellor, BIT Mesra and Prof. V.R Gupta, H.O.D ECE, BIT Mesra, for providing continuous inspiration and encouragement.

References

1. General PLL Description: T.H. Lee, Chap. 15. Gray and Meyer, 10.4 Clock generation: B. Razavi, Design of Analog CMOS Integrated Circuits, Chap. 15, and McGraw-Hill, 2001
2. Hobaral H, Kayanol Y, Inouel H (2012) A study on design of PLL for low phase-noise characteristics. In: SICE annual conference 2012, Akita University, Akita, Japan, 20–23 August 2012
3. Yaniv O, Raphaeli D (2001) Near-optimal PLL design for decision-feedback carrier and timing recovery. In: IEEE transactions on communications, vol 49, no. 9, Sept 2001
4. Suresh B, Vishwanathan V, Krishnan RS, Jamadagni HS (2005) Application of alpha power law models to PLL design methodology. In: Proceedings of the 18th international conference on VLSI design held jointly with 4th international conference on embedded systems design (VLSID'05) 1063-9667/05 $20.00 © 2005 IEEE
5. Janardhan H, Wagdy MF (2006) Design of a 1 GHz digital PLL using 0.18 μm CMOS technology. In: Proceedings of the third international conference on information technology: new generations (ITNG'06) 0-7695-2497-4/06 $20.00 © 2006 IEEE
6. Lakshmikumar KR (2009) Analog PLL design with ring oscillators at low-gigahertz frequencies in nanometer CMOS: challenges and solutions. In: IEEE transactions on circuits and systems—II: express briefs, vol 56, no. 5, May 2009
7. Jiao D, Kim B, Kim CH (2012) Design, modeling, and test of a programmable adaptive phase-shifting PLL for enhancing clock data compensation. IEEE J Solid-state Circ 47(10)
8. Patil RK, Nasre VG (2012) A performance comparison of current starved VCO and source coupled VCO for PLL in 0.18 μm CMOS process. Int J Eng Innovative Technol (IJEIT) 1(2)

Chapter 31
Evaluating the Performance of Dye-Sensitized Solar Cell with Various Key Components such as Electrodes, Dyes, and Electrolytes

Rishi Sharma, Kumar Amit, P.K. Barhai and R.L. Boxmann

31.1 Introduction

The demand for green renewable sources of energy is growing every year. Recently, the main renewable energy sources are solar, wind, geothermal, biological, and hydropower. Conventional fossil energy sources, e.g., petroleum, natural gas, and coal, will not be able to meet the growing global energy demand, which is expected to increase by about 40% over the next two decades. By 2030, petroleum, coal, and natural gas usage is expected to rise by 23, 44, and 37%, respectively [1]. Furthermore, the need to reduce CO_2 emissions will increase the demand for alternative renewable energy sources. Although nuclear energy and biofuels may play some role, both are problematical—e.g., a long-term solution for disposing nuclear waste has not been found, and biofuel production produces considerable waste. The least expensive alternative energy source is the sun, and hence, the demand for solar energy is growing. Solar energy is divided into solar thermal and photovoltaics (PV). Solar thermal refers to schemes in which solar radiation is used to heat a fluid. Photovoltaics use cells, which convert solar photons to electrical power directly. The currently available solar cells are fabricated mainly from silicon. Single crystal, polycrystalline microcrystalline, or amorphous silicon is used for the fabrication of these cells. Single crystal photovoltaic cells fabricated from these wafers have not only the highest laboratory efficiency ($\sim$24%), but also the highest cost. Less expensive, but less efficient ($\sim$13 to 18%) cells can be fabricated by polycrystalline Si wafers, and amorphous Si thin films [2].

R. Sharma · K. Amit (✉) · P.K. Barhai
Department of Physics, Birla Institute of Technology,
Mesra, Ranchi 835215, India
e-mail: aamit.wraj@gmail.com; engr.kumaramit@gmail.com

R.L. Boxmann
School of Electrical Engineering, Tel Aviv University,
P.O. Box 39040, 69978 Tel Aviv, Israel

V. Nath (ed.), *Proceedings of the International Conference on Nano-electronics, Circuits & Communication Systems*, Lecture Notes in Electrical Engineering 403,
DOI 10.1007/978-981-10-2999-8_31

These PV technologies now find potential contenders with the emergence of a third generation of cells based, for example, on nanocrystalline films, conducting polymers films, and especially by dye-sensitized solar cell (DSSC) [3, 4]. The DSSC provides a technically and economically viable alternative concept to present PV technology. In contrast to the conventional systems where a semiconductor absorbs light and separates both the positive and negative charges, in the DSSC, light is absorbed by a dye which is adsorbed to the surface of a wideband semiconductor. Charges are separated at the interface via photo-induced electron injection from the dye into the conduction band of the solid. DSSCs are far less expensive as low-cost polymer substrates, chemicals, and technologies are used in fabrication of DSSC. However, the efficiency of the DSSC is low [3] but it is ever increasing [5]. As all PV cells require transparent conducting electrodes to work efficiently. Tin-doped indium oxide (ITO) and fluorine-doped tin oxide (FTO) are commonly used transparent conducting electrode [6]. In DSSC, all other components are low cost except the transparent conducting electrode. This electrode is often the dominant cost in DSSCs. Hence, a low-cost transparent conductor is crucial for lowering the cost of DSSC. The research community has made considerable advances in identifying and evaluating potential alternative materials. The most promising materials among these are graphene films [7–9] and aluminum-doped zinc oxide (AZO) [10].

In present investigation, DSSC has been fabricated with different combinations of dyes, electrolytes, and conducting electrodes. Observed efficiencies of these cells have also been reported. Fabrication of graphene and its transparency has also been reported here; however, cell could not be fabricated by using graphene.

31.2 Fabrication of DSSC

Fabrication of the cell was done in various steps ranging from preparation of TiO_2 powder, making of dyes and electrolytes. In present investigation, for preparation of TiO_2 semiconductor layer 2 g of TiO_2 powder has been taken and ball-milled for 24 h at moderate speed. After that, the ball-milled sample was suspended in 8 ml of ethanol and was further ultrasonicated for 30 min. Additionally, 0.3 ml of titanium (IV) tetra isopropoxide was added into suspension and ultrasonicated again. Two different dyes, Ruthenium (Red) and 0.5 g Eosin (Y), have been used in present investigation. Dye solutions are prepared by dissolving 0.5 g Ruthenium (Red) and 0.5 g Eosin (Y) in 20 ml of de-ionized water and ethanol, respectively. Two electrolytes, lithium iodide and potassium iodide, are used in the present work. Solutions of the electrolytes are prepared by adding 0.5 g of lithium iodide and 0.5 g of potassium iodide into acetonitrile and distilled water. Furthermore, iodine was also added and ultrasonicated for nearly an hour.

ITO- and AZO-coated glass are used as conducting electrodes. A mask of $2.5 \times 1.5\ cm^2$ is prepared with the help of scotch tape over the conducting side of ITO/AZO-coated glass. The paste of TiO_2 is then evenly spread with the help of glass

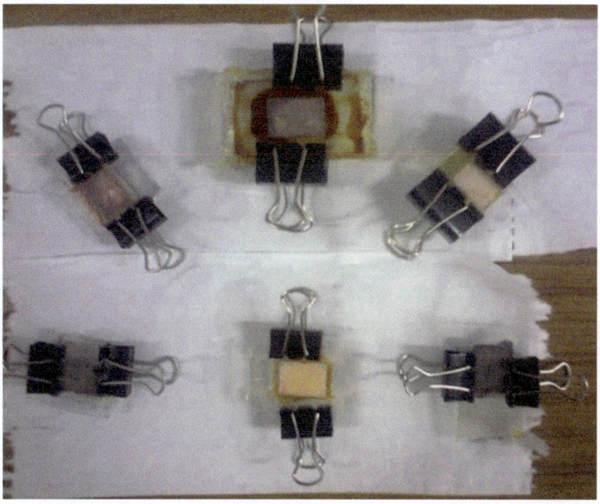

Fig. 31.1 Various types of DSSCs fabricated with different combinations of electrodes, dyes, and electrolytes

rod. After the removal of scotch tape mask, the sample is then annealed over the hot plate for nearly 10 min at 100 °C. The TiO_2 spread ITO/AZO glass is then dipped into the dye solution and soaked for 10 min. The sample is heated mildly for faster absorption and coloration. The sample is then washed away with deionized water or ethanol to remove extra dyes from the electrode surface. A catalytic layer of graphite is spread over counter-electrode. Both the electrodes are sandwiched together with the help of binder clips with conducting side facing each other. It is also ensured that the sides are in little offset condition so that wires can be connected with multimeter. Electrolyte solution is then added from either side and allowed to spread over the sandwiched junction. The cell is now packed properly by silicone gel and tapes. Power can be drawn from DSSC by placing it in solar simulator or in Sun. Fabricated DSSCs are shown in Fig. 31.1.

31.3 Working Principle of DSSC

The working principle of DSSC can be understood in the following steps:

(i) Generation of electron: As the DSSC is illuminated with solar radiation, the absorption of a photon by a dye molecules takes place. After the photon absorption, the dye molecule goes into excited state, S*. Within a very short time, of the order of femtoseconds, the electron is emitted by the dye and it gets oxidized to S^+. Mechanism can be understood through the reaction:

$$\underset{\text{(Dye molecule in ground state)}}{\mathbf{S + Photon}} \longrightarrow \underset{\text{(Dye molecule in excited state)}}{\mathbf{S^*}} \longrightarrow \underset{\text{(Oxidized Dye molecule)}}{\mathbf{S^+}} + \underset{\text{(electron)}}{\mathbf{e^-}}$$

(ii) The excited electron is given off to the conduction band of semiconductor. This electron now flows through the conduction band of the wide band gap semiconductor to the transparent conducting electrode. The electron now then flows through the external load to the counter-electrode.

$$e^- \longrightarrow TiO_2 \longrightarrow TCO \longrightarrow \text{External Load}$$

(iii) The oxidized dye molecules, S^+, are reduced to the original form 'S' by regaining electron from the organic electrolyte solution. The electrolyte solution contains the iodine redox system in which the iodide ions are being oxidized to tri-iodide ions. Tri-iodide reduces to iodine by collecting electron in the presence of catalyst. Catalyst may be platinum or graphite soot.

$$\underset{\text{(Iodide)}}{3/2\,I^-} \xrightarrow{\text{Oxidation}} \underset{\text{(tri-Iodide)}}{I/2I_3^-} + e^-$$

$$\underset{\text{(Oxidized Dye Molecule)}}{S^+} + \underset{\text{(From electrolyte Iodide}(I^-)\text{/tri-Iodide}(I_3^-)\text{ Redox system)}}{e^-} \longrightarrow \underset{\text{(Dye molecule in ground state)}}{S}$$

$$\underset{\text{(tri-Iodide)}}{1/2I_3^-} + \underset{\text{(Electron via external load)}}{e^-} \xrightarrow[\text{Catalyst}]{\text{Reduction}} \underset{\text{(Iodide)}}{3/2I^-}$$

Hence, the device produces potential difference under illumination. The overall voltage produced is derivative of difference between the Fermi level of the electron in the solid and the redox potential of the electrolyte. During the generation of photovoltage, there is no change in chemical composition of electrolyte, dye, and TiO_2.

31.4 Results and Discussion

To study the performance of DSSCs, total 9 cells have been fabricated. Different combinations of conducting transparent electrodes, dyes, and electrolytes have been used for the fabrication of solar cells. Types of conducting electrodes, dyes, semiconductors, electrolytes, cell area, and catalysts on counter-electrode are shown in Table 31.1. Observed short-circuit current (J_{sc}), open-circuit voltage (V_{oc}), fill factor (FF), and efficiency of these cells are shown in Table 31.2.

Tin-doped indium oxide and aluminum-doped zinc oxide have been used as transparent conducting electrodes. Tin-doped indium oxide has single handedly dominated the field transparent conducting electrode for almost four decades. These

Table 31.1 Type of conducting electrodes, dyes, semiconductors, electrolytes, cell area, and catalysts on counter-electrode of different solar cells

Solar cell no.	Conducting electrodes	Dyes	Semiconductor used	Electrolytes	Active area (cm^2)	Catalyst on counter-electrode
Cell-1	ITO	Eosin-Y	TiO_2	KI	1	No catalyst
Cell-2	ITO	Ru	TiO_2	KI	1	Pencil graphite
Cell-3	ITO	Ru	TiO_2–Cu doped	KI	1	Pencil graphite
Cell-4	ITO	Ru	TiO_2-graphene doped	KI	1	Pencil graphite
Cell-5	AZO	Ru	TiO_2	LiI	1	Pencil graphite
Cell-6	AZO	Eosin-Y	TiO_2	LiI	1	Pencil graphite
Cell-7	ITO	Ru	TiO_2	LiI	1	Pencil graphite
Cell-8	ITO	Eosin-Y	TiO_2	LiI	1	Pencil graphite
Cell-9	ITO	Ru	TiO_2	LiI	1	Candle soot

Table 31.2 Open-circuit voltage, short-circuit current, fill factor, and efficiency different solar cells

Solar cell no.	J_{sc} (mA)/cm^2	V_{oc} (mV)	Fill factor	% efficiency
Cell-1	1.2	30	0.1	0.0036
Cell-2	3.6	112	0.2	0.08
Cell-3	4.1	136	0.2	0.11
Cell-4	5.1	200	0.3	0.31
Cell-5	4.6	152	0.2	0.14
Cell-6	8.6	307	0.4	1.05
Cell-7	11.4	657	0.6	4.49
Cell-8	3.2	99	0.2	0.63
Cell-9	9.8	307	0.5	1.50

oxides, however, appear to be increasingly problematic due to (i) the limited availability and high cost of indium, (ii) their instability in the presence of acid or base, (iii) their susceptibility to ion diffusion into polymer layers, and (iv) their limited transparency in the near-infrared region and (v) the current leakage of FTO devices caused by FTO structure defects [11]. AZO is the leading contender as a transparent conducting oxide replacement of ITO [8]. Filtered Vacuum Arc Deposition (FVAD) has been used for the deposition of AZO, in the present investigation. Graphene as an electrode has also been investigated [12, 13]; however, realization of graphene electrode is problematic. The main problem with graphene layers is its electrical conductivity on large area substrates. Transfer of the graphene on transparent substrate is another major problem. It is difficult to obtain continuous single or multilayer graphene by transferring from catalyst-coated Si/SiO_2 substrate to transparent substrates such as Poly(methyl methacrylate) (PMMA) using chemical method. Because of chemical transfer, many impurities are also attached with graphene which reduce the transparency. Graphene samples are grown by using thermal CVD system on Fe, Co, and Ni catalyst-coated Si/SiO_2 substrates. Catalysts coating is done by Anodic Vacuum Arc technique. Raman spectra of graphene samples grown on Ni catalyst at 600 torr are shown in Fig. 31.2.

It shows three main peaks centered at 1350, 1580, and 2700 cm^{-1} known as D peak, G peak, and 2D peak, respectively, which confirm the formation of graphene. Transparency of PMMA, ITO, AZO, and graphene is shown in Figs. 31.3, 31.4, 31.5, and 31.6, respectively.

Both ITO- and AZO-coated glasses are 85–90% transparent in visible range. PMMA sheet gives more than 90% transparency, whereas graphene on PMMA is 70% transparent. It has been observed the transparency of the different samples is in the order of PMMA > ITO > AZO > graphene. However, the transparency of the graphene is still low and it has good potential to replace the conventional

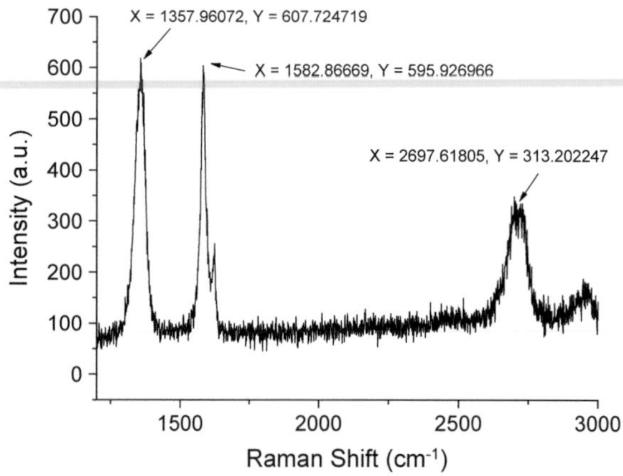

Fig. 31.2 Raman spectra of sample on Ni catalyst at Si/SiO_2 substrate at 600 torr

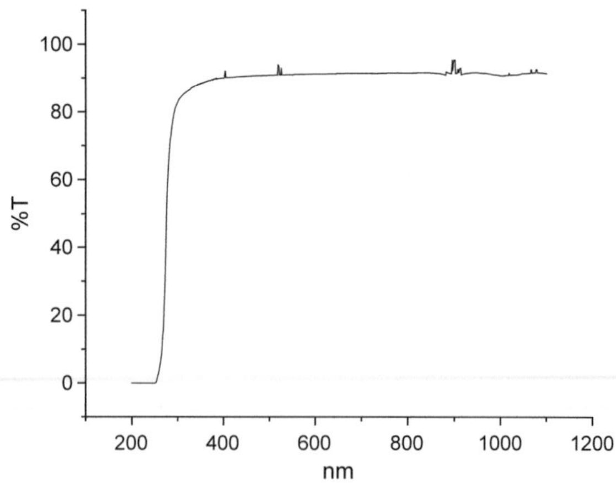

Fig. 31.3 Percentage transmittance of PMMA

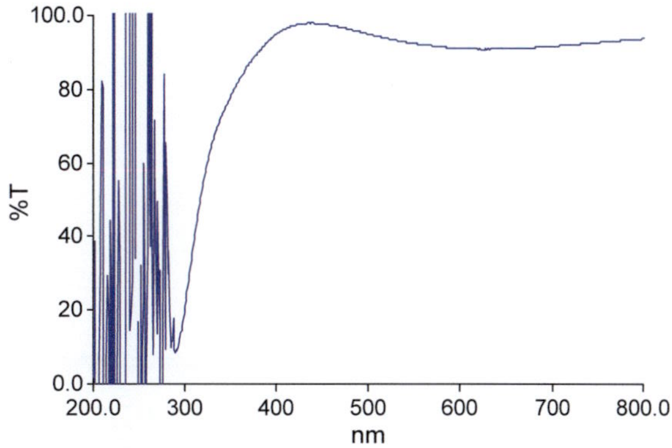

Fig. 31.4 Percentage transmittance of ITO-coated glass

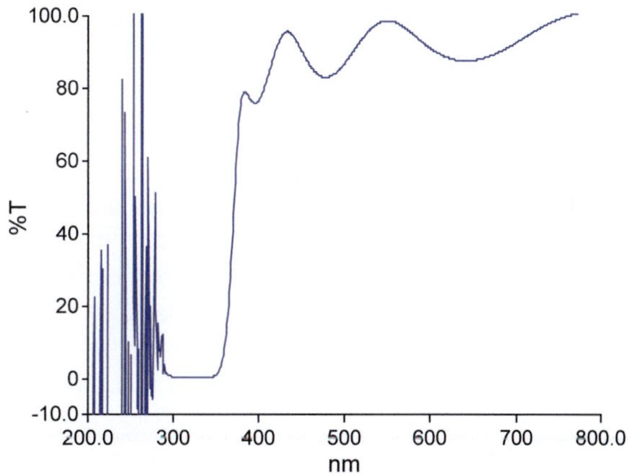

Fig. 31.5 Percentage transmittance of AZO-coated glass

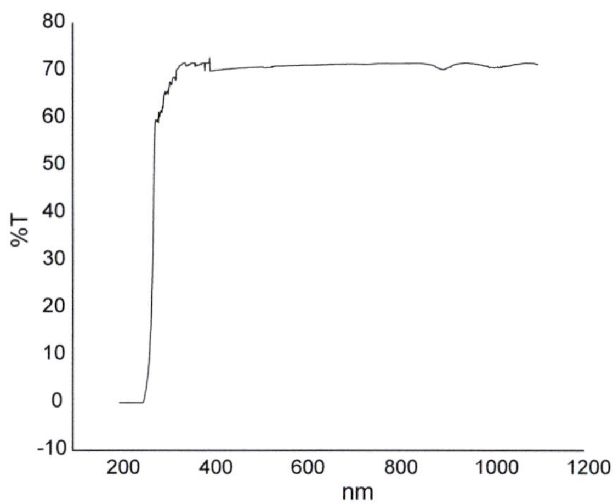

Fig. 31.6 Percentage transmittance of graphene on PMMA

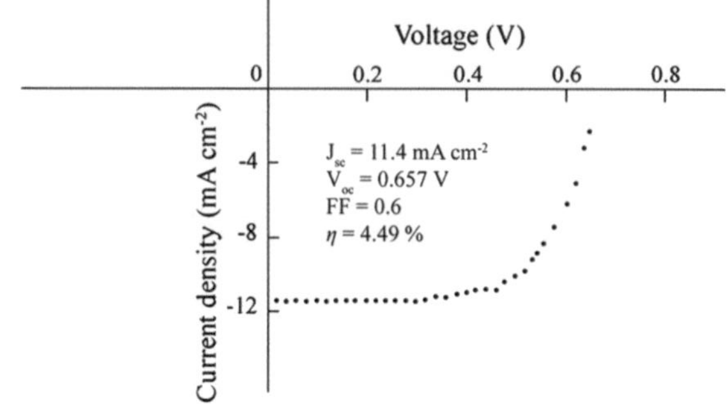

Fig. 31.7 Plot between current density and voltage for the Solar Cell-7

conducting transparent electrodes. Current voltage plot of one of the solar cells (Cell-7) is shown in Fig. 31.7. All measurements are taken under AM 1.5 (1000 W/m^2) with the help of a solar simulator. Current and voltage are measured using digital multimeter of high sensitivity.

Pencil graphite is used as a catalyst on counter-electrode. Cell prepared without any catalyst coated on counter-electrode (Cell-1) shows very poor efficiency. The efficiency of the cell can be increased by using platinum as a catalyst on counter-electrode. It has been observed that the candle soot can also be used as a catalyst (Cell-9) for counter-electrode. Packing of the solar also increases the efficiency. The maximum open-circuit voltage and short-circuit current, and hence highest efficiency (4.49%), have been observed for Cell-7, which correspond to ITO as a conducting electrode, Ru dye and LiI as an electrolyte. The cell fabricated by this combination is costly but have high efficiency. Solar cell fabricated with AZO, Eosin-Y dye, and LiI electrolyte (Cell-6) has low efficiency but relatively cheaper. The cell prepared with ITO electrodes and Eosin-Y dye (Cell-7) has only 0.6% efficiency. Hence, it can be concluded that the AZO electrodes work well with low-cost dye such as Eosin-Y, and it may increase their price to performance ratio. Furthermore, using additives such as Cu or graphene in TiO_2 also increases the performance significantly.

31.5 Conclusions

DSSCs have been fabricated using ITO- and AZO-coated glass electrodes. ITO as a reference sample outperforms AZO, but AZO has full potential to replace ITO. The best efficiency (4.49%) is for the cell prepared with ITO electrode. AZO electrodes have better efficiency with Eosin-Y dye (low cost dye) compared to ITO electrodes.

Thermal CVD-deposited graphene is taken out on PMMA and transparency has been measured. Results show 70% transparency in visible range. However, the transparency of AZO- and ITO-coated glass is better than graphene. It has been observed experimentally that the production of single or few layers graphene is complicated. Uniform layer of graphene on large area transparent substrate cannot be produced/transferred by conventional methods; therefore, it is difficult to produce large area conducting sheets. However, graphene has good potential to replace the conventional conducting transparent electrodes. Besides, the electrical properties many other properties such as magnetic, mechanical, biocompatibility, and chemical are need to be studied to explore the potential of graphene.

Acknowledgments The authors are grateful to the Department of Science and Technology (DST), University Grants Commission for Special Assistance Programme (UGC-SAP), Government of India and the Office of Chief Scientists, Ministry of Science & Technology, Government of Israel for financial support under India-Israel S&T Cooperation Programme.

References

1. Technology review, vol 113, No. 5. MIT, Sept/Oct 2010
2. Solanki CS (2012) Solar photovoltaics: fundamentals, technologies and applications. PHI Learning, Private Limited
3. Grätzel M (2003) Dye-sensitized solar cells. J Photochem Photobiol, C 4(2):145–153
4. Hagfeldt A, Boschloo G, Sun L, Kloo L, Pettersson H (2010) Dye-sensitized solar cells. Chem Rev 110(11):6595–6663
5. Mathew S, Aswani Y, Gao P, Humphry-Baker, Curchod BFE, Ashari-Astani N, Tavernelli I, Rothlisberger U, Nazeeruddin MK, Grätzel M (2014) Dye-sensitized solar cells with 13% efficiency achieved through the molecular engineering of porphyrin sensitizers. Nat Chem 6 (3):242–247
6. Edward PP, Porch A, Jones MO, Morgan DV, Perks RM (2004) Dalton Trans 2995–3002
7. Eda G, Fanchini G, Chhowalla M (2008) Mater Nat Nanotechnol 3:270–274
8. Tung VC, Chen L, Allen MJ, Wassei JK, Nelson K, Kaner RB, Yang Y (2009) Nano Lett 9:1949–1955
9. Wang X, Zhi L, Müllen K (2008) Nano Lett 8:323–327
10. Mercaldo LV, Addonizio ML, Della Noce M, Veneri PD, Scognamiglio A, Privato C (2009) Thin film silicon photovoltaics: architectural perspectives and technological issues. Appl Energy 86:1836–1844
11. Wang X, Zhi L, Müllen K (2008) Nano Lett 8:323
12. Jung S-M, Choi IT, Lim K, Ko J, Kim JC, Lee J-J, Ju MJ, Kim HK, Baek J-B (2014) B-doped graphene as an electrochemically superior metal-free cathode material as compared to Pt over a Co (II)/Co (III) electrolyte for dye-sensitized solar cell. Chem Mater 26(11):3586–3591
13. Roy-Mayhew JD, Aksay IA (2014) Graphene materials and their use in dye-sensitized solar cells. Chem Rev 114(12):6323–6348

Chapter 32
Design of Comparator in Sigma-Delta ADC Using 45 nm CMOS Technology

Varun Kumar, Krishan Kumar Singh, Abhishek Pandey and Vijay Nath

32.1 Introduction

Recent studies on radio frequency (RF) communication focus on higher assimilation, to meet the market demand for a low-cost, very low-power customized personal communication device. Minimization of power can be achieved by inclination towards miniaturization. The performance of today's manufactured devices is highly affected with the varying processes and other non-idealities. Analog-to-digital converter (ADC) is one of the applications where high resolution and speed, minimum offset voltage, and power consumption are required for mobile and portable devices. Gain amplifiers and comparators [1–5] are the limiting factors for the performance in such ADCs. Oversampling analog-to-digital conversion in wireless application has recently become popular because of its increased performance and flexibility. Oversampling incorporates a sampling rate of much greater value than the bandwidth of the signal of interest. Further digital signal processing is applied to the signal for filtering and down sampling. Sigma-delta ADC integrates low bandwidth, low power, low cost, and high resolution. This ADC has different applications in instrumentation, data acquisition, and digital signal processing [6–10]. It came into existence during initial phases of development of pulse

---

V. Kumar (✉) · K.K. Singh · A. Pandey · V. Nath
Electronics and Communication Engineering Department, Birla Institute of Technology, Mesra, Ranchi 835215, Jharkhand, India
e-mail: vk10292011@gmail.com

K.K. Singh
e-mail: singh.krishan.kr@gmail.com

A. Pandey
e-mail: a.p.bitmesra@gmail.com

V. Nath (ed.), *Proceedings of the International Conference on Nano-electronics, Circuits & Communication Systems*, Lecture Notes in Electrical Engineering 403,
DOI 10.1007/978-981-10-2999-8_32

code modulation (PCM) systems. Although this technique has been available for many years, the recent advances in technology made this device practical with widespread usage. The sigma-delta ADC is now a choice for modern voice band, audio, and high-resolution industrial measurement applications.

Sigma-delta ADC comes under the category of oversampling ADC, which samples the signal at an oversampled frequency of $fN = K2F$ where k is the oversampling factor, which is given by below equation [11].

$$K = fn/F \tag{32.1}$$

The sigma-delta ADC is basically divided into two parts: one is the modulator part, and the other is the decimator part. The input analog signal is given to the modulator part where the oversampling, quantization, and noise shaping process are carried out and the result in a pulse stream of digital output. This is given to the decimation filter which decimates higher frequency oversampled signal from the modulator to lower frequency signal; it filters out the noise above the cut-off frequency range [12–15]. The sigma-delta modulator consists of differential amplifier, integrator, and 1-bit comparator, which generate 1-bit output. The decimator is made of N-bit counter. It generates N-bit digital output, where N is the resolution of ADC and is dependent on the oversampling ratio. The modulator and decimator are operated at same oversampling clock.

This chapter has been organized as follows. The designing of opamp is given in Sect. 32.2. Section 32.3 describes the methodology and architecture of the proposed comparator circuit. The simulation result and discussion have been summarized in Sect. 32.4. Finally, the conclusion of the overall paper is shown in Sect. 32.5.

32.2 Designing of Operational Amplifier

One the most commonly used devices in analog world is operational amplifier (opamp). It is a device whose output can be easily related to input in terms of some known mathematical equations and operations. By using some active and passive elements such as resistors and capacitors in integrator and differentiator circuits, this can be achieved. An opamp can be characterized by high input impedance, high gain margin, low output impedance, and high band width. It has also the ability to amplify the differential mode signal to a high extent and attenuates or rejects the common mode signal. Basically, an opamp mainly consists of four different functional blocks, namely the differential gain stage, the single-ended conversion stage, the level shifting stage, and finally the buffer stage. The first stage, that is the differential gain stage, independently amplifies the average and common mode signals. It is one of the most critical and important stages of an opamp, and it decides some of the most important parameters such as common mode rejection ratio (CMRR), input common mode range (ICMR), and the input noise [16, 17].

The second stage that is the single-ended conversion stage which is present after the differential stage is responsible for giving a single-ended output which is referred to the ground. The next is the level shifter stage which is introduced after the single-ended conversion stage and finally the buffer or the second gain stage which provides the necessary gain to the amplifier.

32.3 Designing of Comparator

A comparator is a circuit that performs the operation of comparing two analog input signals and decoding the difference into a single digital output signal.

The comparator is one of the core parts of any analog-to-digital converter. Depending upon their size and structure, they can have severe affects on the performance of a device. The accuracy of the sigma-delta ADC depends on the accuracy of the comparator. The delay, input signal, and input offset voltage directly impact on the resolution and speed of sigma-delta ADC [5]. Comparators are classified depending upon their nature, functionality, and inputs such as voltage and current. Some of the core application of comparators is signal processing, ADC conversion, neural systems, and function generation. The comparator takes the analog input and gives a binary or digital output [18–21].

A comparator circuit compares one signal with another to detect the difference between the input and reference signal. In analog circuit, the comparator is widely used component after operational amplifier (opamp). They have widespread use in A/D converters, data transmission, and switching power regulators. Here, we have used a two-stage opamp for a comparator.

The ideal response of comparator is shown in Fig. 32.1. It shows input response and output response with positive and negative reference voltage (Vref). The proposed comparator is shown in Fig. 32.2, and it consists of two-stage opamp. The capacitor, CL, is used to rectify the amplified signal in pulse form.

32.4 Simulation Results and Discussion

The simulated result of two-stage opamp is shown in Fig. 32.3, which is summarized at different process corner of library. The opamp shows 89 dB gain, 2.3 GHz UGB, and 65° phase margin at Monte Carlo corner simulation.

Figure 32.4 shows the output of unoptimized comparator. The comparator designed was optimized, and its output wave form is shown in Fig. 32.5. We can see that there is considerable gain of about 6 dB in the configured comparator w.r.t the unconfigured one and the output pulses are more feasible.

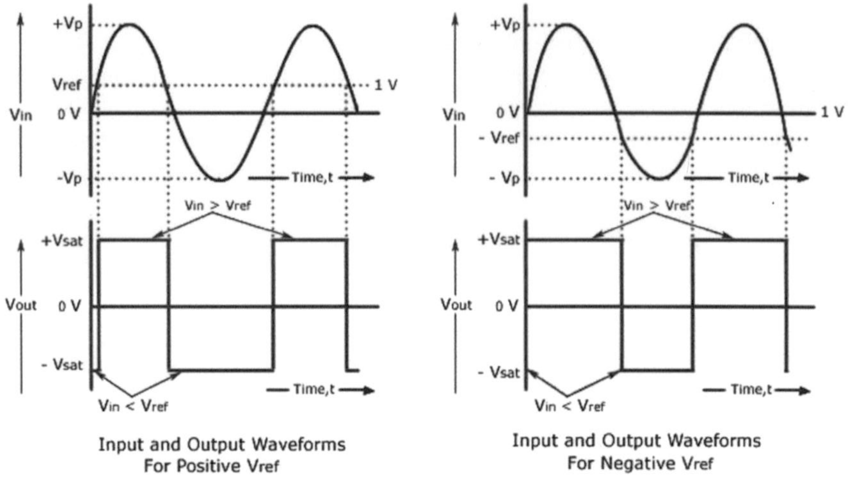

Fig. 32.1 Comparator ideal response

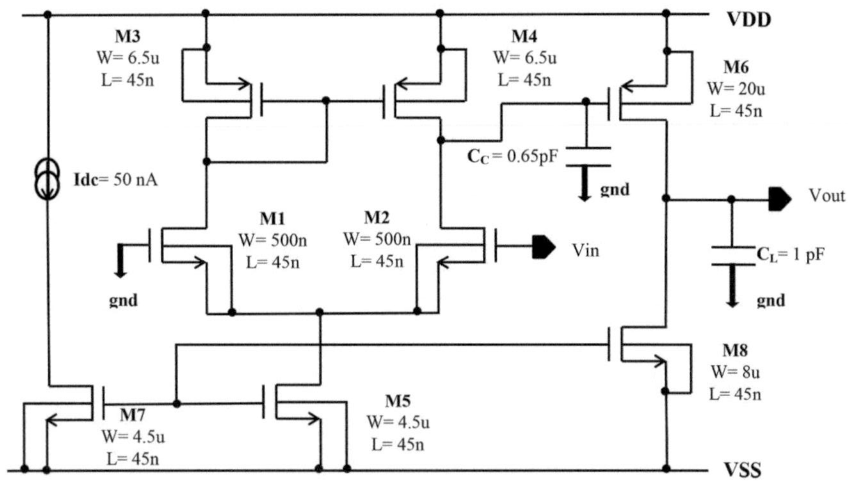

Fig. 32.2 Schematic design of comparator

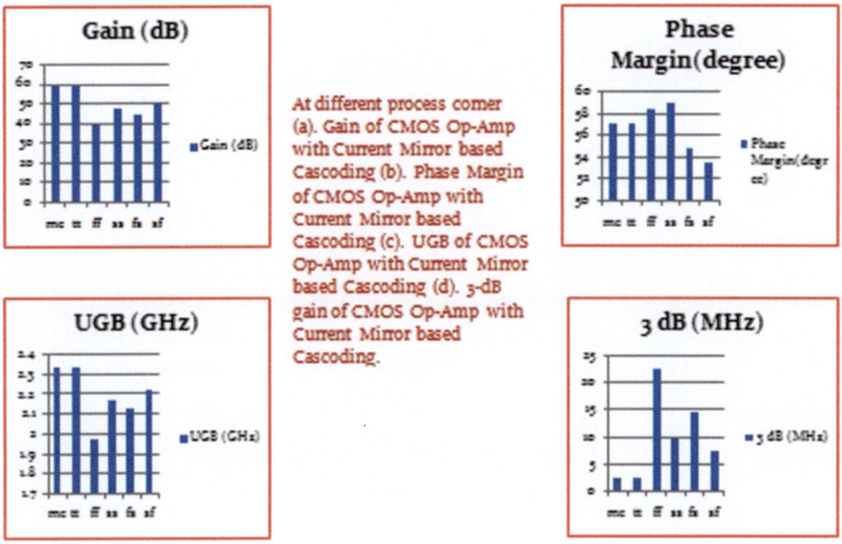

Fig. 32.3 Gain, phase margin, UGB, and 3 dB bandwidth at different model library

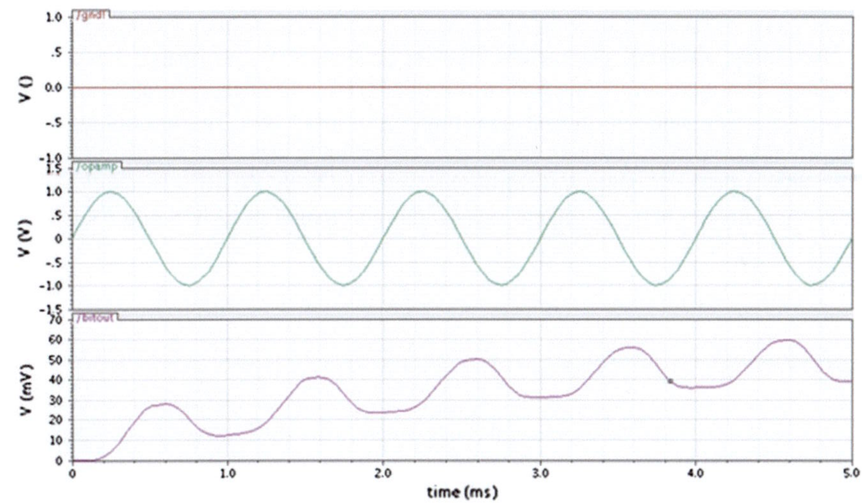

Fig. 32.4 Output of unoptimized comparator

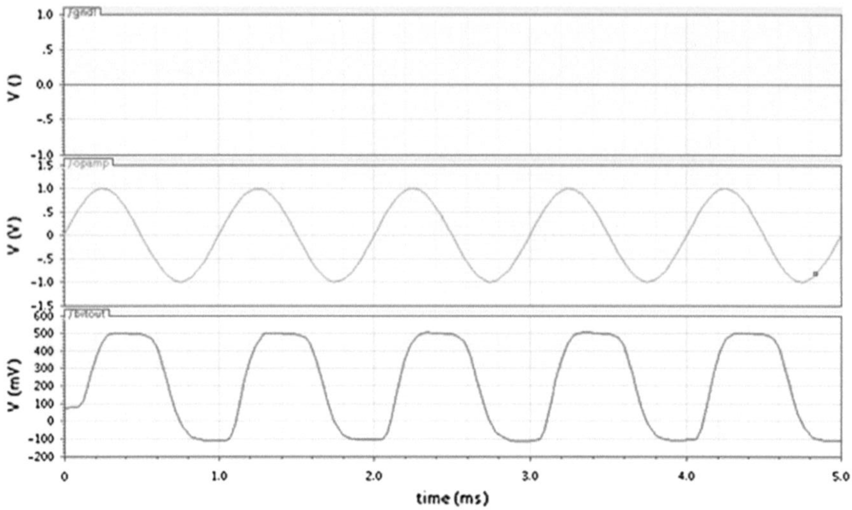

Fig. 32.5 Output of configured/optimized comparator

32.5 Conclusion

In this study, different parameters after optimization of the opamp have been obtained. These are unity gain bandwidth (UGB) of 2.30 GHz, gain of 29 dB, and 991.6 V/μs slew rate with power consumption of 1.02 mW. The technology used was 45 nm CMOS technology. Similarly, comparator was designed using 1 kHz sinusoidal input, and pulse waveform was obtained with a 6 dB gain using Fermi farad capacitances and μm dimensions (width and length of nMOS and pMOS) in designing.

References

1. Baker RJ (2008) CMOS: mixed-signal circuit design. Wiley-IEEE. ISBN978–0470-29026-2. http://CMOSedu.com/
2. Amin N, Guan GC, Ahmad I An efficient first order sigma delta modulator design
3. Ya ML, Khan S, Nordin AN, Alam AZ A low-cost first-order sigma-delta converter design and analysis
4. Boser BE (1988) The design of sigma-delta modulation analog-to-digital converters. IEEE J Solid-State Circ 23(6):001
5. Norsworthy SR, Schreier R, Temes (1998) GC Delta-sigma data converters: theory, design, and simulation. IEEE Press
6. Aziz PM, Sorensen HV, Spiegel JV (1996) An overview of sigma-delta converters. IEEE Signal Process Mag 13:61–84
7. Kim J-I, Sung B-R-S, Kim W, Ryu S-T (2013) A 6-b 4.1-GS/s Flash ADC With Time-Domain Latch Interpolation in 90-nm CMOS. IEEE J Solid-State Circ 48(6):1429–1441

8. Converter for high speed and low power applications. Int J VLSI Design Commun Syst (VLSICS) 3(2)
9. Jincheol Y, Daegyu L, Kyusun C, Tangel A (2001) Future-ready ultrafast 6 Bit CMOS ADC for system-on-chip applications. In: Proceedings of 14th international conference on annual IEEE ASIC/SOC, pp 455–459
10. Yoo J, Choi K, Tangel A (2001) A 1-GSPS CMOS flash A/D converter for system-on-chip applications. In: Proceedings of IEEE computer society workshop on VLSI, pp. 135–139
11. Dalton D, Spalding GJ, Reyhani H, Murphy T, Deevy K, Walsh M, Griffin P (1998) A 200-MSPS 6-Bit Flash ADC in 0.6 μm CMOS. IEEE Trans Circ Syst 45(11):1433–1444
12. Wang Y-T, Razavi B (2000) An 6-bit 150-MHz CMOS A/D Converter. IEEE J Solid-State Circ 35(3):308–317
13. Lee D, Yoo J, Choi K, Ghaznavi J (2002) Fat tree encoder design for ultra-high speed flash A/D converters. In: Proceedings of 45th symposium on circuits and systems MWSCAS, Vol. 2, pp. II–87
14. Wang M, Chen CIH, Radhakrishnan S (2007) Low-power 4-b 2.5-GSPS pipelined flash analog-to-digital converter in 130-nm CMOS. IEEE Trans Instrum Meas 1:287–292
15. He J, Zhan S, Chen D, Geiger RL (2009) Analyses of static and dynamic random offset voltages in dynamic comparators. IEEE Trans Circ Syst I: Regul Papers 56(5):001
16. Padoan S, Boni A, Morandi C, Venturi F (1998) A novel coding schemes for the ROM of parallel ADCs, featuring reduced conversion noise in the case of single bubbles in the thermometer code. IEEE Int Conf Circ Syst 2:271–274
17. Jiang X, Wang Z, Chang MF (2003) A 2GS/s 6-b ADC in 0.18 μm CMOS. IEEE Int Solid-State Circ Conf 1:9–13
18. Hauser MW (1991) Principles of oversampling A/D conversion. J Audio Eng Soc 39(1/2):3–26
19. Deloraine EM, Van Mierlo S, Derjavitch B (1946) Methode et systéme de transmission par impulsions. French Patent, 932, 140
20. Derjavitch B, Edmond MD (1953) Communication System Utilizing Constant Amplitude Pulses of Opposite Polarities. US Patent No. 2,629,857
21. Morizio JC et al (2000) 14-bit 2.2-MS/s sigma-delta ADC. IEEE J Solid-State Circ 35 (7):968–976

Chapter 33
A 21nW CMOS Operational Amplifier for Biomedical Application

Sarika Tyagi, Swapnil Saurav, Abhishek Pandey, Padma Priyadarshini, Madhu Ray, B.B. Pal and Vijay Nath

33.1 Introduction

The operational amplifier has become one of the most versatile and important building blocks in analog electronics. Op-amp plays an important role in analog circuit design as logic gate plays in digital circuitry. Since biomedical application devices are battery operated, so the circuit must consume less power.

The increasing demand of high data rate needs larger bandwidth with high SNR. The current demands of any IC technology are low cost, high performance, high integration, high package density, high noise immunity, portability, robustness, flexibility etc. For this purpose, CMOS technology is a good candidate. The CMOS technology provides the flexibility of device scaling. The advantages of scaling MOS transistors are increased device packing density, improved frequency response, and improved current driving capability. There are also some adverse effects of reducing the channel length, for example, reducing the channel length increases the bandwidth and a good gain becomes difficult to find.

A low voltage CMOS operational amplifier [1–4] is operating at low voltage with good gain but lower unity gain bandwidth. It has been found in literature [5–7] of op-amp that the gain is considerably higher, but slew rate, unity gain bandwidth, and static power dissipation performance is not suitable for wireless communication application. This is the basic problem of wireless devices.

Since the main problem is that the bandwidth of op-amp is directly proportional to the biasing current results increase in power consumption.

S. Tyagi (✉) · S. Saurav · A. Pandey · P. Priyadarshini · M. Ray · B.B. Pal · V. Nath
VLSI Design Group, Department of Electronics and Communication Engineering, Birla Institute of Technology, Mesra, Ranchi 835215, Jharkhand, India
e-mail: sarikatyagi72@gmail.com

A. Pandey
e-mail: a.p.bitmesra@gmail.com

V. Nath (ed.), *Proceedings of the International Conference on Nano-electronics, Circuits & Communication Systems*, Lecture Notes in Electrical Engineering 403,
DOI 10.1007/978-981-10-2999-8_33

The objective of the proposed design is achieved by high gain with low power consumption.

The rest of this report has been prepared as follows. The characteristic of MOSFET operating in subthreshold region is given in Sect. 2. Section 3 depicts the methodology and architecture of the proposed op-amp circuit. The simulation result and discussion has been summarized in Sect. 4. Finally, the conclusion of the overall paper is illustrated in Sect. 5.

33.2 MOSFET in Subthreshold Region

Sometimes the MOSFETs are considered in switched off condition below cut-off, but in reality, the current through MOSFETs decrease exponentially below cut-off. The general equations for MOSFETs are designed for operating in the linear and saturation region. The equation of MOSFET in subthreshold is quite different from the equations for linear and saturation region. For a NMOS operating in sub-threshold region, the current flowing through it can be given by [3]

$$I_D = W/L I_0 \mathrm{e}^{\frac{V_{GS}-V_{th}}{mV_T}}\left[1 - \mathrm{e}^{-\frac{V_{DS}}{V_T}}\right] \cong W/L I_0 \mathrm{e}^{\frac{V_{GS}-V_{th}}{mV_T}} \quad (33.1)$$

where, $V_T = k_B T/q$.

W Width of the MOSFET
L Length of the MOSFET
V_{GS} Gate to source voltage applied to the transistor
V_{DS} Drain to source voltage applied to the transistor
I_o Process dependent parameter
V_{th} Threshold voltage
m Subthreshold slope parameter
k_B Boltzmann constant
q Elementary charge
T Absolute temperature

Again, the V_{th} can be written as

$$V_{th} = V_{th0} - 2V_T \ln\left(\frac{N_A}{n_i}\right) \quad (33.2)$$

where,

V_{th0} Threshold voltage at zero body to source voltage
N_A Carrier concentration
n_i Intrinsic carrier concentration

When the threshold voltage V_{th} is considered, another important phenomenon comes into play called the DIBL (drain-induced barrier lowering) effect. At high drain voltages, a reduced threshold voltage is encountered due to this effect. Using the above written current equation, the transconductance and drain-source resistance of NMOS in subthreshold region can be given as.

$$g_m = \frac{\partial I_D}{\partial V_{GS}} = \frac{I_D}{mV_T} \tag{33.3}$$

$$r_d = \left(\frac{\partial I_D}{\partial V_{DS}}\right) = \frac{mV_T}{\lambda_D I_D} \tag{33.4}$$

where,

λ_D DIBM coefficient.

33.3 Proposed Op-Amp Circuit

When a MOS transistor is configured as an active resistor, if its small signal model is considered, the resistance exhibited by the transistor is $1/g_m$. And while operating in subthreshold region, the output resistance of the MOSFET can be written as

$$r_o = \frac{1}{g_m} = \frac{mV_T}{I_D} = \frac{m}{I_D} \times \frac{k_B T}{q} \tag{33.5}$$

From the above equation, it can be concluded that the output resistance of a MOSFET configured as an active resistor is directly related to its temperature.

A. *Differential Stage*

The first stage of operational amplifier circuit is known as differential stage (Fig. 33.1). The transistors M7–M12 construct the input stage of the differential stage ($V \rightarrow I$), and the transistors M1–M6 make the current mirror for the differential stage ($I \rightarrow V$). The purpose of M13 is to supply biasing current to the differential stage. It consists of cascaded voltage-to-current (transconductance stages) and current-to-voltage stages (load stages) [9–10].

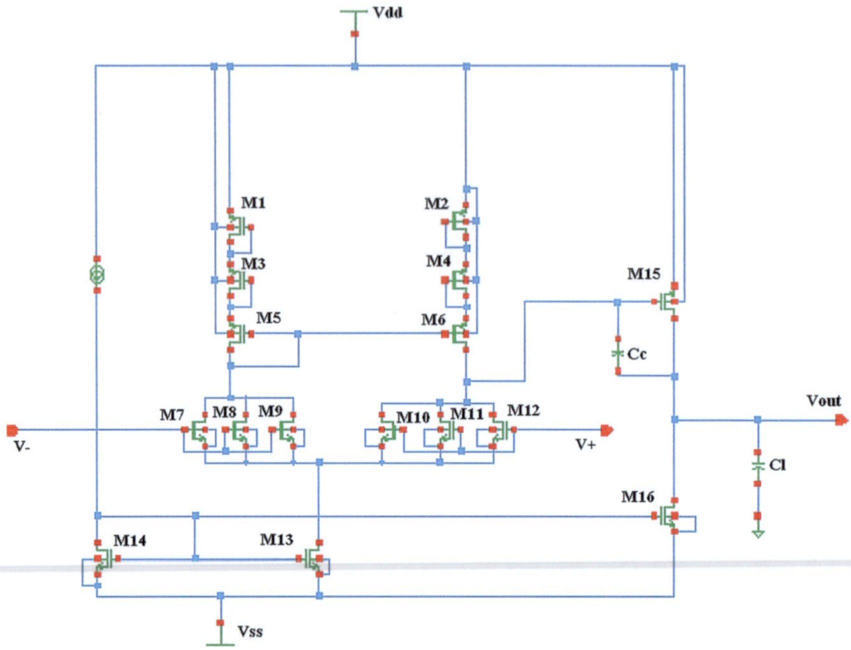

Fig. 33.1 Proposed CMOS Op-amp circuit

B. *Gain Stage*

The second stage of operational amplifier is known as gain stage. It consists of a common source MOSFET ($V \rightarrow I$) (M15 and M16) and a current sink load ($I \rightarrow V$) (M10). Usually, a MOSFET and a current source are used in a buffer circuit, but here the current source has been replaced with the PMOSFET [6, 7].

$$A_{VI} = \frac{g_{m2}}{g_{m4}} = \frac{g_{m1}}{g_{m3}} \tag{33.6}$$

$$A_{VII} = \left(\frac{g_{m6} + g_{m8}}{2}\right) R_{II} \tag{33.7}$$

$$R_{II} = (g_{m7} r_{ds7} r_{ds6}) || (g_{m12} r_{ds12} r_{ds11}) \tag{33.8}$$

$$A_V = A_{V1} A_{VII} = \frac{g_{m2}}{g_{m4}} \left(\frac{g_{m6} + g_{m8}}{2}\right) R_{II} \tag{33.9}$$

33.4 Simulation Results and Discussion

In this part, result analysis of the given op-amp is described. The presented op-amp has been designed in umc 90 nm CMOS process with 500 fF load capacitor and analysis in cadence analog environment system.

The transient response of circuit output and input response is depicted in Fig. 33.2.

The AC response of the presented op-amp is depicted in Fig. 33.3 for 0.5 pF load capacitor. The given op-amp achieves a gain 48 dB and UGB 29 kHz. The variation in temperature on the performance of the op-amp is represented in Fig. 33.4. The ac gain and phase of Monte Carlo simulation result for 10 mV variation in threshold voltage of given op-amp is depicted in Figs. 33.5 and 33.6.

The overall resultant power dissipation of the presented op-amp is 21 nW.

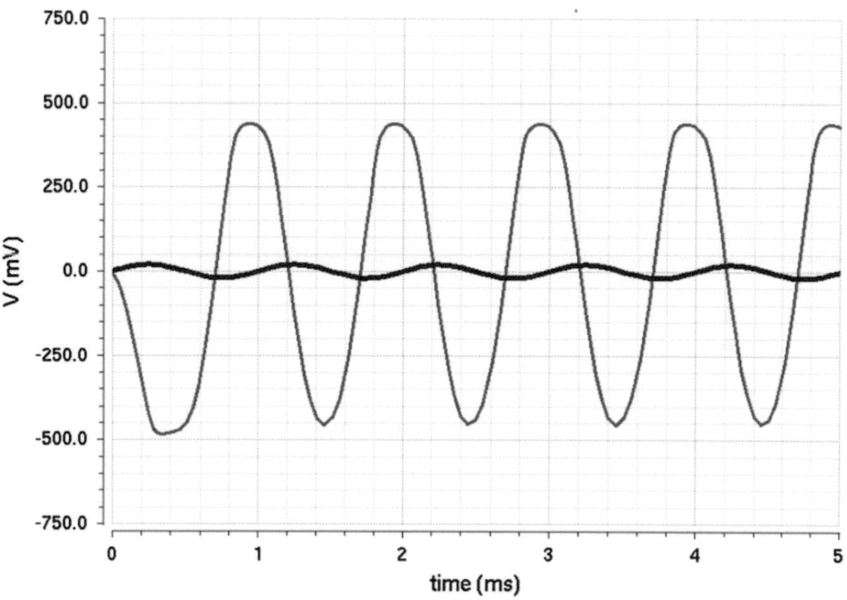

Fig. 33.2 Transient response of proposed op-amp

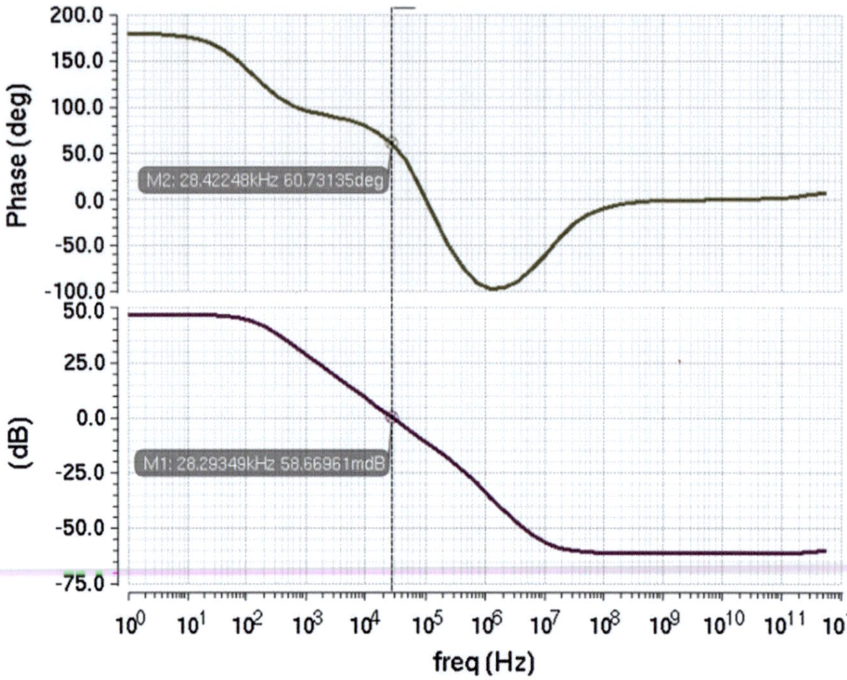

Fig. 33.3 AC response of op-amp

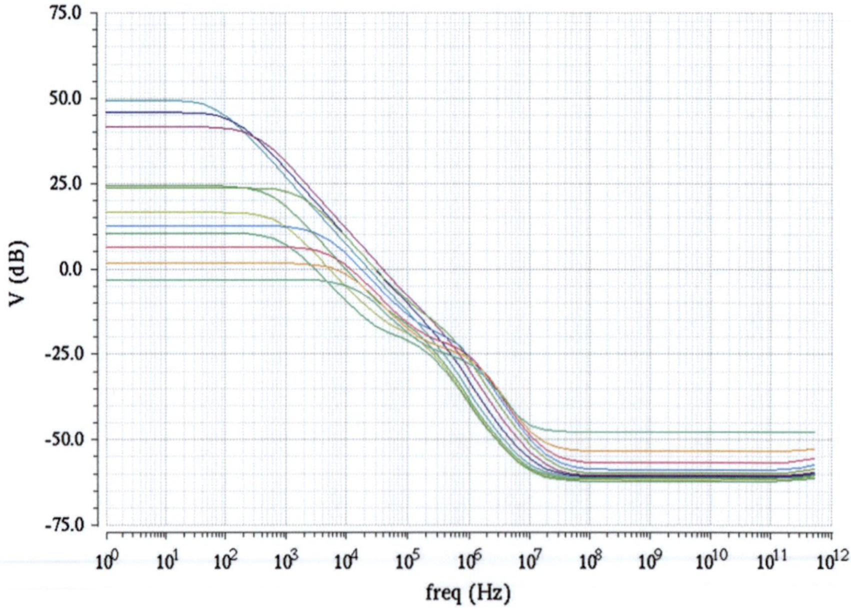

Fig. 33.4 AC response at different temperature −50 to 150 °C

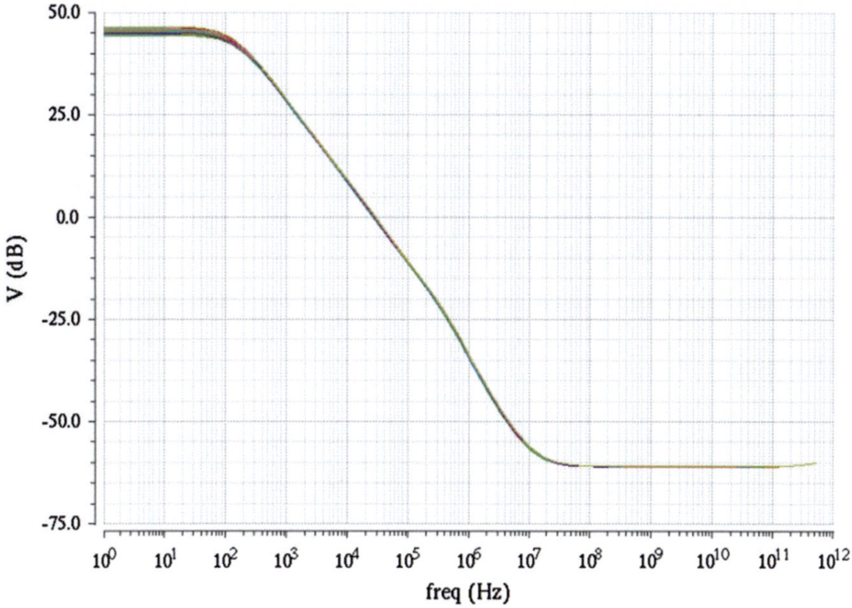

Fig. 33.5 AC gain plot for threshold voltage variation of 10 mV

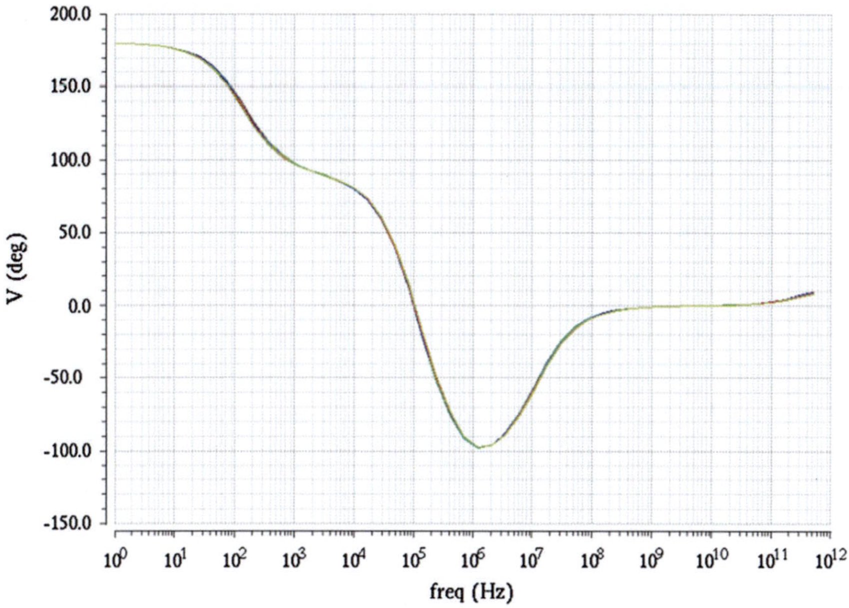

Fig. 33.6 AC Phase margin for 10 mV variation in threshold voltage

33.5 Conclusion

The Op-Amp design presented in this paper produces ultra low power consumption of just 21 nW. Though the gain and UGB are on the lower side with general point of view but if the supply current of the amplifier is considered, then it can be visualized that the results produced are quite good. This paper describes the design of a high speed and low power CMOS two stage operational amplifier with a bandwidth of 29 kHz and gain of 48 dB.

References

1. Guo C, Zhu S, Hu J, Zou JJ, Sun H, Lv X (2011) A low voltage CMOS rail to rail operational amplifier based on flipped differential pairs. In: IEEE microwave, antenna, propagation and EMC technologies for wireless communications (MAPE) 217–220
2. Bhattacharyya TK, Haiti MK (2012) A power efficient and constant-gm 1.8 V CMOS operational transconductance amplifier with rail to rail input and output ranges for charge pump in phase locked loop. In: International conference on device, circuit and systems, pp 38–43
3. Hu Y, Sawan M (2003) A low power 900 mV rail-to- rail class-AB operational amplifier. IEEE J Electr Comput Eng 1:171–174
4. Bahmani F, Fakhraie SM, Khakifirooz A (2000) A rail- to-rail, constant—Gm, 1-volt CMOS OPAMP. IEEE J Circ Syst 2:669–672
5. Uhrmann H, Schlogl F, Schweiger K, Zimmermann H (2009) A 1 GHz-GBW operational amplifier for DVB-H receivers in 65 nm CMOS. In: IEEE design and diagnostic of electronic circuits and systems, pp 182–185
6. Gangopadhyay D, Bhattacharyya TK (2010) A 2.3 GHz gm- boosted high swing class AB ultra wide bandwidth operational amplifier in 0.18um CMOS. In: IEEE 53rd international midwest symposium on circuit and systems (MWSCAS), pp 713–716
7. Chen H et.al (2013) Ultra high bandwidth fully differential three stage operational amplifier in 40 nm Digital CMOS. In: Design and diagnostics of electronic systems (DDECS), IEEE 16th international symposium on digital object identifier, pp 76–81
8. Razavi B (2002) Design of analog CMOS integrated circuits, 1st edn. Tata McGraw-Hill, New York
9. Lloyd J, Lee H-S (1994) A CMOS OP-AMP with fully-differential-gain-enhancement. IEEE Trans Circ Syst-II 41:241–243
10. Usmani FA, Hasan M (2009) Design and parametric analysis of 32 nm OPAMP in CMOS and CNFET technologies for optimum performance. In: IEEE micro nanoelectronics conference, pp 87–92
11. Zaghloul ME, Wang S, (Neil) Tsai Y (2007) CMOS operational amplifier. The George Washington University, Department of Electrical and Computer Engineering, 14 May 2007
12. Phillip E (2004) Allen and Douglas R, CMOS analog circuit design. Oxford University Press, Holberg

Chapter 34
Prediction of Depth-Averaged Velocity and Boundary Shear Stress Distribution in a Single-Stage Channel by Lateral Distribution Method

Bhabani Shankar Das, Kishanjit K. Khatua and Kamalini Devi

34.1 Introduction

Rivers play a very important role for every civilization, as they provide many contributions to human well-being like water for household consumptions, irrigation, industry, sustainable energy, navigation, and valuable wild-life habitat. Human mind is easily enthralled by their power, sometimes peaceful in scenic landscapes, sometimes calamitous when flooding. Therefore, estimation of accurate flow has been a challenging task for river engineers. To achieve this, river engineer adopts a proper depth-averaged velocity measurements or modeling procedures. The prediction of depth-averaged velocity along the width of a channel helps in flood prediction and management, finding the resistance relationships and for many other purposes. Tominaga and Ezaki [1] carried out experiments in simple rectangular channels to know the effects of aspect ratio on secondary currents. Tominaga et al. [2] studied the velocity vectors and observed that the secondary currents flow toward the side wall along a horizontal plane that was located approximately at $0.6H$ (H = depth of flow). Tingsanchali and Maheswaran [3] performed an experiment to compute depth-averaged velocity and bottom shear stress distributions in a rectangular channel near a groyne by using a 2D depth-averaged model by hybrid finite difference scheme. Wilkerson et al. [4] developed and evaluated two models for predicting depth-averaged velocity distributions in straight trapezoidal channels. An analytical technique for finding the depth-averaged velocity in an open channel

B.S. Das (✉) · K.K. Khatua · K. Devi
Department of Civil Engineering, N.I.T. Rourkela, Rourkela, India
e-mail: bsdas7190@gmail.com

K.K. Khatua
e-mail: kkkhatua@yahoo.com

K. Devi
e-mail: kamalinidevi1@gmail.com

V. Nath (ed.), *Proceedings of the International Conference on Nano-electronics, Circuits & Communication Systems*, Lecture Notes in Electrical Engineering 403,
DOI 10.1007/978-981-10-2999-8_34

has been presented by Maghrebi and Ball [5]. They compared their result applied to the River Severn with the best analytical results obtained by the depth-averaged Navier–Stokes equations. Tang and Knight [6] presented two-dimensional analytical solution and compared with the conventional solution for three simple channel shapes and one trapezoidal compound channel. They also presented an analytical solution to the depth-integrated Navier–Stokes equation that includes the effects of bed friction, lateral turbulence, and secondary flow. Yang et al. [7] investigated the influence of flow geometry on the depth-averaged velocity and shear stress in a open-channel flow and also established a theoretical relationship between the boundary shear stress and depth-averaged apparent shear stress.

34.2 LDM for Simple Channel

The Lateral Distribution Method (LDM) is derived from a depth averaging of the Navier–Stokes momentum-conservation equation in the streamwise direction:

$$\begin{aligned}\rho\left(\frac{\partial \bar{u}}{\partial t}+\frac{\bar{u}(\partial \bar{u})}{\partial x}+\frac{\bar{v}(\partial \bar{u})}{\partial y}+\frac{\bar{w}(\partial \bar{u})}{\partial z}\right) &= \rho F_x \\ -\frac{\partial \bar{p}}{\partial x}+\mu\Delta\bar{u}-\rho\left(\frac{\partial \overline{u'u'}}{\partial x}+\frac{\partial \overline{u'v'}}{\partial y}+\frac{\partial \overline{u'w'}}{\partial z}\right)\end{aligned} \tag{34.1}$$

where $(\overline{u}, \overline{v}, \overline{w})$ are Reynolds averaged local velocity components, respectively, in the x- (stream-wise, parallel to the bed), y- (lateral), and z- (normal to bed) directions; ρ is the density of water; F_x is the x-wise component of gravitational force which equals the longitudinal bed slope S_0 time the gravity constant g; p is the pressure; μ is the molecular viscosity; and $\left(\overline{u'u'}, \overline{u'v'}, \overline{u'w'}\right)$ are the Reynolds turbulent shear stresses. From this momentum equation, the LDM equation is derived by taking various assumptions.

Various assumptions made for the derivation of LDM for simple channel are as follows: (1) a permanent $(\partial/\partial t = 0)$ and uniform $(\partial/\partial x = 0)$ flow. (2) Viscous friction with regard to the Reynolds stresses is neglected. (3) There is no secondary current in the simple channel. (4) Water level is assumed to be horizontal $(\partial z_w/\partial y = 0)$ in the transverse direction as a consequence of a one-dimensional flow hypothesis.

The RANS equation in x-direction (longitudinal flow direction) taking all the assumption can be simplified as

$$\rho\left[\frac{\partial \bar{u}\bar{v}}{\partial y}+\frac{\partial \bar{u}\bar{w}}{\partial z}\right] = 0 = \rho g S_0 + \frac{\partial}{\partial y}\left(-\rho\overline{u'v'}\right)+\frac{\partial}{\partial z}\left(-\rho\overline{u'w'}\right) \tag{34.2}$$

where u', v' and w' are the fluctuations of the velocity component. Here the over bar represents a time-averaged parameters. The simplification of (34.2) is done in LDM method. In Eq. (34.2), the first term is the secondary flow term consisting of lateral and vertical components of the velocity. The second term represents the weight component of water. The third and fourth terms account for the apparent shear or Reynolds shear stresses in vertical and horizontal planes, respectively. Considering the mean velocity component in z-direction is very negligible then $\bar{w}$ is equal to zero and $\tau_{yx} = -\rho\overline{u'v'}$, $\tau_{zx} = -\rho\overline{u'w'}$ and Integrating (34.2) over the total flow depth H, we have

$$\int_0^H \rho\left[\frac{\partial\bar{u}\bar{v}}{\partial y} + \frac{\partial\bar{u}\bar{w}}{\partial z}\right]\mathrm{d}z = \rho\frac{\partial H(\bar{u}\bar{v})_d}{\partial y} \tag{34.3}$$

If there is no secondary current, Eq. (34.3) becomes zero.

$$\int_0^H \rho g S_0 \mathrm{d}z = \rho g H S_0 \tag{34.4}$$

$$\int_0^H \frac{\partial}{\partial y}\left(-\rho\overline{u'v'}\right)\mathrm{d}z = \int_0^H \frac{\partial\tau_{xy}}{\partial y}\mathrm{d}z \tag{34.5}$$

$$\int_0^H \frac{\partial}{\partial z}\left(-\rho\overline{u'w'}\right)\mathrm{d}z = \int_0^H \frac{\partial\tau_{zx}}{\partial z}\mathrm{d}z = \tau_b\sqrt{1+S_{0y}^2} \tag{34.6}$$

τ_b = bed shear stress, S_{0y} = lateral slope along the channel.

34.3 Bed Friction and Shear Modeling

Knight et al. [8] used a Boussinesq eddy viscosity model for τ_{xy} and assumed an eddy viscosity ϑt proportional to the water depth H and to the shear velocity U^*.

$$\vartheta t = \lambda H U^*$$

where λ is the coefficient of eddy viscosity. Hence, the Darcy–Weisbach friction factor,

$$f = \frac{8\tau_b}{\rho U^2}, \quad \text{then } \vartheta t = \lambda H U^* = \lambda H\left(\frac{f}{8}\right)^{\frac{1}{2}} U$$

From the eddy viscosity theory, we can take

$$\tau_{xy} = \rho \vartheta t \frac{\partial U}{\partial y} = \rho \lambda H \sqrt{\frac{f}{8}} U \frac{\partial U}{\partial y} \tag{34.7}$$

Now Eq. (34.5) can be written as

$$\int_0^H \frac{\partial \tau_{xy}}{\partial y} \mathrm{d}z = \frac{\partial}{\partial y}\left(\rho \lambda H^2 \sqrt{\frac{f}{8}} U \frac{\partial U}{\partial y}\right) \tag{34.8}$$

And Eq. (34.6) is written as

$$\int_0^H \frac{\partial \tau_{zx}}{\partial z} \mathrm{d}z = \tau_b \sqrt{1 + S_{0y}^2} = \frac{f}{8} \rho U^2 \sqrt{1 + S_{0y}^2} \tag{34.9}$$

Neglecting the secondary current term for simple channel, finally Eq. (34.2) is simplified to

$$\rho g H S_{0x} + \frac{\partial}{\partial y}\left(\rho \lambda H^2 \sqrt{\frac{f}{8}} U \frac{\partial U}{\partial y}\right) - \rho \frac{f}{8} U^2 \sqrt{1 + S_{0y}^2} = 0 \tag{34.10}$$

Using Manning's n for the Darcy's friction factor f, Eq. (34.10) becomes

$$\rho g H S_{0x} + \frac{\partial}{\partial y}\left(\rho \lambda H^2 \sqrt{\frac{g n^2}{H^{1/3}}} U \frac{\partial U}{\partial y}\right) - \frac{\rho g n^2}{H^{1/3}} U^2 \sqrt{1 + S_{0y}^2} = 0 \tag{34.11}$$

34.4 Numerical Solution of LDM

In the present research work, the LDM numerical solution is obtained by writing the differential equation in a discrete form, using the finite differences method (Fig. 34.1). The finite difference method is carried out by central difference method by taking small grid size of 0.001 m for appropriate prediction of the hydraulic parameter like depth-averaged velocity and boundary shear stress. This reduces the LDM equation to a set of quadratic algebraic equations linking together velocities U_i at each node of the mesh. At the boundaries, a no-slip condition is used by setting the velocity equal to zero along the walls. The so-defined set of equation is solved either by a Newton method, or by any other appropriate method.

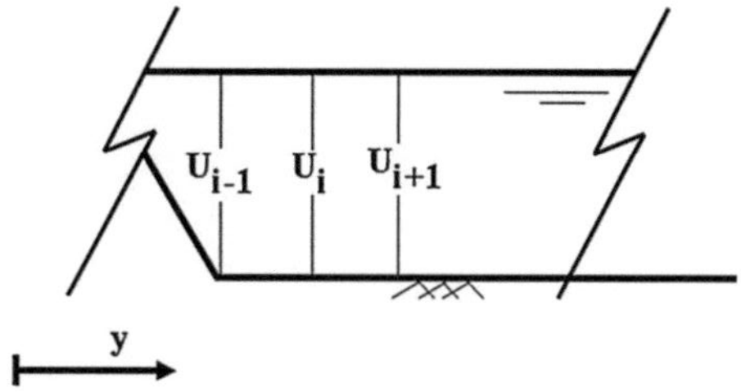

Fig. 34.1 Discrete mesh for LDM numerical solving

As the friction term is also proportional to the square of U, all the equation can be rewritten as a function of U^2 rather of U. When expressed in discrete form, the LDM reduces now to a set of linear equations linking the square of velocities U^2. We know that

$$\frac{\partial}{\partial y}\left(U\frac{\partial U}{\partial y}\right) = \frac{1}{2}\frac{\partial}{\partial y}\left(\frac{\partial U^2}{\partial y}\right) = \frac{\partial^2(U^2)}{\partial y^2}$$

So, the LDM equation may be written as

$$\rho HgS_{0x} + \frac{1}{2}\rho\lambda\sqrt{\frac{gn^2}{H^{\frac{1}{3}}}}H^2\frac{\partial^2(U^2)}{\partial y^2} - \frac{\rho gn^2}{H^{1/3}}(U^2)\sqrt{1+S_{0y}^2} = 0 \tag{34.12}$$

Now Eq. (34.12) reduce to

$$A\frac{\partial^2(U^2)}{\partial y^2} + B\left(U^2\right) + C = 0 \tag{34.13}$$

where

$$A = \frac{1}{2}\rho\lambda\sqrt{\frac{gn^2}{H^{1/3}}}H^2, \quad B = \frac{\rho gn^2}{H^{1/3}}\sqrt{1+\frac{1}{s_{0y}^2}}, \quad C = \rho gHS_{0x}$$

By substituting $U^2 = t$, we have

$$A\frac{\partial^2 t}{\partial y^2} + Bt + C = 0 \tag{34.14}$$

Now by solving it by finite difference method using central difference scheme;

$$A\left(\frac{t_{i+1}-2t_i+t_{i-1}}{\Delta y^2}\right)+B(t_i)+C=0$$
$$\Rightarrow \left(\frac{t_{i+1}-2t_i+t_{i-1}}{\Delta y^2}\right)+\frac{B}{A}(t_i)+\frac{C}{A}=0$$
$$\Rightarrow t_{i+1}-2t_i+t_{i-1}=-\frac{B}{A}(t_i)\Delta y^2-\frac{C}{A}\Delta y^2$$
$$\Rightarrow -t_{i-1}+2t_i-t_{i+1}-\frac{B}{A}(t_i)\Delta y^2=\frac{C}{A}\Delta y^2$$
$$\Rightarrow -t_{i-1}+\left(2-\frac{B\Delta y^2}{A}\right)t_i-t_{i+1}=\frac{C}{A}\Delta y^2$$

If we assume $\left(2-\frac{B\Delta y^2}{A}\right)=P$, $\frac{C}{A}\Delta y^2=Q$, then we have

$$-t_{i-1}+Pt_i-t_{i+1}=Q \tag{34.15}$$

Now for i = 1–N Eq. (34.14) is written below

$$\begin{array}{ll} \text{for } i=1, & -t_0+Pt_1-t_2=Q \\ \text{for } i=2, & -t_1+Pt_2-t_3=Q \\ \text{for } i=3, & -t_2+Pt_3-t_4=Q \\ \text{for } i=4, & -t_3+Pt_4-t_5=Q \\ \vdots & \vdots \\ \text{for } i=n-1, & -t_{n-2}+Pt_{n-1}-t_n=Q \\ \text{for } i=n, & -t_{n-1}+Pt_n-t_{n+1}=Q \end{array} \tag{34.16}$$

For solving this system of equations, we have to use matrix form.

$$AX=B,$$

where

$$A=\begin{bmatrix} P & -1 & 0 & \cdot & \cdot & \cdot & 0 \\ -1 & P & -1 & 0 & \cdot & \cdot & \cdot \\ 0 & -1 & P & -1 & 0 & \cdot & \cdot \\ \cdot & \cdot & \cdot & \cdot & \cdot & \cdot & \cdot \\ \cdot & \cdot & \cdot & \cdot & \cdot & \cdot & 0 \\ \cdot & \cdot & \cdot & \cdot & -1 & P & -1 \\ 0 & \cdot & \cdot & \cdot & 0 & -1 & P \end{bmatrix}$$

$$X=\begin{bmatrix} t_1 & t_2 & t_3 & \cdot & \cdot & t_{n-1} & t_n \end{bmatrix}^{\mathrm{T}}$$

$$B = [Q \quad Q \quad Q \quad \cdot \quad \cdot \quad Q \quad Q]^{\mathrm{T}}$$

The matrix corresponding to this set of equation is tridiagonal. Crout factorization technique is used for solving the matrix writing code in MATLAB tool. After solving the matrix, it gives $U_d = \sqrt{t}$; from the value of depth-averaged velocity (U_d), the boundary shear stress (τ_b) is found out by the relation

$$\tau_b = U_d^2 \frac{gn^2}{H^{1/3}} \tag{34.17}$$

34.5 Experimental Datasets

In this current research work, the experiment dataset is collected from the *Flood Channel Facility data* (*FCF Data*)-*University of Birmingham* (Fig. 34.2). The link for the FCF data is http://www.birmingham.ac.uk/research/activity/civil-engineering/archive/short-term/floods/flowdata/fcf-data/fcf-phase-a.aspx. From those datasets, we consider only in-bank flow experiments for research having experiment ID as follows: IB0501_2, IB7501_1, IB1002_3, and IB1502_2. The bed slope of the channel is 0.001027, and the semi-base width of the main channel is b = 0.75 m, with main channel depth h = 0.150 m, thus giving an aspect ratio of 10 (=$2b/h$) for the main channel at the bank full stage. For the present analysis, the λ value is taken as 0.05 [9]. The four sets of data are available for different in-bank depths. The manning's n and experimental values of boundary shear stress and depth-averaged velocity are given in Table 34.1.

Conveyance Afflux and Estimation system (CES) is commercially available software based on Shiono and Knight Method (SKM). We use this to compare the results obtained from LDM numerical solution and the experimental datasets.

34.6 Results and Discussion

The boundary shear stress (τ_b) and depth-averaged velocity (U_d) are found from numerical solution of the LDM equation. Also using CES, for the four different experiments ID: IB0501_2, IB7501_1, IB1002_3, and IB1502_2 the boundary shear stress (τ_b) and depth-averaged velocity (U_d) value are obtained.

The comparison of the experimental values, LDM, and CES is shown in Figs. 34.3 and 34.4. The averaged value of the depth-averaged velocity and boundary shear stress is entered in Table 34.2. From Fig. 34.3, it is clearly shown that depth-averaged velocity found from LDM numerical method has good

Table 34.1 Experimental datasets

Experiment ID	Manning's n	Water depth (H)	Averaged value	
			U_d (m/s)	τ_b (N/m^2)
IB0501_2	0.01036	0.04863	0.39715	0.46281
IB7501_1	0.01007	0.07553	0.53782	0.69884
IB1002_3	0.01008	0.10089	0.64081	0.91040
IB1502_2	0.00985	0.14875	0.82487	1.28488

Fig. 34.2 General view of the flood channel facility at H.R. Wallingford [10]

agreement with the experimental value obtained for different depths. In Fig. 34.3a, b, CES value for depth-averaged velocity is over estimated and for Fig. 34.3c, d, both CES and LDM numerical method have good agreement with the experimental value for the depth-averaged velocity. From Fig. 34.4, it is clearly shown that for boundary shear stress both LDM numerical method and CES have good agreement with experimental data.

34.7 Conclusions

- For higher flow depths, CES is found to give good depth-averaged velocity results; however, LDM provides better results for all depth of flow.
- For the evaluation of boundary shear stress distribution, both CES and LDM provide good results for all experimental channels and for all flow depths.

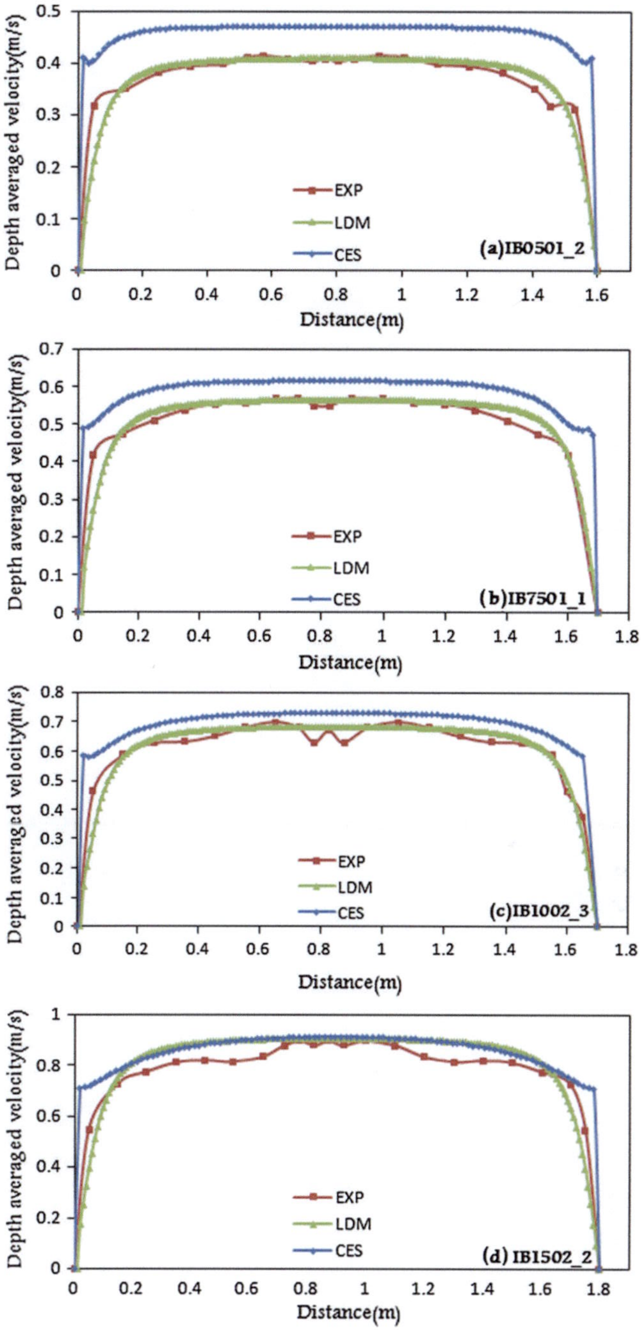

Fig. 34.3 a–d Comparison of depth-averaged velocity (U_d) in (m/s) for the LDM, CES, and with experimental data for different depths

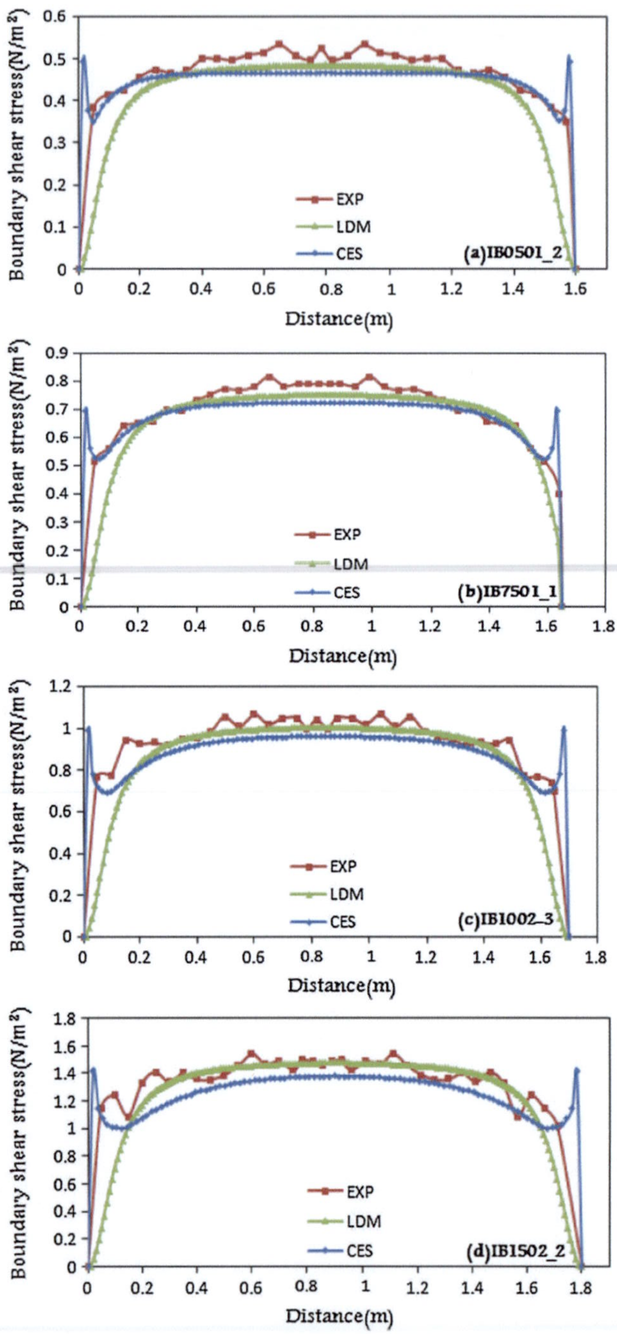

Fig. 34.4 **a–d** Comparison of boundary shear stress (τ_b) in (N/m^2) for the LDM, CES, and with experimental data for different depths

Table 34.2 Comparison of results (experimental values, LDM, and CES)

Experiment ID	Manning's n	Water depth (H) (m)	$(U_d)_{avg}$ (m/s)			$(\tau_b)_{avg}$ (N/m^2)		
			EXP	LDM	CES	EXP	LDM	CES
IB0501_2	0.01036	0.04863	0.39715	0.409	0.459	0.46281	0.458	0.451
IB7501_1	0.01007	0.07553	0.53782	0.563	0.613	0.69884	0.687	0.675
IB1002_3	0.01008	0.10089	0.64081	0.684	0.694	0.91040	0.891	0.887
IB1502_2	0.00985	0.14875	0.82487	0.822	0.855	1.28488	1.262	1.258

- The eddy viscosity coefficient λ needs to be calibrated properly to improve the results from LDM.
- From the current research, it was found that the LDM numerical method is well developed and some modification can be done to predict the depth-averaged velocity and boundary shear stress distribution more accurately for simple and compound channel.

References

1. Tominaga A, Ezaki K (1985) An experimental study on three-dimensional turbulent structure in a rectangular open channel flow (in Japanese). Dobaku Gakkai Ronbunshuu 357(II-3): 81–88
2. Tominaga A, Nezu I, Ezaki K, Nakagawa H (1989) Three-dimensional turbulent structure in straight open channel flows. J Hydraul Res 27(1):149–173
3. Tingsanchali T, Maheswaran S (1990) 2-D depth-averaged flow computation near groyne. J Hydraul Eng 116(1):71–86
4. Wilkerson GV, McGahan JL (2005) Depth-averaged velocity distribution in straight trapezoidal channels. J Hydraul Eng 131(6):509–512
5. Maghrebi MF, Ball JE (2006) New method for estimation of discharge. J Hydraul Eng 132 (10):1044–1051
6. Tang X, Knight DW (2008) A general model of lateral depth-averaged velocity distributions for open channel flows. Adv Water Resour 31(5):846–857
7. Yang S-Q (2010) Depth-averaged shear stress and velocity in open-channel flows. J Hydraul Eng 136(11):952–958
8. Knight DW, Shiono K, Pirt J (1989) Prediction of depth mean velocity and discharge in natural rivers with overbank flow. In: Proceedings of international conference on hydraulic and environmental modelling of coastal, estuarine and river waters. Gower Technical, University of Bradford
9. Chlebek J (2009) PhD thesis on "Modelling of simple prismatic channels with varying roughness using the SKM and a study of flows in smooth non-prismatic channels with skewed floodplains". The University of Birmingham
10. Knight DW, Shiono K (1990) Turbulence measurements in a shear layer region of a compound channel. J Hydraul Res IAHR 28(2):141–156

Chapter 35
Flow Computation in Symmetric and Asymmetric Compound Channels Using Conveyance Estimation System

Kamalini Devi, Kishanjit K. Khatua and Jnana Ranjan Khuntia

35.1 Introduction

Natural rivers are often flanked by one/two floodplains which may be symmetric or asymmetric. During high stage or floods, these floodplains act as additional storage capacity. However, accurate prediction of stage discharge relationship in such two-stage channels is very essential for flood warning, flood defense, or flood mitigation and also for the environmental disbalance. So it is a vital work for a hydraulic engineer to estimate discharge through the channel after simulating or measuring stage in compound channel. It is a task to hydraulic engineers to predict the discharge accurately for inbank as well as over bank conditions. The distribution of flow to different community is also an important work during flood. Due to the secondary flow formation at the junction between main river channel and floodplain, the flow structure becomes more complex in overbank case. So this affects in predicting the discharge [4]. Asymmetric compound channels of different configurations, i.e., rectangular and trapezoidal cross sections, have been studied, and suitable equations for predicting discharge have been developed [1]. Uniform flow formulas such as Chezy's or Manning's are used to calculate the flow in simple channel. But having the complex geometrical cross section, the SCM, divided channel method (DCM), and MDCM methods are widely used for discharge prediction in overbank sections due to the exchange of momentum at junction. The 1D software package, commonly known as Conveyance Estimation

K. Devi (✉) · K.K. Khatua · J.R. Khuntia
Department of Civil Engineering, N.I.T, Rourkela, India
e-mail: kamalinidevi1@gmail.com

K.K. Khatua
e-mail: kkkhatua@yahoo.com

J.R. Khuntia
e-mail: jnanaranjan444@gmail.com

V. Nath (ed.), *Proceedings of the International Conference on Nano-electronics, Circuits & Communication Systems*, Lecture Notes in Electrical Engineering 403,
DOI 10.1007/978-981-10-2999-8_35

System (CES), is capable of predicting various flow features such as stage–discharge relationship, depth averaged velocity, and boundary shear distribution across the flow cross sections. Some methods are also developed by researchers by dividing the section vertically, horizontally, or diagonally which are considered in this study for assessing the flow in compound channel. After identifying the need to reduce the uncertainty associated with the prediction of flood, the numerical software CES is developed for predicting the stage–discharge relationship in open-channel flow by solving the Shiono–Knight method. Here, three traditional methods such as VDM, HDM, and IDM along with this CES software are used for finding out the discharge and suitability of these methods and for predicting the discharge evaluated for both symmetric and asymmetric cases. Then, some best suitable methods are recommended for specific ranges of data sets as none of these methods give the best results for every individual set of data. The evaluation of CES for computing discharge for both symmetric and asymmetric compound channels is presented herein.

35.2 Theory of Application

In straight compound channels, the longitudinal velocity in main river section is usually faster compared to its floodplain especially for lower relative depth. This causes a shear layer at junction between the main channel and the floodplain [6]. Division planes, i.e., the three vertical, horizontal, and inclined planes, are generally used for dividing asymmetrical and symmetric compound channel into main channel and floodplain as suggested by Wormleaton et al. [7, 8]. Vertical, horizontal, and inclined division planes are used for dividing the compound channel to estimate the capacity. The most chosen location of that subdivision is shown in Figs. 35.1 and 35.2, where B is the width of total compound channel, b is the

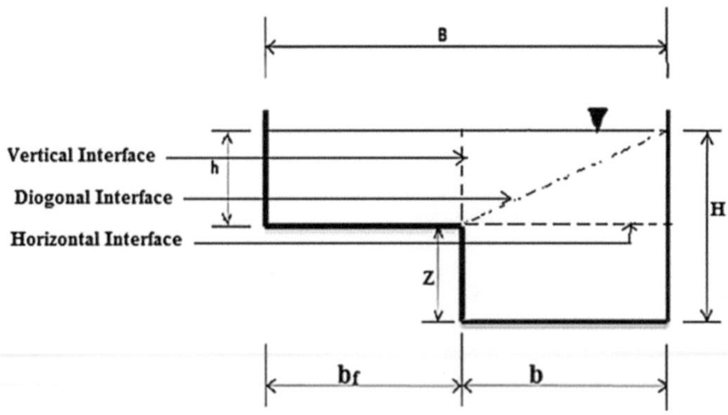

Fig. 35.1 Asymmetric compound channel with interface

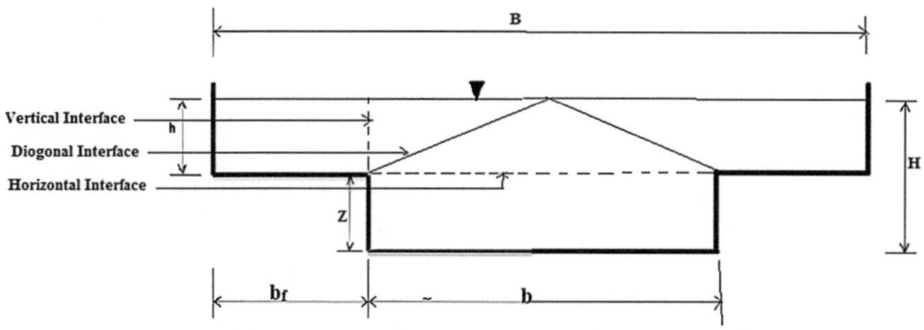

Fig. 35.2 Symmetric compound channel with interface

bottom width of main channel, *bf* is the width of floodplain, *H* is the total flow depth of main channel, and *Z* is bankfull depth. The uniform flow equation, i.e., Manning's equation, is adopted for finding out the flow rate in every subdivisional cross section, and the equation for main channel and floodplain are as follows:

$$Q_{\mathrm{mc}} = \frac{A_{\mathrm{mc}}}{n_{\mathrm{mc}}} R_{\mathrm{mc}}^{\frac{2}{3}} S_0^{\frac{1}{2}}$$

$$Q_{\mathrm{f}} = \frac{A_{\mathrm{f}}}{n_{\mathrm{f}}} R_{\mathrm{f}}^{\frac{2}{3}} S_0^{\frac{1}{2}}$$

where Q_{mc} is the rate of flow of the main channel, A_{mc} is the area of the main channel, R_{mc} is the hydraulic radius of the main channel, n_{mc} is the Manning's roughness coefficient of the main channel, n_f is the Manning's roughness coefficient of the floodplain, Q_{f} is the rate of flow of the floodplain, A_{f} is the area of the floodplain, $R_{\mathrm{f}} = A_{\mathrm{f}}/P_{\mathrm{f}}$ is the hydraulic radius of the floodplain, and S_0 is the bed slope.

Then, the total discharge capacity for an asymmetric compound channel is found out by adding the flow through main channel and floodplain.

$$Q = Q_{\mathrm{mc}} + Q_{\mathrm{f}} \tag{35.1}$$

A. Traditional channel method

When bankfull condition occurs, it has been observed that the single channel method underestimates the discharge as it considers the whole cross section as one unit. Sometimes, it is seen that the Manning's equation overestimates the discharge. The reduction in hydraulic radius underestimates the carrying capacity as it takes the whole cross section as a unit. So three DCMs are used as a conventional flow predicting model in which the symmetric and asymmetric compound channels of different width ratios are simply divided by vertical or horizontal or inclined interface planes as shown in Figs. 35.1 and 35.2. Then, the individual flow capacity for main channel section and floodplain sections is computed considering any interface plane. The summation of main channel and floodplain discharge provides the total discharge. To perform the analysis, three theoretical discharge computation methods have been used as follows: VDM, HDM, and IDM. From the notation of name, *V*, *H*, and *I* represent vertical interface, horizontal interface, and inclined interface, respectively. DM stands for division method. For the application in the numerical model, CES which has recently been adopted by the environment agency (EA) is used for the estimation of discharge of such channels. CES was built by the joint Agency/DEFRA research program on flood defense, with contributions from the Scottish Executive and the Northern Ireland Rivers Agency. It was carried out under a New Research Framework Agreement between HR Wallingford and the Environment Agency. This Software is recommended for reliable prediction of conveyance. In DCM, the bed resistance is chosen on the basis of Manning's *n* or Chezy's *C*, where *n* is the Manning roughness coefficient, and the Chezy's coefficient *C* is related to Manning's *n*. CES is developed to overcome these shortcomings by adopting the Reynolds-averaged Navier–Stokes (RANS) approach, and the solution estimates the conveyance. Outputs are compared with the previous division methods and discussed.

35.3 Calculation

For asymmetric channel VDM, $\text{Amc} = bH$, $A_f = bf\ h$, $\text{Pmc} = (H + b + Z)$, $P_f = bf + h$; for HDM, $\text{Amc} = bZ$, $A_f = Bh$, $\text{Pmc} = (b + 2Z)$, and $P_f = bf + 2h$; for IDM, $\text{Amc} = b(Z + H)/2$, $A_f = h(bf + B)/2$, $\text{Pmc} = (H + b + Z)$, and $P_f = bf + \text{h}$;

For symmetric channel VDM, $\text{Amc} = bH$, $A_f = 2bf\ h$, $\text{Pmc} = (Z + b + Z)$, $P_f = 2(bf + h)$; for HDM, $\text{Amc} = bZ$, $A_f = Bh$, $\text{Pmc} = (b + 2Z)$, and $P_f = 2bf + 2h$; for IDM, $\text{Amc} = Bz + (bh)/2$, $A_f = h(b_f + 0.5B)$, $\text{Pmc} = (Z + b + Z)$, and $P_f = 2(bf + h)$.

35.4 Experimental Apparatus and Procedure

In this present paper, 4 series of experiments were considered for the analysis. Two series of the asymmetric experimental data (width ratio 1.5 and 3) are taken which are conducted on the flume at the fluid mechanics laboratory, Mechanical Engineering Department, Birzeit University [2]. This asymmetric glass-walled horizontal laboratory flume is 7.5 m long, 0.30 m wide and a height of 0.3 m with a bottom slope of 0.0025. The Manning's *n* is assumed as 0.015. For symmetric cross section of width ratio 1.8, the data are taken from [3] in which experiments had been conducted in prismatic compound channels with non-prismatic case after half length of the channel. This channel is built inside a concrete flume measuring 15 m long × 0.95 m width × 0.55 m depth with a roughness coefficient of 0.011. One series of data (width ratio = 3.66) have been taken from model of symmetric rectangular compound cross

Table 35.1 Geometrical parameters of experimental channels

Channel type	*b* (m)	*z* (m)	*Bf* (m)	*B* (m)	*B*/*Z*	*B*/*b*	*bf*/*b*
Asymmetric (A)	0.2	0.06	0.1	0.3	5	1.5	0.5
Symmetric (A)	0.5	0.1	0.2	0.9	9.47	1.8	0.4
Asymmetric (B)	0.1	0.06	0.2	0.3	5	3	2
Symmetric (B)	0.12	0.12	0.16	0.44	3.66	3.66	1.33

sections [5]. This straight experimental compound channel is fabricated inside tilting flumes in the Hydraulics Engineering laboratory at the National Institute of Technology, Rourkela, India. The straight compound channel has equal floodplain of at both sides of the main channel. From the series of data, four data set series are considered for the analysis. The required experimental values with dimensions of the models used in the experiments are given in Table 35.1. The values of varying width ratio 1.5, 1.8, 3 and 3.66 are taken for calculation.

35.5 Results and Discussions

In the subsequent sections, the variations of relative depth and discharges (measured and calculated) of different symmetric and asymmetric series of data of varying non-dimensional parameter are graphically shown in Figs. 35.3, 35.4, 35.5,

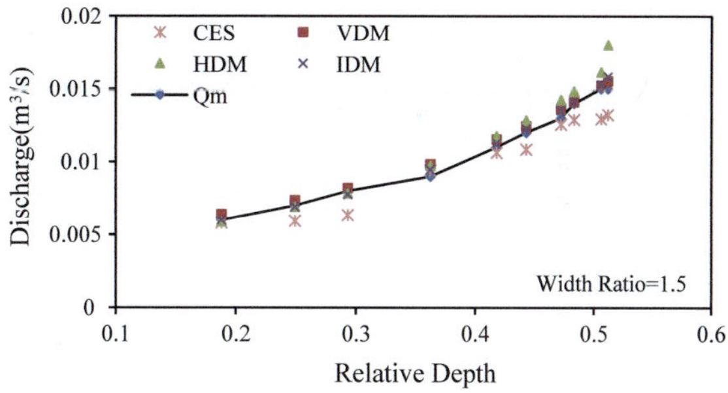

Fig. 35.3 Comparison of methods in asymmetric channel

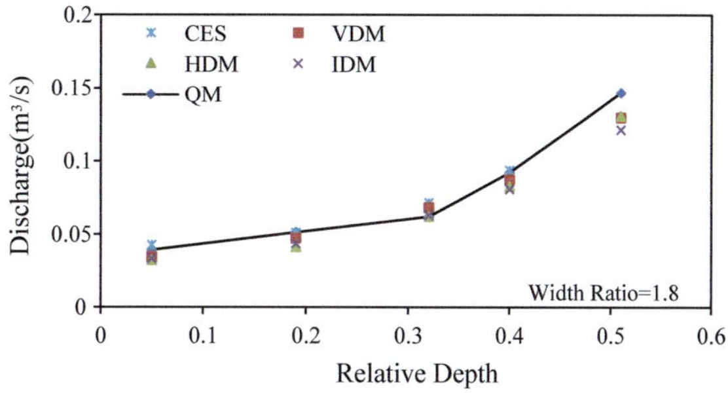

Fig. 35.4 Comparison of methods in symmetric channel

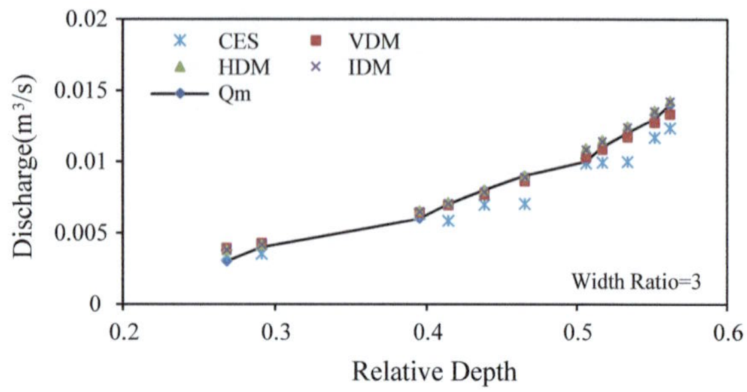

Fig. 35.5 Comparison of methods in asymmetric channel

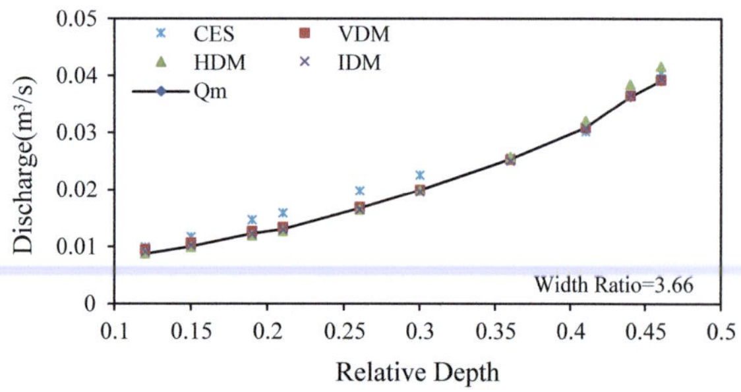

Fig. 35.6 Comparison of methods in symmetric channel

Table 35.2 Error analysis and best discharge computation methods for mode

Channel type	Width ratio *B*/*b*	*b*/*Z*	Relative depth *h*/*H*	Percentage error analysis				Suitable method	The best prediction method
				VDM	HDM	IDM	CES		
Asymmetric (A)	1.50	3.33	0.1880–0.512	4.300	7.000	2.600	9.370	IDM	IDM
Symmetric (A)	1.80	5.26	0.05–0.51	9.340	11.500	12.100	7.640	VDM, CES	CES
Asymmetric (B)	3.00	1.66	0.26–0.562	6.100	6.100	5.700	12.030	IDM, HDM	IDM
Symmetric (B)	3.66	1.00	0.12–0.46	2.300	2.600	1.400	10.998	VDM, IDM	IDM

and 35.6. The relative depths in abscissa with discharge value in ordinate are taken in each figure. The ranges of width ratio (α), depth ratio (β), b/z of different compound channels along with the error analysis are presented in Table 35.2. From those graphs, it is clearly seen that IDM method gives good agreement with the measured discharges for symmetric and asymmetric channels. But for symmetric channel of lower width ratio 1.8, the software CES gives good result and IDM gives the maximum error. For this asymmetric channel of width ratio 3, all the VDM, HDM, and IDM methods give better results. For symmetric channel of width ratio 3.66 (higher width ratio), the software CES overestimates the flow and IDM gives

minimum error to estimate the flow. But the error comes minimum for all method. It is observed that for asymmetric channels, CES underestimates the discharge. From the error analysis, it is analyzed that all methods give good agreement for asymmetric channel of width ratio 1.5. It is observed that for asymmetric channels, CES underestimates the discharge.

35.6 Conclusions

Experimental data sets have been taken to study the effect of flow variables in symmetric and asymmetric compound channels. In this study, a series of laboratory experiments of four models of different width ratios were considered for the analysis. Flow characteristics in each model were observed, and corresponding error analysis is done for a wide range of discharge. From the error analysis of the results, it was observed that selected methods of the DCMs should be used to estimate discharges. One should keep in mind that for the specified range of relative value (h/H) value, this best predicted method should be applied accurately. The best calculation methods with the validity ranges of h/H ratios are given in Table 35.2. It is seen that for asymmetric case, inclined methods predict the flow well, whereas for symmetric cases, the selected method should be preferred carefully.

References

1. Al-Khatib IA, Dweik AA, Gogus M (2012) Evaluation of separate channel methods for discharge computation in asymmetric compound channels. Flow Meas Instrum 24:19–25
2. Al-Khatib IA, Hassan HA, Abaza KA (2013) Application and validation of regression analysis in the prediction of discharge in asymmetric compound channels. J Irrig Drainage Eng 139(7):542–550
3. Naik B, Khatua KK, Sahoo R, Satapathy SS (2014) Flow analysis for a converging compound channel. Int J Appl Eng Res 9(2):133–138
4. Hin LS, Bessaih N, Ling LP, Ghani AA, Seng MY (2008) Determination of apparent and composite friction factors for flooded equatorial natural rivers. Int J River Basin Manage 6(1):3–12
5. Khatua KK, Patra KC (2007) Interaction of flow and estimation of discharge in two stage meandering compound channels. Interaction
6. Knight DW, Hamed ME (1984) Boundary shear in symmetric compound channels. J Hydraul Eng 110(10):1412–1430
7. Wormleaton PR, Allen J, Hadjipanos P (1982) Discharge assessment in compound channel flow. J Hydraul Div 108(9):975–994
8. Wormleaton PR, Hadjipanos P (1985) Flow distribution in compound channels. J Hydraul Eng 111(2):357–361

MIX
Papier aus verantwortungsvollen Quellen
Paper from responsible sources
FSC® C105338

If you have any concerns about our products,
you can contact us on
ProductSafety@springernature.com

In case Publisher is established outside the EU,
the EU authorized representative is:
Springer Nature Customer Service Center GmbH
Europaplatz 3, 69115 Heidelberg, Germany

Printed by Libri Plureos GmbH
in Hamburg, Germany